Zeolites as Catalysts: Applications in Chemical Engineering, Energy Sources and Environmental Protection

Zeolites as Catalysts: Applications in Chemical Engineering, Energy Sources and Environmental Protection

Editors

De Fang
Yun Zheng

Basel • Beijing • Wuhan • Barcelona • Belgrade • Novi Sad • Cluj • Manchester

Editors

De Fang
Center for Materials
Research and Analysis
Wuhan University of
Technology
Wuhan
China

Yun Zheng
Key Laboratory of
Optoelectronic Chemical
Materials and Devices
Jianghan University
Wuhan
China

Editorial Office
MDPI
St. Alban-Anlage 66
4052 Basel, Switzerland

This is a reprint of articles from the Special Issue published online in the open access journal *Catalysts* (ISSN 2073-4344) (available at: www.mdpi.com/journal/catalysts/special_issues/RD0HY99HP8).

For citation purposes, cite each article independently as indicated on the article page online and as indicated below:

Lastname, A.A.; Lastname, B.B. Article Title. *Journal Name* **Year**, *Volume Number*, Page Range.

ISBN 978-3-0365-9659-4 (Hbk)
ISBN 978-3-0365-9658-7 (PDF)
doi.org/10.3390/books978-3-0365-9658-7

© 2023 by the authors. Articles in this book are Open Access and distributed under the Creative Commons Attribution (CC BY) license. The book as a whole is distributed by MDPI under the terms and conditions of the Creative Commons Attribution-NonCommercial-NoDerivs (CC BY-NC-ND) license.

Contents

Wangxiang Pan, Jingping He, Guanlin Huang, Wentao Zhang and De Fang
Research Progress of the Selective Catalytic Reduction with NH_3 over ZSM-5 Zeolite Catalysts for NO_x Removal
Reprinted from: *Catalysts* **2023**, *13*, 1381, doi:10.3390/catal13101381 1

Shu Ren, Fan Yang, Chao Tian, Yinghong Yue, Wei Zou, Weiming Hua and Zi Gao
Selective Alkylation of Benzene with Methanol to Toluene and Xylene over H-ZSM-5 Zeolites: Impact of Framework Al Spatial Distribution
Reprinted from: *Catalysts* **2023**, *13*, 1295, doi:10.3390/catal13091295 21

Lin Chen, Shan Ren, Tao Chen, Xiaodi Li, Zhichao Chen, Mingming Wang, et al.
Effect of Different Zinc Species on Mn-Ce/CuX Catalyst for Low-Temperature NH_3-SCR Reaction: Comparison of $ZnCl_2$, $Zn(NO_3)_2$, $ZnSO_4$ and $ZnCO_3$
Reprinted from: *Catalysts* **2023**, *13*, 1219, doi:10.3390/catal13081219 39

Ruiqi Wu, Ning Liu, Chengna Dai, Ruinian Xu, Gangqiang Yu, Ning Wang and Biaohua Chen
Mechanistic Insight into the Propane Oxidation Dehydrogenation by N_2O over Cu-BEA Zeolite with Diverse Active Site Structures
Reprinted from: *Catalysts* **2023**, *13*, 1212, doi:10.3390/catal13081212 57

Wentao Zhang, De Fang, Guanlin Huang, Da Li and Yun Zheng
Research and Application Development of Catalytic Redox Technology for Zeolite-Based Catalysts
Reprinted from: *Catalysts* **2023**, *13*, 1197, doi:10.3390/catal13081197 71

Muhammad Usman and Munzir H. Suliman
Silver-Doped Zeolitic Imidazolate Framework (Ag@ZIF-8): An Efficient Electrocatalyst for CO_2 Conversion to Syngas
Reprinted from: *Catalysts* **2023**, *13*, 867, doi:10.3390/catal13050867 91

Kai Qi, Lili Gao, Xuelian Li and Feng He
Research Progress in Gas Separation and Purification Based on Zeolitic Materials
Reprinted from: *Catalysts* **2023**, *13*, 855, doi:10.3390/catal13050855 103

Yuqian Liang, Rui Li, Ruicong Liang, Zhanhong Li, Xiangqiong Jiang and Jiuxing Jiang
Positive Effect of Ce Modification on Low-Temperature NH_3-SCR Performance and Hydrothermal Stability over Cu-SSZ-16 Catalysts
Reprinted from: *Catalysts* **2023**, *13*, 742, doi:10.3390/catal13040742 117

Svitlana Orlyk, Pavlo Kyriienko, Andriy Kapran, Valeriy Chedryk, Dmytro Balakin, Jacek Gurgul, et al.
CO_2-Assisted Dehydrogenation of Propane to Propene over Zn-BEA Zeolites: Impact of Acid–Base Characteristics on Catalytic Performance
Reprinted from: *Catalysts* **2023**, *13*, 681, doi:10.3390/catal13040681 133

Yik-Ken Ma, Taghrid S. Alomar, Najla AlMasoud, Zeinhom M. El-Bahy, Stephen Chia, T. Jean Daou, et al.
Effects of Synthesis Variables on SAPO-34 Crystallization Templated Using Pyridinium Supramolecule and Its Catalytic Activity in Microwave Esterification Synthesis of Propyl Levulinate
Reprinted from: *Catalysts* **2023**, *13*, 680, doi:10.3390/catal13040680 153

Norberto J. Abreu, Héctor Valdés, Claudio A. Zaror, Tatianne Ferreira de Oliveira, Federico Azzolina-Jury and Frédéric Thibault-Starzyk
Evidence of Synergy Effects between Zinc and Copper Oxides with Acidic Sites on Natural Zeolite during Photocatalytic Oxidation of Ethylene Using Operando DRIFTS Studies
Reprinted from: *Catalysts* **2023**, *13*, 610, doi:10.3390/catal13030610 . . . . **171**

Mohamed Adel Sayed, Jamaan S. Ajarem, Ahmed A. Allam, Mostafa R. Abukhadra, Jianmin Luo, Chuanyi Wang and Stefano Bellucci
Catalytic Characterization of Synthetic K^+ and Na^+ Sodalite Phases by Low Temperature Alkali Fusion of Kaolinite during the Transesterification of Spent Cooking Oil: Kinetic and Thermodynamic Properties
Reprinted from: *Catalysts* **2023**, *13*, 462, doi:10.3390/catal13030462 . . . . **193**

Amal A. Atran, Fatma A. Ibrahim, Nasser S. Awwad, Mohd Shkir and Mohamed S. Hamdy
Facial One-Pot Synthesis, Characterization, and Photocatalytic Performance of Porous Ceria
Reprinted from: *Catalysts* **2023**, *13*, 240, doi:10.3390/catal13020240 . . . . **213**

Review

Research Progress of the Selective Catalytic Reduction with NH₃ over ZSM-5 Zeolite Catalysts for NO$_x$ Removal

Wangxiang Pan [1,2,†], Jingping He [1,2,†], Guanlin Huang [1,2], Wentao Zhang [3] and De Fang [1,2,4,*]

1. State Key Laboratory of Silicate Materials for Architectures, Wuhan University of Technology, Wuhan 430070, China
2. School of Materials Science and Engineering, Wuhan University of Technology, Wuhan 430070, China
3. Aircraft Maintenance Engineering College, Guangzhou Civil Aviation College, Guangzhou 510403, China
4. Center for Materials Research and Analysis, Wuhan University of Technology, Wuhan 430070, China
* Correspondence: fangde@whut.edu.cn; Tel./Fax: +86-27-87651843
† These authors contributed equally to this work.

Abstract: Nitrogen oxides (NO$_x$) are very common air pollutants that are harmful to the environment and human bodies. Selective catalytic reduction with ammonia (NH$_3$-SCR) is considered an effective means to remove NO$_x$ emissions due to its good environmental adaptability, high catalytic activity, and remarkable selectivity. In this paper, the preparation methods, types, advantages, and challenges of ZSM-5 catalysts are reviewed. Special attention is paid to the catalytic properties and influence factors of ZSM-5 catalysts for NH$_3$-SCR. The SCR performances of ZSM-5 catalysts doped with single or multiple metal ions are also reviewed. In addition, the environmental adaptabilities (sulfur resistance, alkali resistance, water resistance, and hydrothermal stability) of ZSM-5 catalysts are discussed, and the development of ZSM-5 catalysts in denitrification is summarized.

Keywords: NH$_3$-SCR; ZSM-5; catalyst; NO$_x$ removal

1. Introduction

1.1. Denitration and Its Necessity

Nitrogen oxides, including nitric oxide, nitrous oxide, nitrogen dioxide, and other compounds, are very common air pollutants. They cause a series of problems such as the greenhouse effect, acid rain, acid deposition, and soil acidification, and may induce a series of diseases in humans such as cough, chest tightness, and pulmonary edema. With the improvement in people's living standards, the popularity of cars has increased, and, thus, has the emission of nitrogen oxides. In addition, the boilers that are used in most industrial installations, such as power plants and factories, are also an unavoidable source of nitrogen oxides. From the currently available "China Ecological and Environmental Statistics Annual Report 2021", the nitrogen oxide emission data obtained in 2021 in China, only within the scope of emission sources, was 9.884 million tons. In 2019, China's annual emissions of nitrogen oxides were second only to the United States, and the annual emissions have remained above the level of 10 million tons. The treatment of nitrogen oxides can prevent acid rain and the greenhouse effect, and all of society can better obtain sustainable development with a long-term solution for NO$_x$ removal [1,2].

There are many methods for treating nitrogen oxides. The emergence and maturity of denitrification make this method widely used throughout the world. Denitration technology can be divided into wet denitration (acid absorption, alkali absorption, oxidation absorption, etc.) and dry denitration (selective catalytic reduction flue gas denitration and selective non-catalytic reduction denitration). A total of 90% of the flue gas from the combustion is nitric oxide, whose nature makes it difficult to dissolve in water, so the dry denitration method receives more attention in academic and industrial circles.

1.2. NH$_3$-SCR Technology

Selective catalytic reduction technology is also named SCR, and its essence is to add a reducing agent to flue gas to remove the nitrogen oxides with the help of a suitable catalyst. The catalyst can convert the nitrogen oxides into non-toxic and harmless nitrogen and water within a certain temperature range [3–6]; if the reducing agent is ammonia, the technology is named NH$_3$-SCR. There are serval advantages of SCR technology for the flue gas denitration method. Firstly, the product is non-toxic and harmless nitrogen which can be directly put into the atmosphere without subsequent recovery operations and will not cause secondary pollution. Moreover, denitration efficiency is terribly remarkable and can reach more than 85%. Secondly, the SCR technology is relatively mature and reliable in popular use.

SCR can be divided into three types according to its reaction equation, standard SCR, NO$_2$ SCR, and fast SCR. Most of the SCR catalytic reactions are based on the standard SCR reaction (1) because the ratio of NO and NO$_2$ in the exhaust is generally more than 9:1. When the concentration of NO$_2$ is larger than NO (NO$_2$/NO ratio is more than 1), the NO$_2$ can react via the NO$_2$ SCR routes (2). The fast SCR reaction (3) involves an equimolar mixture of NO and NO$_2$ at lower temperatures, and its reaction rate is over 10 times that of the standard SCR reaction. Furthermore, the oxidation process of NO to NO$_2$ by O$_2$ is the rate-limiting step that separates standard SCR from fast SCR [7,8].

$$4NO + 4NH_3 + O_2 \rightarrow 4N_2 + 6H_2O \tag{1}$$

$$6NO_2 + 8NH_3 \rightarrow 7N_2 + 12H_2O \tag{2}$$

$$2NO + 4NH_3 + 2NO_2 \rightarrow 4N_2 + 6H_2O \tag{3}$$

The key to SCR technology is the catalyst, and there are many kinds of catalysts for NH$_3$-SCR. Precious metal catalysts were investigated earlier than others. Generally, platinum and palladium are used as active ingredients, and aluminum oxide, titanium oxide, and molecular sieves are used as carriers. Metal oxide catalysts are widely used at present, mostly vanadium, tungsten, cerium, manganese, iron, copper, and others. Metal oxide catalysts can promote the reaction activity of ammonia, and their denitrification rates are considerable. For molecular sieve catalysts, most of them have certain adsorption properties and exhibit excellent NH$_3$-SCR activity and N$_2$ selectivity.

1.3. Classification of SCR Catalysts

In general, SCR catalysts for commercial use can be divided into several types, such as precious metal catalysts, metal oxide catalysts, metal salt catalysts, molecular sieve catalysts, and carbon-based catalysts [9–13].

1.3.1. Precious Metal Catalysts

The earliest SCR catalysts appeared in the 1970s, and the active ingredients were basically precious metal catalysts. Precious metal catalysts are made by dispersing small amounts of precious metals, such as Pt [14] and Ag [15], on the supports (such as TiO$_2$, Al$_2$O$_3$, etc.). The denitrification efficiency of this kind of catalyst is high, but the reaction temperature needs to be accurately controlled. Due to the expensive price and limited production of precious metals, the related catalysts cannot be used widely in industry, so partial or complete replacement of the precious metals has become a major area of research.

1.3.2. Metal Oxide Catalysts

There are many kinds of metal oxide catalysts, such as vanadium oxides [16], manganese oxides [17], copper oxides [18], iron oxides [19], cerium oxides [20], nickel oxides, cobalt oxides, chromium oxides [6], samarium oxides [21], etc. The advantages of metal oxide catalysts are their high catalytic efficiency, good stability, strong renewable ability, and low cost. However, at the same time, they have a certain toxicity, poor selectivity, and limited service life. These disadvantages also limit their use. At present, researchers are mainly investigating the synergistic effect of composite oxides on SCR, since the activity of catalysts with composite oxides was found to be significantly higher compared with that of single oxide catalysts.

1.3.3. Metal Salt Catalysts

Among the metal salt catalysts, ferric sulfate, cerium phosphate, and cupric sulfate catalysts are the main ones [22–25]. The $Fe_2(SO_4)_3/TiO_2$ catalyst showed about 90% NO_x conversion in the temperature range between 350 and 450 °C, and was synthesized by a solid-state impregnation method [22]. Zr, Co, Mo, or Cu-doped $CePO_4$ catalysts were prepared via the co-precipitation method for NH_3-SCR for NO_x removal. Mo-doped $CePO_4$ catalysts exhibited about 99% NO_x conversion at 300 °C, illustrating excellent low-temperature activity, N_2 selectivity, long stability, and SO_2 and H_2O resistances [23]. The catalyst with a 10 wt% $CuSO_4$ displayed excellent NH_3-SCR activity, a broad temperature window, and remarkable SO_2 and H_2O tolerance, suggesting great potential for NH_3-SCR at low temperatures [24]. The 0.3 Mn-FeV catalyst showed a superior NH_3-SCR performance (100%) at 250 °C. However, the crystal size, acidity, and redox of the catalyst could be improved and the V species with higher valence states could also be obtained due to the addition of Mn [25]. Metal salt catalysts have a wide range of applications and they are not as susceptible to toxic inactivation as metal oxides.

1.3.4. Carbon-Based Catalysts

At present, the main research scope of carbon-based catalysts includes activated carbon [26], graphene [27], carbon nanotubes [28], and other carbon materials. Carbon-based catalysts have remarkable and beneficial properties for the NH_3-SCR process, which are large specific surface areas, developed pore structures, oxygen-containing functional groups, and good adsorption capacity. Meanwhile, carbon-based catalysts are ideal for use in industrial production because they are low-cost due to their rich storage and wide sources. However, there are some unavoidable disadvantages for carbon-based catalysts during the NH_3-SCR process, such as low strength and combustibility.

1.3.5. Molecular Sieve Catalysts

A molecular sieve catalyst, also known as a zeolite catalyst, is a ceramic-based catalyst made of water with alkaline ions and a porous crystalline substance of aluminum silicate in the form of a pellet or honeycomb. There are various classification methods for molecular sieve catalysts, such as 3A (potassium type A), 4A (sodium type A), 5A (calcium type A), 10Z (calcium type Z), 13Z (sodium type Z), Y (sodium type Y), sodium mercerite type, etc., according to the differences in the molecular ratio of SiO_2 and Al_2O_3. Molecular sieve catalysts demonstrate excellent NH_3-SCR performance and good hydrothermal stability, using ZSM-5, SSZ-13, SAPO, ZK-5, KFI, and CHA molecular sieves as the supports.

A molecular sieve is a kind of artificially synthesized hydrated silicoaluminate (bubble zeolite) or natural zeolite with the function of screening molecules. It has many pores with a uniform size that are neatly arranged in the structure; molecular sieves with different pore sizes can separate molecules of different sizes and shapes. A molecular sieve has the characteristics of a large specific surface area, rich pore structure, and strong stability. The thermal stability of a molecular sieve catalyst can be optimized by changing the type and proportion of added materials when constructing the core–shell structure. By controlling the content of the supported metal elements, the temperature window of the catalyst can be adjusted to achieve denitrification at different temperatures [29]. The pore size of molecular sieves with small grains is shorter than that of large-grain molecular sieves, and its pore utilization rate and catalytic activity are greatly improved [30]. The molecular sieve catalyst has good denitrification performance and N_2 selectivity [31,32], but the pore size of the molecular sieve can cause certain restrictions on the diffusion of some molecules, thus affecting its pore utilization rate and catalytic performance.

In different molecular sieve catalyst systems, the active species of metal ions or oxides show different physical and chemical properties, affecting the complexity of the redox reaction mechanism in NH_3-SCR. For Cu-based molecular sieve catalysts, the Si/Al ratio in zeolite and the metal loadings are the key factors for the distribution of active metal species, while the main active sites for the standard SCR reaction are Cu^{2+} species at low temperatures. For the standard NH_3-SCR process over a Fe/SSZ-13 catalyst, isolated Fe^{3+} species are the primary active sites at low temperatures (<300 °C) while $[HO-Fe-O-Fe-OH]^{2+}$ species are the primary active sites at high temperatures (≥300 °C). NO oxidation was caused by dinuclear Fe^{3+} species and NH_3 oxidation was caused by highly aggregated Fe_xO_y species [33,34]. For the NH_3-SCR process over Fe/ZSM-5 zeolites, the main active sites for standard SCR were Fe^{3+} species at the α site, while the fast SCR reaction was promoted by Fe^{3+} species at the β sites and γ sites [35]. Moreover, due to the presence of rich metal species with higher value states and oxygen vacancies over the surfaces of catalysts, the oxidation of NO to NO_2 can be accelerated to promote the fast SCR reaction, causing higher SCR activity at low temperatures.

Due to the high Si–Al ratio and the structurally stable five-membered ring in its skeleton, ZSM-5 zeolite has high thermal stability. Its acid resistance is also very good, so it can resist a variety of acids other than hydrofluoric acid corrosion. At the same time, ZSM-5 does not easily accumulate ash, because there is no large cavity in the ZSM-5 skeleton, limiting the formation and accumulation of large condensates. ZSM-5 has good stability in water steam while the structures of other zeolites are destroyed and irreversibly deactivated when they are subjected to water steam and heat. Therefore, the ZSM-5 molecular sieve catalyst for NH_3-SCR has been a hot topic in recent years [36,37].

2. Preparation Method of ZSM-5 Catalysts

2.1. Catalyst Prepared with the Impregnation Method

The impregnation method is a common method for the preparation of solid catalysts. For example, the Mg/ZSM-5 catalyst was prepared by impregnation [38]. The carrier ZSM-5 was placed in contact with an aqueous solution of a certain metal salt (MgX) as the carrier of the active component so the metal salt solution could be adsorbed or stored in the capillary tube of the carrier. After filtrating, drying, grinding, activation, and immersing in the solution containing the co-catalyst precursor, the catalyst was finally obtained after drying.

The principle of the impregnation method is that the active component is impregnated on the porous carrier in the form of a salt solution and permeates into the inner surface to form a high-efficiency catalyst. Most of the active components are permeated to the inner surface, and only a small number of active components are loaded on the carrier surfaces. The salt of the active component is left on the inner surface of the carrier when the water evaporates and escapes. The impregnated metal and metal oxide salts are evenly distributed in the pores of the carrier, and then the highly dispersed carrier catalyst can be obtained after heating decomposition and activation. The specific surface, pore size, and strength of the molecular sieve play a significant role in the properties of catalysts [39], and different active components can also affect the properties of catalysts. For example, Ebrahim Mohiuddin et al. [40] found that the total acidity of Fe-ZSM-5 was higher than that of Cr-ZSM-5, which improved the selectivity of the catalyst.

Impregnation methods are mainly divided into four kinds, namely, excessive impregnation, equal amount overflow, multiple impregnation, and impregnation precipitation methods. Ricardo H. Gil-Horan et al. [41] used multiple impregnation methods to reduce the interference between different components when preparing P, Fe, and Ni-reinforced ZSM-5 zeolite catalysts. Compared with the excessive impregnation method, the impregnation solution enters the void of the carrier with the impregnation precipitation method, and the precipitating agent is added to precipitate the active component on the inner hole and surface of the carrier.

Impregnation methods do not require special equipment and complex processes, reducing the preparation cost and preparation difficulty. By adjusting the concentration of the solution, impregnation time, and temperature, the morphology, size, and distribution of catalyst particles can be accurately controlled. However, the impregnation method does not easily form a uniform distribution for large-scale production.

2.2. Catalyst Prepared with the Ion Exchange Method

Since most of the cations that need to be exchanged can exist in the cation (simple or complex) state in an aqueous solution, the aqueous ion exchange method is commonly used. When the pH value of an aqueous solution is adjusted to the extent that the crystal structure of the molecular sieve cannot be destroyed and the molecular sieve contacts an aqueous solution of a metal salt, the metal cation in the aqueous solution will exchange with the original cation of the molecular sieve (usually sodium ions), and the anionic skeleton of the molecular sieve will be recombined. For example, a Fe-ZSM-5 catalyst was prepared with the aqueous solution exchange method by Yu Shen et al. [42]. The Fe^{3+} in ferric chloride hexahydrate is exchanged with the Na^+ ions in a molecular sieve to obtain the catalyst. If a higher degree of exchange is required in the ion exchange process, intermittent multiple exchange methods or continuous exchange methods could be used.

The multiple exchange method filters and washes the molecular sieve after the first exchange. Then, the above molecular sieve is treated with a second exchange, filtration, and washing. Moreover, the above steps are repeated until the exchange degree reaches the desired requirements. For multiple exchanges, the exchange degree and exchange rate can be improved by alternating ion exchange and high-temperature roasting. A variety of cations can be exchanged into molecular sieves at the same time, and the molecular sieve after the exchange often has better properties. The ratio of each cation exchange capacity can be better controlled by using the mixed solution or by successive exchanges, according to the strength of the selectivity for various cation exchanges.

Regarding molten salt solution technology, the interference of the solvent effect can be eliminated. Molten salts with high ionization, such as alkali metal halides, sulfates, or nitrates, can be used to provide a cation exchange molten salt solution, but the melting point of the molten salt solution must be lower than the destruction temperature of the zeolite molecular sieve structure. According to a previous study, Houda Jouini prepared Fe-Cu-ZSM-5 from CuCl molten salt [43]. In addition to the cation exchange reaction in the molten salt solution, there were some salts contained in the molecular sieve cage, causing the special properties of zeolites. At the same time, external assistance (such as microwaves) could also promote ion exchange. Somayeh Taghavi [44] heated a mixture of copper chloride and zeolite in a microwave oven with a power of 700 W (about 235 °C) for 10 min to assist the molten salt exchange to produce Cu-ZSM-5.

Some salts can be sublimed into a gas at low temperatures, and the ion exchange of zeolites can be carried out in this gaseous environment. For example, ammonium chloride is sublimed into gas at 300 °C, and Na^+ in a molecular sieve can be exchanged with ammonium chloride vapor. Cuprous chloride can also sublimate above 300 °C, and the non-skeleton cation of the molecular sieve can exchange with it to overcome its insolubility in water and instability in aqueous solution.

No special equipment is required for the preparation of the ion exchange method, so the production cost is relatively low. The preparation process is simple, and the operation is convenient. However, the service life of the prepared catalyst is relatively short, and it needs to be replaced frequently. At the same time, its stability is poor, and it is easily affected by environmental conditions and operating factors, which leads to the decline of its catalytic efficiency.

2.3. Catalyst Prepared with the Chemical Vapor Deposition Method

Chemical vapor deposition (CVD) uses gaseous or vapor substances to react at the gas phase or gas–solid interface to form solid deposits. For example, a manganese (II) acetylacetone precursor was used to deposit manganese oxide on a clay-bonded SiC support by an aerosol-assisted metal–organic chemical vapor deposition (AA-MOCVD) technique [45]. The preparation process of chemical vapor deposition is mainly divided into three steps.

First, the reaction gas is diffused on the surface of the body. Second, the reaction gas is adsorbed on the surface of the matrix. Finally, the solid sediment and gaseous by-products formed on the surface of the matrix are separated from the surface of the matrix.

The advantages of CVD are that a smooth deposition on the surface can be obtained, the mean free path of molecules (atoms) is relatively larger, and the spatial distribution of molecules is more uniform. For example, Milad Rasouli introduced ZnO into the HZSM-5 molecular sieve by chemical vapor deposition in their experiment [46]. In addition, it has excellent product performance, but there are some problems of pollution and insufficient precision.

2.4. Catalyst Prepared with the Sublimation Method

Under a certain atmospheric pressure, the vapor pressure of a solid substance is equal to the external pressure, and sublimation occurs not only on the surface of the solid but also in its interior. In related studies, El-Malki et al. [47] sublimed volatile compounds onto HZSM-5 to produce Z/ZSM-5 molecular sieve catalysts with Ga, Fe, and Zn. In addition, the catalytic activity and durability of the Fe/ZSM-5 catalyst prepared by Haiying Chen in a water-rich environment were promoted by sublimation [48]. The advantage of the sublimation method is that it does not use a solvent and the product purity is high. Its disadvantage is that the product loss is larger, and it is generally used for the preparation of a small number of catalysts.

2.5. Catalyst Prepared with the Hydrothermal Method

The hydrothermal method is adopted to prepare materials in a sealed pressure vessel after a powder is dissolved and recrystallized with water as a solvent. A certain form of the precursor is put into an autoclaved aqueous solution, a hydrothermal reaction under high-temperature and high-pressure conditions is performed, and then separation, washing, drying, and other operations are carried out.

The advantage of the hydrothermal method is that the powder prepared with it has complete grain development, a small particle size, and uniform distribution, and it is convenient to design and characterize its stoichiometry and crystal type. However, it usually depends on the equipment because the hydrothermal method requires high temperature and high pressure. At present, some researchers have synthesized ZSM-5 composite membranes by the hydrothermal method [49].

In short, the hydrothermal method can promote the growth of crystals. The catalyst that is prepared by this method has a smaller crystal size and its catalytic activity is very high. Meanwhile, a wide variety of catalysts can be synthesized. However, the reaction rate is slow and there are high temperature and high pressure steps, and a high-precision reactor is needed to ensure that the reaction conditions are well controlled.

According to the above information, ZSM-5 catalysts can be prepared with the impregnation method, ion exchange method, chemical vapor deposition method, sublimation method, hydrothermal method, and so on. The preparation method can have a significant effect on the surface enrichment of active compositions, BET surface area, uniform distribution, acid sites, redox properties, and so on. Therefore, NO conversions especially are dependent on the preparation methods for ZSM-5 catalysts. Among them, the catalysts that are prepared by the impregnation method and ion exchange method have a low cost. The product with the sublimation method has higher purity, but the method is only suitable for the preparation of a small number of catalysts. The quality of the product that is produced by the chemical vapor deposition method is low and the product that is prepared by the hydrothermal method has good SCR performance, but the cost is high.

3. Effect of Single Metal Ion on ZSM-5 Catalysts

3.1. ZSM-5-Cu Catalysts

Marzieh Hamidzadeh et al. proved that the hydrothermal impregnation of copper chloride and nitrate on HZSM-5 enhanced its denitrification catalytic activity, although this method could cause the loss of zeolite skeleton. The Cu (Cl)/ZSM-5 catalyst had the highest activity among these catalysts. Its activity increased with temperature, remaining constant until about 300 °C, and then weakened with a small slope [50]. At the same time, Wang's team [51] found that an excellent Cu/ZSM-5 catalyst was prepared by a polymer-assisted deposition method. Its excellent catalytic performance could be correlated with the encapsulation and confinement dispersion of Cu-based active species within the ZSM-5 molecular sieve channels, which dramatically optimized the catalytic properties and structural stability of the solid. The catalytic conversions are shown in Figure 1.

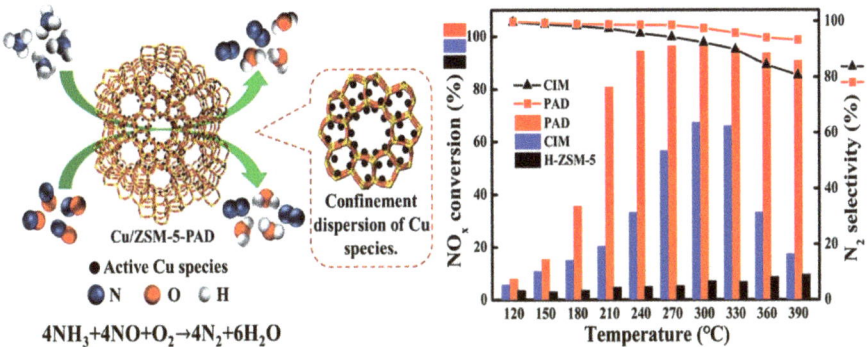

Figure 1. High-silica Cu/ZSM-5 with confinement encapsulated Cu-based active species for NH_3-SCR [51].

3.2. ZSM-5-Fe Catalysts

Fe-based zeolites are widely used in the NH_3-SCR of NO_x at high temperatures, and the Fe species play significant roles in the NO_x conversions, illustrating good sulfur resistance and SCR activity. Therefore, the relationships between SCR performances and Fe species over Fe-ZSM-5 were systematically investigated. According to the previous study [52], the oligomeric Fe and Fe_xO_y species could cause weakened high-temperature acid sites and stronger NH_3 oxidation, and the NH_3-SCR activity over Fe-ZSM-5 with a high Fe loading decreased at high temperatures. Due to nitrate decomposition, the formation of N_2O was found to be over 1.0 wt% Fe-ZSM-5. The calculated reaction pathways of NO_2 and NH_3 depended on the formation rate of [Fe-NO_2]-1/[FeO-NO], based on the DFT results. The catalytic activity and the calculated reaction pathway of the Fe-ZSM-5 catalyst are shown in Figure 2.

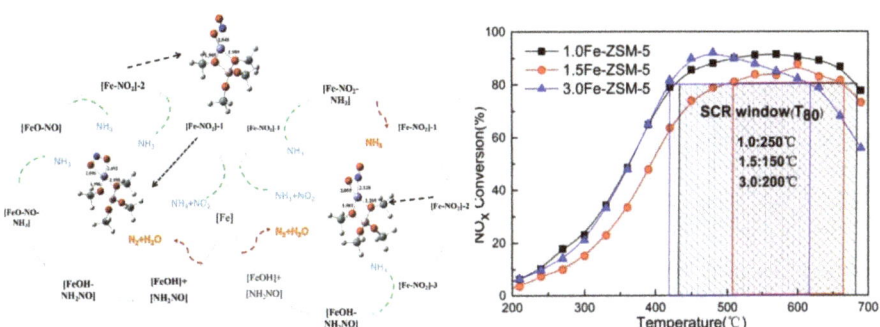

Figure 2. Catalytic activity and calculated reaction pathway of the Fe-ZSM-5 catalyst [52].

In Yu's study [53], ZSM-5 zeolite catalysts doped with Fe were prepared with the solution ion-exchange method. The active Fe species, rather than acidity, played a decisive role in the SCR activity of Fe-ZSM-5, and the Fe-oxo-related Lewis sites were important for the low-temperature SCR reaction. Compared to the dimeric $[OH-Fe-O-Fe-OH]^{2+}$, the isolated Fe^{3+} species in Fe-ZSM-5 were shown more active during the SCR process at a middle temperature of 500 °C.

According to Wang's research [54], a series of Fe/ZSM-5 catalysts with different iron contents via the impregnation method were prepared to study the effect of iron loading on the performance and structure of the Fe/ZSM-5 catalyst for NH_3-SCR of NO. The optimal loading capacity of the Fe active component was 10 wt%. In the experiment, the activities of other catalyst samples under different temperatures were characterized. When the temperature increased from 350 to 450 °C, the conversion rate of the 10 wt% Fe/ZSM-5 catalyst was more than 80% and reached the highest conversion rate (96.91%) at 431 °C. The crystallization of ZSM-5 was not affected by the Fe contents. The more dispersed nanoparticles as active components on the surface of the carrier, the more favorable the NH_3-SCR reaction was. The redox performance of the 10 wt% Fe/ZSM-5 catalyst was significant. More Fe^{3+} species and easier formation of Lewis acid sites could improve the conversion efficiency of NO_x.

3.3. ZSM-5-Co Catalysts

Marzieh Hamidzadeh proved that the hydrothermal impregnation of cobalt chloride and nitrate on HZSM-5 enhanced its denitrification catalytic activity. The Co (Cl)/ZSM-5 catalyst showed the highest activity around 300 °C [50]. However, the related research is very limited since many previous works focus on the NO-SCR reaction over Co-containing H-ZSM-5 catalysts with a C_xH_y reductant instead of NH_3 [55].

3.4. ZSM-5-Mn Catalysts

Most Mn in Mn/ZSM-5 exists in the form of oxides. Mn/ZSM-5 molecular sieve catalysts were prepared by Xue Hongyan's team with the impregnation method [56]. Below 200 °C, the conversion of NH_3 with the Mn/ZSM-5 catalyst is negligible. The conversion rates of NH_3 on the Mn/ZSM-5 catalyst increase sharply in the range of 250–300 °C, reaching almost 100% at 300 °C. In Xiao's research [57], he found that the Mn/ZSM-5 catalyst had weak sulfur resistance. The addition of Fe significantly improved the conversion rate of SO_2-SO_3, and the presence of appropriate Fe promoted the formation of SO_3 and enhanced the conversion of NO_x.

As shown in Figure 3, this work reveals the relationship between the location of MnO_x nanoparticles and the SCR properties, which can provide a new idea for designing low-temperature zeolite catalysts. MnO_x-ZSM-5 catalysts with different MnO_x loadings were explored for the NH_3-SCR reaction. The catalyst exhibited superior low-temperature SCR activity when MnO_x was dispersed on the outer surface of ZSM-5. Thus, the confinement effect could influence the redox properties of the catalysts [58].

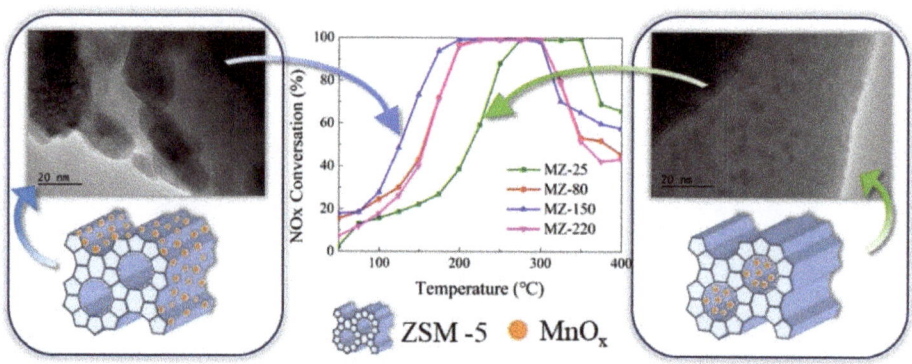

Figure 3. ZSM-5 catalyst with a superior low-temperature SCR activity [58].

3.5. ZSM-5-Ni Catalysts

The Ni in Ni/ZSM-5 exists in the form of NiO crystals (3–50 nm) and charge-compensated Ni^{2+} cations. After the introduction of Ni by the initial wet impregnation method, Vu's team [59] largely preserved the structure and texture properties of ZSM-5. It was pointed out that the ZSM-5 molecular sieve functionalized with cheap and abundant transition metal Ni (Ni/ZSM-5) was a highly active and cost-effective catalyst. In addition, the excellent (water) thermal stability under harsh operating conditions (e.g., temperature up to 873 K and pressure up to 60 bar) made this silicon-rich zeolite bifocal catalyst more attractive, especially for industrially relevant gas applications.

According to Marzieh Hamidzadeh's study [50], Ni-ZSM5 catalysts were prepared with the hydrothermal impregnation method, using nickel chloride and nitrate as the precursors. The catalyst prepared with the chloride precursor system showed higher activity and higher metal dispersion. However, compared with nickel, copper and cobalt improved the conversion rate of SCR reaction more significantly.

As shown in Figure 4, double active sites are verified by redox ability: Cu/ZSM-5 > Mn/ZSM-5 > Ni/ZSM-5. Copper, manganese, and nickel were introduced into H/ZSM-5 zeolites with the incipient wetness impregnation method. The modified catalysts were evaluated for the NH_3-SCR of nitric oxides, suggesting that the NH_3-SCR route over catalysts was through the Langmuir–Hinshelwood mechanism. Metal clusters and bulk oxides were the active sites used to adsorb NO and NH_3 was adsorbed on the acid sites. Nonselective NH_3 oxidation at higher temperatures appeared at higher temperatures due to the double active sites [60].

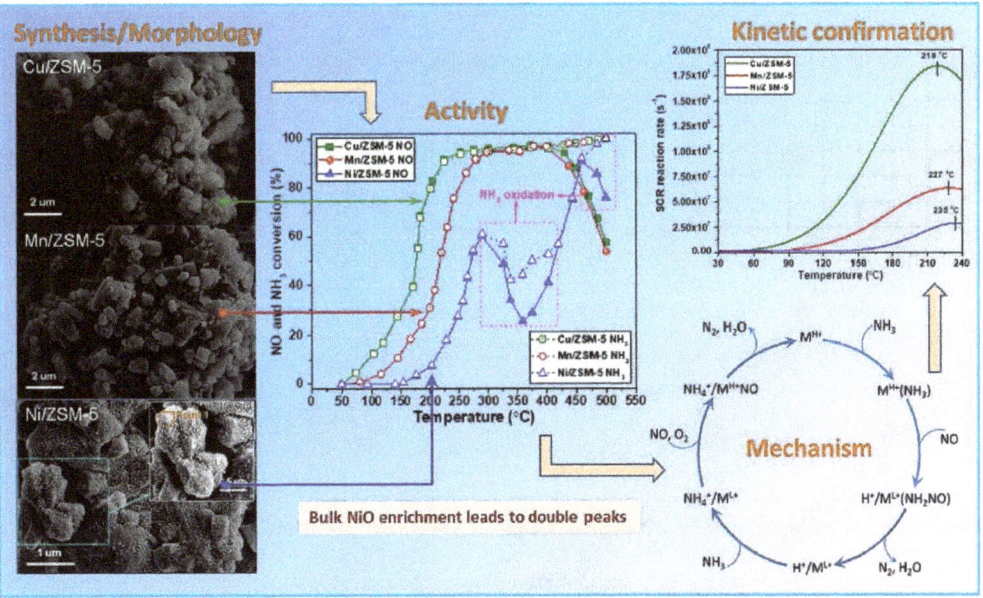

Figure 4. Copper, manganese, and nickel/ZSM-5 catalytic mechanisms for nitric oxide selective reduction with ammonia [60].

According to Table 1, it can be seen that the molecular sieve catalysts Cu(Cl)/ZSM-5, Cu-ZSM-5, Co(Cl)/ZSM-5, Mn/ZSM-5, MnO_x-ZSM-5, Cu(Cl)/ZSM-5, CO (Cl)/ZSM-5, and MnO_x-ZSM-5 can achieve a catalytic efficiency of 90% at a low temperature (<300 °C). The Ni(Cl)/ZSM-5 molecular sieve catalyst requires a relatively high reaction temperature (>300 °C) for catalysis, but its conversion is slightly lower than that of other catalysts at lower temperatures. Meanwhile, the distribution and species of metal elements play important roles in the SCR reaction and NO_x removal.

Table 1. The effect of single metal ions on the performances of ZSM-5 catalysts.

Catalysts	Method	Conditions	Conversion	Ref.
Cu(Cl)/ZSM-5	hydrothermal impregnation method	[NO] = [NH$_3$] = 350 ppm, [O$_2$] = 2.6%, GHSV = 30,000 h^{-1}	>90% (300 °C)	[50]
Cu-ZSM-5	polymer-assisted deposition method	[NO] = [NH$_3$] = 600 ppm, [O$_2$] = 5%, GHSV = 45,000 h^{-1}	>90% (240–390 °C)	[51]
Fe-ZSM-5	wet incipient impregnation method	[NO] = [NH$_3$] = 500 ppm, [O$_2$] = 8%, GHSV = 50,000 h^{-1}	>80% (425–680 °C)	[52]
Fe-ZSM-5	ion exchange method	[NO] = [NH$_3$] = 500 ppm, [O$_2$] = 8%, GHSV = 50,000 h^{-1}	>90% (550 °C)	[53]
Fe/ZSM-5	impregnation method	[NO] = [NH$_3$] = 3%, [O$_2$] = 21%, GHSV = 24,000 h^{-1}	96.91% (431 °C)	[54]
Co(Cl)/ZSM-5	hydrothermal impregnation	[NO] = [NH$_3$] = 350 ppm, [O$_2$] = 2.6%, GHSV = 30,000 h^{-1}	>90% (300 °C)	[50]
Mn/ZSM-5	Mn (NO) impregnation method	[NO] = [NH$_3$] = 500 ppm, [O$_2$] = 5.0%, GHSV = 80,000 h^{-1}	>90% (250 °C)	[55]
MnO$_x$-ZSM-5	impregnation method T = 150 °C	[NO] = [NH$_3$] = 500 ppm, [O$_2$] = 4%, GHSV = 30,000 h^{-1}	>80% (157–315 °C)	[58]
Ni/ZSM-5	incipient wetness impregnation method	[NO] = [NH$_3$] = 500 ppm, [O$_2$] = 5%	90.9% (460 °C)	[60]
Ni(Cl)/ZSM-5	hydrothermal impregnation	[NO] = [NH$_3$] = 350 ppm, [O$_2$] = 2.6%, GHSV = 30,000 h^{-1}	>80% (400 °C)	[50]

Ordinary ZSM-5 catalysts cannot meet the needs of denitration in all environments, which is why scholars have repeatedly studied the significance of introducing different metal cations to ZSM-5 catalysts. The introduction of Cu and Co can enhance the denitrification activity over 300 °C. Fe enhances the redox capacity of the catalyst and obtains the maximum conversion rate of nitrogen oxides over 400 °C, Mn expands the range of applications due to its remarkably low temperature (below 300 °C) SCR activity, and the Ni-ZSM-5 catalyst shows excellent SCR activity over 400 °C. Moreover, Fe and Ce can enhance sulfur resistance. Sometimes, adding a single metal ion is not enough to meet the complex conditions of realistic denitrification, so it is necessary to study the effect of multiple metal ions on the performance of a ZSM-5 catalyst.

4. Effect of Multiple Metal Ions on ZSM-5 Catalysts

Houda Jouini's team [43] utilized the solid-state ion exchange method to prepare Ce-Fe-Cu-ZSM-5 and Fe-Cu-ZSM-5 catalysts. The Fe-Cu-ZSM-5 catalytic system demonstrated its efficiency in NO_x emission reduction at a wide temperature window (180 °C–550 °C). The addition of cerium promoted the oxygen transport capacity, and the combination of the three significantly enhanced the low-temperature SCR activity. Cerium, acting as an oxygen storage promoter, facilitated the oxidation of NO to NO_2, which was a crucial step in NH_3-SCR. While reducing NO emissions in the presence of O_2, it also facilitated the oxidation of ammonia. Therefore, the addition of cerium promoted the catalyst's activity at high temperatures.

Furthermore, the catalyst (WZZ-SG700) that was prepared by the sol-gel method had a higher surface enrichment when compared with the impregnation method and grinding method methods, illustrating the highest NO conversion (as shown in Figure 5) [61]. The SCR performance at high temperatures was improved by its remarkable surface acidity and redox property due to the significant interaction between WO_3 and ZrO_2. Meanwhile, the high resistance against SO_2 and H_2O and high N_2 selectivity were enhanced for the WZZ-SG700 catalyst. The main reaction intermediates of monodentate nitrite, ad-NO_2 species, NH_4^+, surface-adsorbed NH_3, and amide (-NH_2) were found during the reaction by the in situ DRIFTS tests.

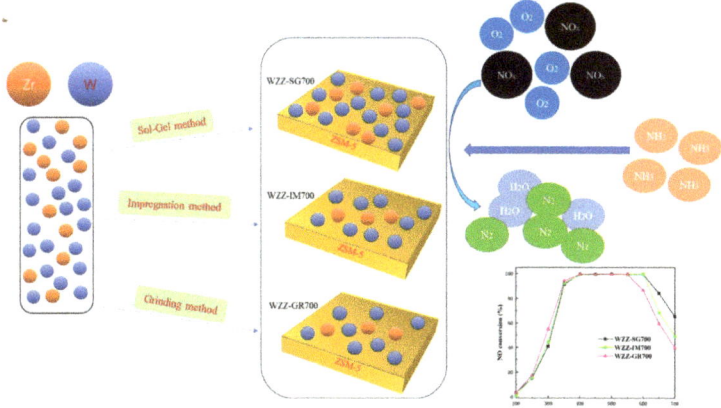

Figure 5. The high activity of NH$_3$-SCR at high temperatures compared to W-Zr/ZSM-5 [61].

As shown in Figure 6, Xue's team [62] employed a combination of ion exchange and impregnation methods to prepare a series of MnCo/Cu-ZSM-5 (MnCo/Cu-Z) catalysts with varying Mn/Co ratios for the NH$_3$ selective catalytic reduction of NO. It was found that the reason for the increase in SCR activity over the MnCo catalyst was the improvement of the high-price metal ions on the catalyst surface and the reduction of metal oxides. Below 200 °C, MnCo/Cu-Z catalysts exhibited higher NH$_3$-SCR activity for NO$_x$ compared to Cu-Z catalysts, with the highest activity observed for Mn$_1$Co$_2$/Cu-Z catalysts.

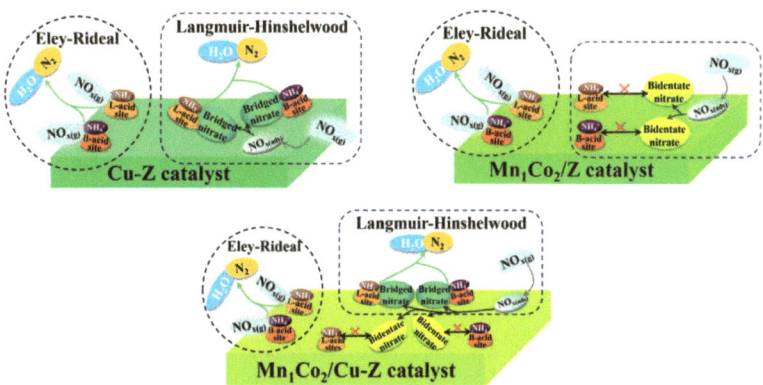

Figure 6. Proposed reaction mechanism of the NH$_3$-SCR of NO$_x$ compared to Cu-Z, Mn$_1$Co$_2$/Z, and Mn$_1$Co$_2$/Cu-Z catalysts at 150 °C [62].

ZSM-5 with Co and Mn was prepared by Zhang's team with the co-precipitation method and they tested their catalysts for the low-temperature NH$_3$ selective catalytic reduction of NO$_x$. The optimal catalyst was found to be 5 wt% Co 10 wt% Mn/ZSM-5(2), showing a high NO$_x$ conversion (98.8%) and N$_2$ selectivity (90.7%) at 150 °C [63].

Wen's team [64] proposed a four-step catalytic mechanism for SCR and fast SCR using Mn/Co-Al/Ce-doped ZSM-5 and conducted a detailed study on the catalytic mechanism using quantum chemical methods. It was concluded that when Ce/Al was doped into ZSM5-Mn/Co and replaced the active center Si, the activation energy for fast SCR was further reduced to about 40–60 kJ/mol. The catalysts for SCR and fast SCR were not only related to the supported transition metals such as Mn and Co but also highly correlated with active centers such as Al and Ce.

As shown in Figure 7, Wentao Mu [65] discovered that the Fe content and metal ion ratio of Fe-Mn/ZSM-5 catalysts must be suitable for optimal catalytic performance at low temperatures. A series of catalysts (Fe-Mn/ZSM-10) were prepared with a different iron content and constant manganese content (5 wt%) by conventional co-precipitation. The addition of Fe may bring about a strong synergistic effect between Fe and Mn to produce a bimetallic active site and establish a multifunctional electron transfer bridge to improve the catalytic performance of Fe-Mn/ZSM-5 catalyst. However, the bridge transfer rate depends on the relative contents of iron and manganese ions, so the content of Fe should be moderate.

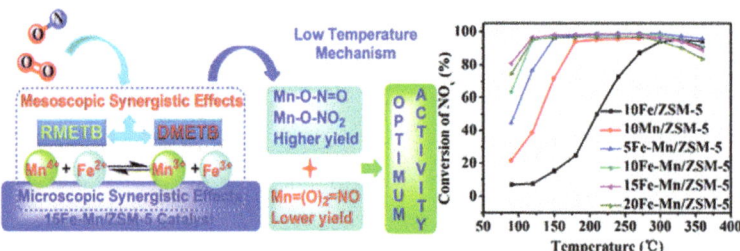

Figure 7. Fe-Mn/ZSM-5 catalyst for the NH_3-SCR of NO_x at low temperatures [65].

The effects of different La loadings on Ce-Cu/ZSM-5 catalysts were investigated, and their performances on the NH_3-SCR of NO_x were evaluated. The Ce-Cu/ZSM-5 catalyst doped with 2 wt% La illustrated the best NH_3-SCR performance at 200–500 °C, suggesting the promotion of La doping. The redox cycle between Cu and Ce could promote the oxidation of NO, resulting in a "Fast SCR". The surface acidic and redox properties could be improved by the addition of La, causing an increase in NH_3-SCR performance [66]. The related results are shown in Figure 8.

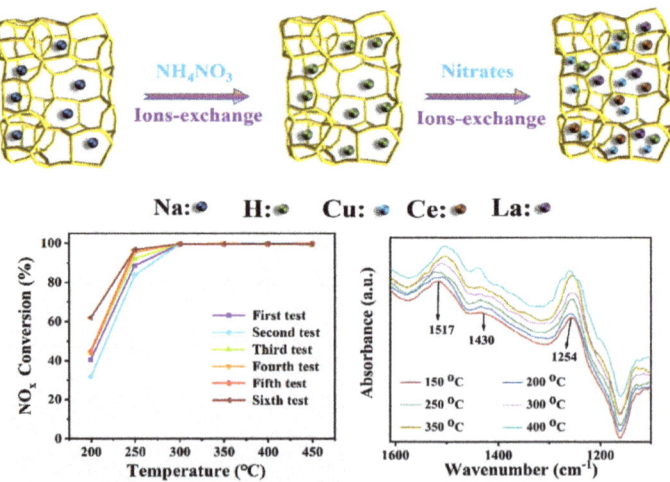

Figure 8. Effect of the V, Cu, and Ce-loading on the simultaneous removal of SO_2 and NO [66].

Ma's team [67] investigated the structure and physicochemical properties of Mn-Ce species-loaded mesoporous ZSM-5 (MZ) for the selective catalytic reduction of NO with ammonia. Compared to traditional ZSM-5 and SBA-15, Cu-Ce/MZ exhibited numerous mesopores and more accessible active sites, facilitating diffusion and improving the internal mass transfer during the denitrification process. Cu-Ce/MZ displayed a high NO conversion of 55.9% at

150 °C. At 200 °C, the Cu-Ce/MZ catalyst showed higher activity than Cu-Ce/ZSM-5 and Cu-Ce/SBA-15 for fast SCR, maintaining NO_x conversion rates of about 92.7%.

According to Xue's study [68], Ca/ZrCu/ZSM-5 catalysts were synthesized with different Zr contents using the initial wet impregnation method. It was concluded that the addition of Zr enhanced the resistance of Cu/ZSM-5 catalysts to calcium poisoning during the SCR process. Additionally, the acidity of Ca/Cu/ZSM-5 did not change significantly after the addition of Zr. The optimal Zr content was found to be 0.10%, resulting in the highest catalytic activity.

MnO_x-CeO_2/ZSM-5 catalysts were prepared by Yang's team [69] with the impregnation method. Compared to MnO_x/ZSM-5, the catalytic performances for the fast SCR reaction and NO oxidation were significantly improved with the addition of CeO_2. The doping of CeO_2 facilitated the absorption and oxidation of NO and therefore promoted the conversion of NO to NO_2. MnCe(0.39)/ZSM-5 exhibited the best performance, maintaining fast SCR NO_x conversion rates above 95% within the temperature range of 150–350 °C. The NOx conversion reached 95.4% at 150 °C.

As mentioned above, the introduction of a single metal cation greatly improves the SCR performance of a catalyst. The complex reality often requires all aspects of the properties of the catalyst, and the introduction of multiple ions will improve the many catalytic properties of the catalysts. According to Table 2, it is common to introduce two or more ions into catalysts, such as Ce, Fe, Cu, Mo, La, Zr, W, and Mn, and the impregnation method and ion exchange method are commonly used for their preparation. Fe-Mn/ZSM-5, 5% Co 10% Mn/ZSM-5, Cu-ZSM-5 with MN-Co, Fe-Mn/ZSM-5, CuCe-La/ZSM-5, Cu-Ce /ZSM-5, Ca/ZrCu/ZSM-5, MnO_x-CeO_2/ZSM-5, and $K_{0.5}$/Cu/ZSM-5 showed high activity at low temperature (<300 °C), while (W-Zr-ZSM-5 (WZZ)), W-Zr-ZSM-5 (WZZ), $K_{0.5}$/Cu/ZSM-5 (WZZ), and Ce-Fe-Cu-ZSM-5 showed better catalytic performance at high temperatures (>300 °C). The temperature window is not only an important issue to be considered in the laboratory stage, but also a remarkable factor from the perspective of energy consumption and economy. Moreover, the interaction of multiple ions over the ZSM-5 catalysts usually can change the catalytic active sites and active centers, generating active species. Fe and Mn enhance redox properties, Mn and Ce improve redox activity, Cu and Ce exhibit numerous mesopores and more accessible active sites, and WO_3 and ZrO_2 improve remarkable surface acidity and redox properties.

Table 2. Multiple metal ions on the performances of ZSM-5 catalysts.

Catalysts	Method	Conditions	Conversion	Ref.
Ce-Fe-Cu-ZSM-5	solid-state ion exchange method	[NO] = [NH$_3$] = 1000 ppm, [O$_2$] = 8%, GHSV = 333,333 h^{-1}	63% (550 °C)	[43]
Fe-Cu-ZSM-5	solid-state ion exchange method	[NO] = [NH$_3$] = 1000 ppm, [O$_2$] = 8%, GHSV = 333,333 h^{-1}	72% (550 °C)	[43]
Fe-Mn/ZSM-5	wet impregnation method	[NO$_2$] = [NO] = 250 ppm, [O$_2$] = 5%, GHSV = 37,500 h^{-1}	99.44% (250 °C)	[57]
W-Zr-ZSM-5 (WZZ)	sol-gel method, impregnation method, and grinding method	[NO$_x$] = [NH$_3$] = 500 ppm, [O$_2$] = 5% GHSV = 15,000 h^{-1}	>90% (300–650 °C)	[61]
Cu-ZSM-5 with Mn-Co	ion-exchange technique	[NO$_x$] = [NH$_3$] = 500 ppm, [O$_2$] = 5%	>90% (185–470 °C)	[62]
5%Co10%Mn/ZSM-5	impregnation method	[NO$_x$] = [NH$_3$] = 500 ppm, [O$_2$] = 5%	98.8% (150 °C)	[63]
Fe-Mn/ZSM-5	conventional co-precipitation method	[NO$_x$] = [NH$_3$] = 600 ppm, [O$_2$] = 5%	100% (90 °C)	[65]
CuCe-La/ZSM-5	ion exchange method	[NO] = [NH$_3$] = 500 ppm, [O$_2$] = 5%, GHSV = 17,000 h^{-1}	99.5% (300 °C)	[66]
Cu-Ce/ZSM-5	ion exchange method	[NO] = [NH$_3$] = 500 ppm, [O$_2$] = 5%,	>92.7% (200 °C)	[67]
Ca/ZrCu/ZSM-5	impregnation method	[NO] = [NH$_3$] = 500 ppm, [O$_2$] = 5%, GHSV = 80,000 h^{-1}	>90% (200–250 °C)	[68]
MnO$_x$-CeO$_2$/ZSM-5	impregnation method	[NO] = [NO$_2$] = 500 ppm, [NH$_3$] = 1000 ppm, [O$_2$] = 3%, GHSV = 30,000 h^{-1}	95.4% (150 °C)	[69]
K$_{0.5}$/Cu/ZSM-5	impregnation method	[NO] = [NH$_3$] = 500 ppm, [O$_2$] = 5%, 80,000 mL/g·h	>90% (200–250 °C)	[70]

5. Environmental Adaptability and Future Prospect of ZSM-5 Catalyst

5.1. Sulfur Resistance

The sulfur resistance of a catalyst refers to its ability to maintain high denitrification activity in the presence of sulfides (such as hydrogen sulfide or sulfur oxides). The toxicity mechanism of SO_2 on catalysts can be divided into three ways. The first way is that the competitive adsorption between SO_2 and reactants (NO and NH_3) reduces the adsorption sites, causing poor SC activity. The second way is that the formation of SO_3 from SO_2 reacts with NH_3 to generate the ammonium sulfate salt, covering the active sites. The third way is that the active metal species are consumed by SO_2 to form metal sulfates or sulfites, reducing the reducibility of metal ions.

ZSM-5 is a commonly used catalyst with excellent denitrification properties. This kind of catalyst illustrates remarkable sulfur resistance due to its special distribution and species of metal ions and its characteristic pore structures. Moreover, the introduction of other specific metal ions or oxides can enhance the sulfur resistance of ZSM-5 [71–73]. Improving the sulfur resistance can prolong the service life of a catalyst by slowing down the reaction between the sulfide and the active site on the catalyst surface. It can also reduce the adsorption of sulfide and improve the selective removal efficiency of the catalyst for nitrogen oxides. The adsorption of sulfide is then reduced and the selective removal efficiency of nitrogen oxides is improved. Wang [54] discovered that Fe-ZSM-5 catalysts with different iron contents exhibited high catalytic activity in SCR, along with good sulfur and water resistance.

5.2. Alkali Resistance

Fuel and lubricant additives may contain a large amount of alkali metals, and their deposition could significantly reduce the performance of SCR catalysts. Alkaline substances will competitively adsorb on the active site of ZSM-5, causing the active site to be blocked and reducing the denitrification performance. If ZSM-5 has good alkaline resistance, it can reduce the blocking of the active site, improve the selective removal efficiency of the catalyst for nitrogen oxides, slow down the alkali vapor deactivation rate of the catalyst, and extend the service life of the catalyst.

Hongyan Xue and his team [70] found that the introduction of potassium reduced the exposure of active CuO on ZSM-5. To some extent, it inhibited the adsorption of NH_3 and NO_x, reduced the formation of active bridging nitrates, and promoted the formation of inactive bidentate nitrates. As shown in Figure 9, the alkali metal deposition on the SCR catalyst reduces the specific surface area and pore volume of the catalyst, destroys the active site on the catalyst surface, and eventually leads to catalyst poisoning [74]. The alkali metal compounds present in the exhaust gases of diesel vehicles and certain factories after dust removal could severely poison Fe-Cu-ZSM-5 catalysts and affect their activity [75]. Characterization results revealed that the introduction of alkali metals did not decrease the crystallinity and textural properties (specific surface area and pore volume) of the parent molecular sieve. On the contrary, it induced new mesoporosity in ZSM-5 materials. By comparing the pore size distributions (PSD) of the fresh and alkali-treated catalysts obtained by adsorption branches, the pore size distributions were found to be relatively wide in the poisoned samples.

Meanwhile, the doping of other metal elements, rational design of catalyst structures, and effective use of catalyst supports can effectively improve the anti-poisoning ability of SCR catalysts. Water washing, pickling, and electrophoresis treatment are some common methods for alkaline poisoning catalyst regeneration.

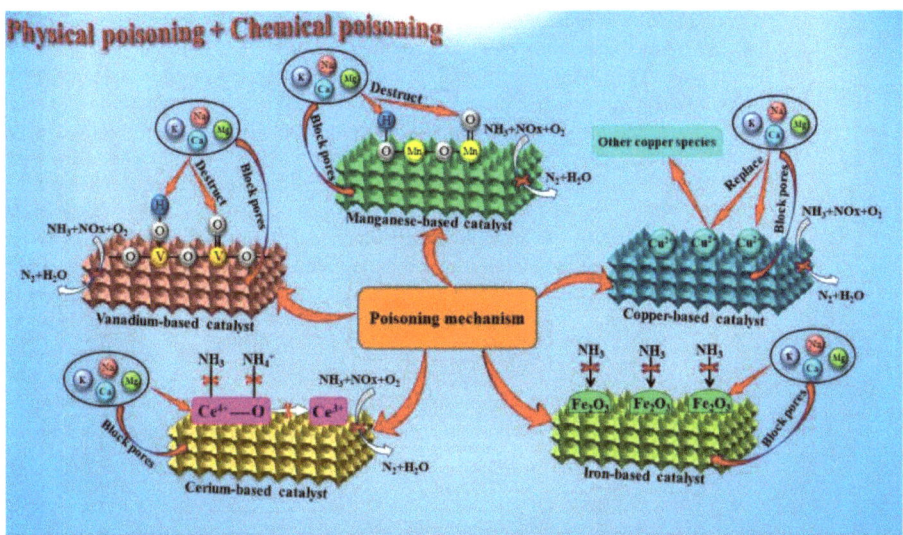

Figure 9. Schematic diagram of the poisoning mechanism of alkali metals on different SCR catalysts [74].

5.3. Water Resistance

Water will competitively adsorb on the active site of ZSM-5, causing the active site to be blocked, thus reducing the denitrification performance. If ZSM-5 has good water resistance, it can reduce the blocking of the active site, improve the selectivity, slow down the hydrothermal deactivation rate of the catalyst, and extend the service life of the catalyst. Guiying Xu and his team [76] focused on the study of different active metals and carriers on the water resistance and sulfur resistance of Mn-based catalysts and proposed that the different physical and chemical properties of different carriers have a certain impact on the performance of catalysts; so, selecting the right carrier can improve the water resistance and sulfur resistance of the catalyst. At the same time, in the existing research, it is difficult to study the durability of catalysts in regard to water resistance and sulfur resistance, so it is one of the important issues that should be further investigated.

When other molecular sieves are exposed to water vapor and heat, their structures are generally destroyed, leading to irreversible deactivation. However, ZSM-5 was used as a catalyst for methanol conversion (with water being one of the main products). This indicated that ZSM-5 had good stability toward water vapor. ZSM-5 molecular sieves had a high silicon-to-aluminum ratio, resulting in lower surface charge density. H_2O, being a highly polar molecule, was not easily absorbed by ZSM-5 molecular sieves. The coreshell structure Mn-Ce mixed oxide catalyst designed by Honggen Peng and his team [77] improved the water resistance of ZSM-5. Additionally, Wang [54] found that a Fe-ZSM-5 (20, 1:1) catalyst exhibited good water resistance under the conditions of a space velocity of 55,510 h^{-1}, 5% (φ) water vapor, and 600×10^{-6} sulfur dioxide. This indicated that the introduction of other metal ions could alter the water resistance of ZSM-5.

5.4. Hydrothermal Stability

Hydrothermal stability is an important property of molecular sieve catalysts that needs to be investigated. Many industrial catalytic reactions have high requirements for hydrothermal stability, which is often the key to determining the catalyst life and reaction process selection. For molecular sieve catalysts, hydrothermal treatment can cause the dealumination of zeolites, reducing the surface area and pore volume, inactivating the active metal species to the aggregation of active metal species, and degrading the structure due to the mobility of metal oxides [78].

In order to obtain molecular sieve catalysts with good hydrothermal stability, several kinds of methods are investigated for the issue. The molecular sieve catalyst can improve the stability of the active center of the catalytic material under water vapor conditions by modifying the active center of the porous material with phospho-oxygen compound assembly and introducing skeleton heteroatoms [79]. Hydrothermal treatment results in chemical and structural changes that are sometimes irreversible, depending on the stability of the zeolite frame. In order to maintain the structural integrity of zeolites and improve the hydrothermal stability of zeolites, high-silicon zeolites have been synthesized. Meanwhile, small-pore zeolites show stronger resistance to dealumination under hydrothermal conditions, preventing the formation of metal oxide species [78,80]. Therefore, appropriate or special molecular sieve catalysts should be selected according to the temperature window of the application situation. Fe-exchanged ZSM-5 catalysts prepared from Na^+ and H^+ forms of ZSM-5, by Xiaoyan Shi [81], and better hydrothermal stability could be obtained when Fe-ZSM-5 catalysts had greater quantities of Fe^{3+} species in ion-exchanged sites and fewer residual protonic sites. Therefore, modifying the active center, preparing high-silicon zeolites, selecting special molecular sieves, and forming suitable metal species can promote the hydrothermal stability of NH_3-SCR.

5.5. Future Prospect

The occurrence of side reactions will consume reactants and produce unwanted products, resulting in a waste of resources and secondary pollution. N_2O as the side reaction during the SCR process over molecular sieve catalysts will bring secondary pollution. Therefore, how to reduce the formation of N_2O is also a significant issue for NO_x removal. It is very efficient to choose highly selective catalysts by introducing suitable metal ions to ensure selectivity.

Metal ions (such as Cu, Fe, Co, Ni, Cr, Zn, Zr, W, Mo, La, Ce, and Mn) are widely used for preparing ZSM-5 catalysts. During the synthesis and scrapping process, the residual metal ions resulting in environmental pollution are also a major problem. This problem can be solved by obtaining a green approach for NO_x removal by designing and selecting an environmentally friendly synthesis and scrapping process. During the synthesis process, we can choose the appropriate ions and control the concentration and temperature to improve the ion exchange efficiency. Meanwhile, using ions with good activity and low toxicity as much as possible can prevent risks to the environment, even if there are metal ions remaining after preparation. For example, between the Fe ion and Cr ion, without considering other factors, the Fe ion with good activity and low toxicity is preferred. Furthermore, in terms of energy, it is important to choose a preparation method with a low cost and low energy consumption. The hydrothermal method is expensive and has high energy consumption, while the lower production cost of the impregnation method and ion exchange method are in line with the theme of green energy. During the scrapping process, catalyst regeneration is an excellent way to solve the problem of residual metal ions in the ZSM-5 catalysts, and acid/alkali washing and thermal treatment are the common regeneration methods for deactivated catalysts [82]. Furthermore, metal ions from waste ZSM-5 catalysts can be extracted and used to prepare new materials, if the catalysts have no regeneration value. In conclusion, the green approach for NO_x removal can be further enhanced by controlling the synthesis and scrapping process.

6. Conclusions

Most of the molecular sieve catalysts were prepared with the immersion method, and some were prepared with the ion exchange method. The more commonly introduced metal ions are Fe, Co, Ni, Mn, Cr, Cu, Zr, W, Ce, La, and so on. Meanwhile, the SCR activity and environmental adaptability of multiple metal ions over the ZSM-5 molecular sieve catalysts were found to be significantly better compared to that of single metal ions, due to the strong interaction of the multiple metal ions. However, ZSM-5 molecular sieve catalysts still face many challenges in environmental adaptability, including the poisoning of toxic substances, competitive adsorption of volatile substances, influence of water vapor, changes in reaction conditions, and long-term stability. Overcoming these challenges requires further research and development to improve catalyst stability and performance.

Denitrification, the process of removing nitrogen oxides from flue gas, is still highly regarded by the public and has been sharply raised as a worldwide problem. SCR denitration technology is the most widely used technology in China, and the molecular sieve catalyst is one of the main catalysts that can meet the needs of fixed source denitration in the future. They demonstrate excellent catalytic activity and selectivity in a variety of fields, such as exhaust gas purification, petrochemical, biomass conversion, olefin separation, and conversion. The ZSM-5 molecular sieve catalyst has a highly controllable pore structure and acid site, and its catalytic performance can be optimized by adjusting its composition, morphology, and active center. In addition, the ZSM-5 molecular sieve catalyst also has good environmental adaptability and can resist adverse factors such as sulfur, alkali, and water, so it has a wide application prospect.

Author Contributions: Conceptualization, W.P., J.H. and D.F.; methodology, W.P., J.H. and D.F.; software, W.P.; validation, G.H. and W.Z.; formal analysis, D.F.; investigation, W.P., J.H., W.Z. and D.F.; resources, G.H. and W.Z.; data curation, G.H., W.Z. and D.F.; writing—original draft preparation, W.P., J.H. and D.F.; writing—review and editing, D.F., J.H. and G.H.; visualization, D.F.; supervision, D.F.; project administration, D.F.; funding acquisition, D.F. All authors have read and agreed to the published version of the manuscript.

Funding: The 2023 program of graduate education and teaching reform at Wuhan University of Technology (construction and research of the teaching case base for the course named Contemporary Analytical Techniques for material characterization).

Data Availability Statement: Data will be made available on request.

Conflicts of Interest: The authors declare no conflict of interest.

References

1. Kumar, M.S.; Alphin, M.S.; Manigandan, S.; Vignesh, S.; Vigneshwaran, S.; Subash, T. A review of comparison between the traditional catalyst and zeolite catalyst for ammonia-selective catalytic reduction of NO_x. *Fuel* **2023**, *344*, 128125. [CrossRef]
2. Khanal, V.; Balayeva, N.O.; Gunnemann, C.; Mamiyev, Z.; Dillert, R.; Bahnemann, D.W.; Subramanian, V. Photocatalytic NO_x removal using tantalum oxide nanoparticles: A benign pathway. *Appl. Catal. B Environ.* **2021**, *291*, 119974. [CrossRef]
3. Busca, G.; Lietti, L.; Ramis, G.; Berti, F. Chemical and mechanistic aspects of the selective catalytic reduction of NO_x by ammonia over oxide catalysts: A review. *Appl. Catal. B Environ.* **1998**, *18*, 1–36. [CrossRef]
4. Chen, H.; Zhang, Y.; He, P.; Li, C.; Liu, L. Facile synthesis of cost-effective iron enhanced hetero-structure activated carbon/geopolymer composite catalyst for NH_3-SCR: Insight into the role of iron species. *Appl. Catal. A Gen.* **2020**, *605*, 117804. [CrossRef]
5. Chen, Z.; Wu, X.; Yang, B.; Huang, Z.; Shen, H.; Jing, G. Submonolayer vanadium and manganese binary metal oxides supported on three-dimensionally ordered mesoporous CeO_2 for efficient low-temperature NH_3-SCR. *Catal. Lett.* **2021**, *151*, 1385–1396. [CrossRef]
6. Fang, D.; He, F.; Mei, D.; Zhang, Z.; Xie, J.; Hu, H. Thermodynamic calculation for the activity and mechanism of Mn/TiO_2 catalyst doped transition metals for SCR at low temperature. *Catal. Commun.* **2014**, *52*, 45–48. [CrossRef]
7. Bendrich, M.; Scheuer, A.; Hayes, R.; Votsmeier, M. Unified mechanistic model for Standard SCR, Fast SCR, and NO_2 SCR over a copper chabazite catalyst. *Appl. Catal. B Environ.* **2018**, *222*, 76–87. [CrossRef]
8. Iwasaki, M. A comparative study of, H.S. "standard", "fast" and "NO_2" SCR reactions over Fe/zeolite catalyst. *Appl. Catal. A Gen.* **2010**, *390*, 71–77. [CrossRef]
9. Su, C.; Zhu, L.; Xu, M.; Zhong, Z.; Wang, X.; Gao, Y.; Zhu, Y. Influences analysis of Sb/Si dopant in TiO_2 on NH_3-SCR activity and low temperature SO_2 resistance of V_2O_5/TiO_2 catalysts. *Appl. Surf. Sci.* **2023**, *637*, 157996. [CrossRef]
10. Song, I.; Youn, S.; Lee, H.; Lee, S.G.; Cho, S.J.; Kim, D.H. Effects of microporous TiO_2 support on the catalytic and structural properties of V_2O_5/microporous TiO_2 for the selective catalytic reduction of NO by NH_3. *Appl. Catal. B Environ.* **2017**, *210*, 421–431. [CrossRef]
11. Park, D.C.; Moon, S.; Song, J.H.; Kim, H.; Lee, E.; Lim, Y.H.; Kim, D.H. Widening the operating window of Pt/ZSM-5 catalysts for efficient NO_x removal in H_2-SCR: Insights from thermal aging. *Catal. Today* **2024**, *425*, 114318. [CrossRef]
12. Liang, Y.; Li, R.; Liang, R.; Li, Z.; Jiang, X.; Jiang, J. Positive effect of Ce modification on low-temperature NH_3-SCR performance and hydrothermal stability over Cu-SSZ-16 catalysts. *Catalysts* **2023**, *13*, 742. [CrossRef]

13. Huang, G.; Yang, J.; Lv, C.; Li, D.; Fang, D. Research progress of NH$_3$-SCR over carbon-based catalysts for NO$_x$ removal. *J. Environ. Chem. Eng.* **2023**, *11*, 110966. [CrossRef]
14. Valtanen, A.; Huuhtanen, M.; Rautio, A.-R.; Kolli, T.; Kordás, K.; Keiski, R.L. Noble Metal/CNT based catalysts in NH$_3$ and EtOH assisted SCR of NO. *Top. Catal.* **2015**, *58*, 984–992. [CrossRef]
15. Zhou, L.; Zhang, M.; Liu, C.; Zhang, Y.; Wanga, H.; Zhangd, Z. Catalytic activity and mechanism of selective catalytic oxidation of ammonia by Ag-CeO$_2$ under different preparation conditions. *RSC Adv.* **2023**, *13*, 10239–10248. [CrossRef] [PubMed]
16. Nuguid, R.J.G.; Ferri, D.; Marberger, A.; Nachtegaal, M.; Krocher, O. Modulated excitation raman spectroscopy of V$_2$O$_5$/TiO$_2$: Mechanistic insights into the selective catalytic reduction of NO with NH$_3$. *ACS Catal.* **2019**, *9*, 6814–6820. [CrossRef]
17. Xu, S.; Chen, J.; Li, Z.; Liu, Z. Highly ordered mesoporous MnO$_x$ catalyst for the NH$_3$-SCR of NO$_x$ at low temperatures. *Appl. Catal. A Gen.* **2023**, *649*, 118966. [CrossRef]
18. Zhang, J.; Tian, H.; Yu, Y.; Jiang, Z.; Ma, M.; He, C. Novel CuO@TiO$_2$ core–shell nanostructure catalyst for selective catalytic reduction of NO$_x$ with NH$_3$. *Catal. Lett.* **2021**, *151*, 2502–2512. [CrossRef]
19. Chang, H.; Zhang, T.; Dang, H.; Chen, X.; You, Y.; Schwank, J.W.; Li, J. Fe$_2$O$_3$@SiTi core–shell catalyst for the selective catalytic reduction of NO$_x$ with NH$_3$: Activity improvement and HCl tolerance. *Catal. Sci. Technol.* **2018**, *8*, 3313–3320. [CrossRef]
20. Wu, X.; Yu, X.; He, X.; Jing, G. Insight into low-temperature catalytic NO reduction with NH$_3$ on Ce-doped manganese oxide octahedral molecular sieves. *J. Phys. Chem. C* **2019**, *123*, 10981–10990. [CrossRef]
21. Xie, J.; Ye, Y.; Li, Q.; Kang, T.; Hou, S.; Jin, Q.; He, F.; Fang, D. Denitrification performance and sulfur resistance mechanism of Sm-Mn catalyst for low temperature NH$_3$-SCR. *Front. Chem. Sci. Eng.* **2023**, *17*, 617–633. [CrossRef]
22. Yu, Y.; Chen, C.; Ma, M.; Douthwaite, M.; He, C.; Miao, J.; Chen, J.; Li, C. SO$_2$ promoted in situ recovery of thermally deactivated Fe$_2$(SO$_4$)$_3$/TiO$_2$ NH$_3$-SCR catalysts: From experimental work to theoretical study. *Chem. Eng. J.* **2019**, *361*, 820–829. [CrossRef]
23. Zhao, R.; Pang, R.; Wang, Y.; Zhao, Z. Effect of metal elements doping on the CePO$_4$ catalysts for selective catalytic reduction of NO with NH$_3$. *Mol. Catal.* **2022**, *530*, 112627. [CrossRef]
24. Yu, Y.; Miao, J.; Wang, J.; He, C.; Chen, J. Facile synthesis of CuSO$_4$/TiO$_2$ catalysts with superior activity and SO$_2$ tolerance for NH$_3$-SCR: Physicochemical properties and reaction mechanism. *Catal. Sci. Technol.* **2017**, *7*, 1590–1601. [CrossRef]
25. Xie, H.; Shu, D.; Chen, T.; Liu, H.; Zou, X.; Wang, C.; Han, Z.; Chen, D. An in-situ DRIFTs study of Mn doped FeVO$_4$ catalyst by one-pot synthesis for low-temperature NH$_3$-SCR. *Fuel* **2022**, *309*, 122108. [CrossRef]
26. Yang, J.; Ren, S.; Su, B.; Wang, M.; Chen, L.; Liu, Q. Understanding the dual-acting of iron and sulfur dioxide over Mn-Fe/AC catalysts for low-temperature SCR of NO. *Mol. Catal.* **2022**, *519*, 112150. [CrossRef]
27. Miao, Y.; Zheng, J.; Liu, Y.; Xiang, N.; Li, Y.; Han, X.; Huang, Z. Relationship between oxygen-containing groups and acidity of graphene oxide supported Mn-based SCR catalysts and the effects on the catalytic activity. *Catal. Lett.* **2020**, *150*, 3243–3255. [CrossRef]
28. Li, P.; Zhang, T.; Sun, H.; Gao, Y.; Zhang, Y.; Liu, Y.; Ge, C.; Chen, H.; Dai, X.; Zhang, X. Cobalt doped Fe-Mn@CNTs catalysts with highly stability for low-temperature selective catalytic reduction of NO$_x$. *Nano Res.* **2022**, *15*, 3001–3009. [CrossRef]
29. Xu, J.; Wang, H.; Guo, F.; Zhanga, C.; Xie, J. Recent advances in supported molecular sieve catalysts with wide temperature range for selective catalytic reduction of NO$_x$ with C$_3$H$_6$. *RSC Adv.* **2019**, *9*, 824–838. [CrossRef]
30. Krishnamurthy, M.; Swaminathan, M. Synthesis of hierarchical micro-mesoporous ZSM-5 zeolite and its catalytic activity in benzylation of mesitylene. *Silicon* **2023**, *15*, 3399–3405. [CrossRef]
31. Okamoto, M. Core-shell structured zeolite catalysts with minimal defects for improvement of shape selectivity. In *Core-Shell and Yolk-Shell Nanocatalysts*; Springer: Singapore, 2021; pp. 187–198.
32. Guo, A.; Liu, H.; Li, Y.; Luo, Y.; Ye, D.; Jiang, J.; Chen, P. Recent progress in novel zeolite catalysts for selective catalytic reduction of nitrogen oxides. *Catal. Today* **2023**, *422*, 114212. [CrossRef]
33. Zhang, W.; Chen, J.; Guo, L.; Zheng, W.; Wang, G.; Zhang, S.; Wu, X. Research progress on NH$_3$-SCR mechanism of metal-supported zeolite catalysts. *J. Fuel Chem. Technol.* **2021**, *49*, 1294–1315. [CrossRef]
34. Gao, F.; Zheng, Y.; Kukkadapu, R.K.; Wang, Y.; Walter, E.D.; Schwenzer, B.; Szanyi, J.; Charles, H.; Peden, F. Iron loading effects in Fe/SSZ-13 NH$_3$-SCR catalysts: Nature of the Fe ions and structure−function relationships. *ACS Catal.* **2016**, *6*, 2939–2954. [CrossRef]
35. Ellmers, I.; Velez, R.P.; Bentrup, U.; Brückner, A.; Grünert, W. Oxidation and selective reduction of NO over Fe-ZSM-5—How related are these reactions? *J. Catal.* **2014**, *311*, 199–211. [CrossRef]
36. Yuan, E.; Li, M.; Yang, M.-H.; Huang, X.; Zhang, K.; Han, W.; Tang, Z.; Liu, Z. Encapsulation of ultra-small Cu–Fe into ZSM-5 zeolites for NH$_3$-SCR with broad reaction-temperature ranges. *Microporous Mesoporous Mater.* **2022**, *331*, 111675. [CrossRef]
37. Ye, T.; Chen, Z.; Chen, Y.; Xie, H.; Zhong, Q.; Qu, H. Green synthesis of ZSM-5 zeolite for selective catalytic reduction of NO via template-free method from tailing residue. *J. Environ. Chem. Eng.* **2022**, *10*, 107766. [CrossRef]
38. Han, S.; Huang, R.; Chen, S.; Wang, Z.; Jiang, N.; Park, S.-E. Hierarchical Mg/ZSM-5 catalysts for methanol-to-propylene reaction via one-step acid treatment. *Res. Chem. Intermed.* **2021**, *47*, 249–268. [CrossRef]
39. Shan, S.; Tian, Y.; Chen, F.; Wu, S.; Zhou, R.; Xie, Y.; Li, N.; Zeng, X.; Lin, C.; Acidity, W.Y. Crystallite size and pore structure as key factors influencing 1,3,5-trimethylbenzene hydrodealkylation performance of NiMoS/ZSM-5. *Catal. Surv. Asia* **2022**, *26*, 35–45. [CrossRef]
40. Mohiuddin, E.; Mdleleni, M.M.; Key, D. Catalytic cracking of naphtha: The effect of Fe and Cr impregnated ZSM-5 on olefin selectivity. *Appl. Petrochem. Res.* **2018**, *8*, 119–129. [CrossRef]

41. Gil-Horan, R.H.; Chavarria-Hernandez, J.C.; Quintana-Owen, P.; Gutierrez-Alejandre, A. Ethanol conversion to short-chain olefins over ZSM-5 zeolite catalysts enhanced with P, Fe, and Ni. *Top. Catal.* **2020**, *63*, 414–427. [CrossRef]
42. Shen, Y.; Wang, F.; Liu, W.; Zhang, X. The preparation of Fe^{3+} ion-exchanged mesopore containing ZSM-5 molecular sieves and its high catalytic activity in the hydroxylation of phenol. *J. Porous Mater.* **2018**, *25*, 1587–1595. [CrossRef]
43. Jouini, H.; Mejri, I.; Rhimi, B.; Mhamdi, M.; Blasco, T.; Delahay, G. Ce-promoted Fe-Cu-ZSM-5 catalyst: SCR-NO activity and hydrothermal stability. *Res. Chem. Intermed.* **2021**, *47*, 2901–2915. [CrossRef]
44. Taghavi, S.; Ghedini, E.; Menegazzo, F.; Giordana, A.; Cerrato, G.; Cruciani, G.; Di Michele, A.; Zendehdel, M.; Signoretto, M. Balanced acidity by microwave-assisted ion-exchange of ZSM-5 zeolite as a catalyst for transformation of glucose to levulinic acid. *Biomass Convers. Biorefinery* **2022**. [CrossRef]
45. Manafi, A.; MirMoghtadaei, G.; Falamaki, C. Aerosol assisted chemical vapor deposition of $Mn(AC)_2$ for MnO_x/(Clay-Bonded SiC) catalyst synthesis for propane-SCR of NO_x. *Russ. J. Inorg. Chem.* **2021**, *66*, 684–695. [CrossRef]
46. Rasouli, M.; Yaghobi, N. Bifunctional ZnO/HZSM-5 catalysts in direct hydrogenation of CO_2 to aromatics; influence of preparation method. *Catal. Lett.* **2023**, *153*, 1450–1463. [CrossRef]
47. El-Malki, E.-M.; Van Santen, R.A.; Sachtler, W.M.H. Introduction of Zn, Ga, and Fe into HZSM-5 Cavities by Sublimation: Identification of Acid Sites. *J. Phys. Chem. B* **1999**, *103*, 4611–4622. [CrossRef]
48. Chen, H.Y.; Sachtler, W.M. Promoted Fe/ZSM-5 catalysts prepared by sublimation: De-NO_x activity and durability in H_2O-rich streams. *Catal. Lett.* **1998**, *50*, 125–130. [CrossRef]
49. Kwak, B.I.; Hyun, S.H.; Kim, G.T. CO_2 separation characteristics of ZSM-5 composite membranes synthesized by the hydrothermal treatment. *J. Mater. Sci. Lett.* **2001**, *20*, 1893–1896. [CrossRef]
50. Hamidzadeh, M.; Ghassemzadeh, M.; Tarlani, A.; Sahebdelfar, S. The effect of hydrothermal impregnation of Ni, Co, and Cu on HZSM-5 in the nitrogen oxide removal. *Int. J. Environ. Sci. Technol.* **2018**, *15*, 93–104. [CrossRef]
51. Wang, Y.; Ji, X.; Meng, H.; Qu, L.; Wu, X. Fabrication of high-silica Cu/ZSM-5 with confinement encapsulated Cu-based active species for NH_3-SCR. *Catal. Commun.* **2020**, *138*, 105969. [CrossRef]
52. Wang, P.; Yu, D.; Zhang, L.; Ren, Y.; Jin, M.; Lei, L. Evolution mechanism of NO_x in NH_3-SCR reaction over Fe-ZSM-5 catalyst: Species-performance relationships. *Appl. Catal. A Gen.* **2020**, *607*, 117806. [CrossRef]
53. Yu, D.; Wang, P.; Li, X.; Zhao, H.; Lv, X. Study on the role of Fe species and acid sites in NH_3-SCR over the Fe-based zeolites. *Fuel* **2023**, *336*, 126759. [CrossRef]
54. Wang, X.; Hu, H.; Zhang, X.; Su, X.; Yang, X. Effect of iron loading on the performance and structure of Fe/ZSM-5 catalyst for the selective catalytic reduction of NO with NH_3. *Environ. Sci. Pollut. Res.* **2019**, *26*, 1706–1715. [CrossRef] [PubMed]
55. Lonyi, F.; Solt, H.E.; Paszti, Z.; Valyon, J. Mechanism of NO-SCR by methane over Co,H-ZSM-5 and Co,H-mordenite catalysts. *Appl. Catal. B Environ.* **2014**, *150–151*, 218–229. [CrossRef]
56. Xue, H.; Guo, X.; Meng, T.; Mao, D.; Ma, Z. NH_3-SCR of NO over M/ZSM-5 (M = Mn, Co, Cu) catalysts: An in-situ DRIFTS study. *Surf. Interfaces* **2022**, *4*, 101722. [CrossRef]
57. Xiao, H.; Dou, C.; Li, J.; Yuan, Z.; Lv, H. Experimental study on SO_2-to-SO_3 conversion over Fe-modified Mn/ZSM-5 catalysts during the catalytic reduction of NO_x. *Catal. Surv. Asia* **2019**, *23*, 332–343. [CrossRef]
58. Qiao, T.; Liu, Z.; Liu, C.; Meng, W.; Sun, H.; Lu, Y. MnO_x location on MnO_x-ZSM-5 to influence the catalytic activity for selective catalytic reduction of NO_x by NH_3. *Appl. Catal. A Gen.* **2021**, *617*, 118128. [CrossRef]
59. Vu, H.; Arcon, I.; de Souza, D.O.; Pollastri, S.; Drazic, G.; Volavsek, J.; Mali, G.; Zabukovec Logar, N.; Novak Tusar, N. Insight into the interdependence of Ni and Al in bifunctional Ni/ZSM-5 catalysts at the nanoscale. *Nanoscale Adv.* **2022**, *4*, 2321–2331. [CrossRef]
60. Liu, C.; Kang, R.; Bin, F.; Wei, X.; Hui, K.N.; Kasipandi, S.; San Hui, K. Insights on copper, manganese, and Nickel/ZSM-5 catalytic mechanisms for nitric oxides selective reduction with ammonia. *Carbon Resour. Convers.* **2022**, *5*, 15–25. [CrossRef]
61. Feng, S.; Li, Z.; Shen, B.; Yuan, P.; Wang, B.; Liu, L.; Ma, J.; Kong, W. High activity of NH_3-SCR at high temperature over W-Zr/ZSM-5 in the exhaust gas of diesel engine. *Fuel* **2022**, *323*, 124337. [CrossRef]
62. Xue, H.; Guo, X.; Meng, T.; Guo, Q.; Mao, D.; Wang, S. Cu-ZSM-5 catalyst impregnated with Mn-Co oxide for the selected catalytic reduction of NO: Physicochemical property-catalytic activity relationship and In Situ DRIFTS study for the reaction mechanism. *ACS Catal.* **2021**, *11*, 7702–7718. [CrossRef]
63. Zhang, S.; Zhang, C.; Wang, Q.; Ahn, W.S. Co- and Mn-coimpregnated ZSM-5 prepared from recycled industrial solid wastes for low-temperature NH_3-SCR. *Ind. Eng. Chem. Res.* **2019**, *58*, 22857–22865. [CrossRef]
64. Wen, Z.; Li, S.; Li, H.; Li, Y.; Wang, G. Quantum chemical study on the reaction mechanism of fast SCR catalyzed by ZSM-5 doped with Mn/Co-Al/Ce. *Arab. J. Sci. Eng.* **2019**, *44*, 5549–5557. [CrossRef]
65. Mu, W.; Zhu, J.; Zhang, S.; Guo, Y.; Su, L.; Lia, X.; Li, Z. Novel proposition on mechanism aspects over Fe-Mn/ZSM-5 catalyst for NH_3-SCR of NO_x at low temperature: Rate and direction of multifunctional electron-transfer-bridge and in situ DRIFTs analysis. *Catal. Sci. Technol.* **2016**, *6*, 7532–7548. [CrossRef]
66. Yang, J.; Li, Z.; Yang, C.; Ma, Y.; Li, Y.; Zhang, Q.; Song, K.; Cui, J. Significant promoting effect of La doping on the wide temperature NH_3-SCR performance of Ce and Cu modified ZSM-5 catalysts. *J. Solid State Chem.* **2022**, *305*, 122700. [CrossRef]
67. Ma, Y.; Liu, Y.; Li, Z.; Geng, C.; Bai, X.; Cao, D. Synthesis of CuCe co-modified mesoporous ZSM-5 zeolite for the selective catalytic reduction of NO by NH_3. *Environ. Sci. Pollut. Res.* **2020**, *27*, 9935–9942. [CrossRef]

68. Xue, H.; Meng, T.; Liu, F.; Guo, X.; Wang, S.; Mao, D. Enhanced resistance to calcium poisoning on Zr-modified Cu/ZSM-5 catalysts for the selective catalytic reduction of NO with NH_3. *RSC Adv.* **2019**, *9*, 38477–38485. [CrossRef]
69. Yang, X.; Xiao, H.; Liu, J.; Wan, Z.; Wang, T.; Sun, B. Influence of Ce-doping on MnO_x-ZSM-5 catalysts for the selective catalytic reduction of NO/NO_2 with NH_3. *React. Kinet. Mech. Catal.* **2018**, *125*, 1071–1084. [CrossRef]
70. Xue, H.; Guo, X.; Meng, T.; Mao, D.; Ma, Z. Poisoning effect of K with respect to Cu/ZSM-5 used for NO reduction. *Colloid Interface Sci. Commun.* **2021**, *44*, 100465. [CrossRef]
71. Guan, J.; Zhou, L.; Li, W.; Hu, D.; Wen, J.; Huang, B. Improving the performance of Gd addition on catalytic activity and SO_2 resistance over MnO_x/ZSM-5 catalysts for low-temperature NH_3-SCR. *Catalysts* **2021**, *11*, 324. [CrossRef]
72. Feng, S.; Kong, W.; Wang, Y.; Xing, Y.; Wang, Z.; Ma, J.; Shen, B.; Chen, L.; Yang, J.; Li, Z.; et al. Mechanistic investigation of Sm doping effects on SO_2 resistance of W-Zr-ZSM-5 catalyst for NH_3-SCR. *Fuel* **2023**, *353*, 129139. [CrossRef]
73. Di, Z.; Wang, H.; Zhang, R.; Chen, H.; Wei, Y.; Jia, J. ZSM-5 core-shell structured catalyst for enhancing low-temperature NH_3-SCR efficiency and poisoning resistance. *Appl. Catal. A Gen.* **2022**, *630*, 118438. [CrossRef]
74. Wu, P.; Tang, X.; He, Z.; Liu, Y.; Wang, Z. Alkali metal poisoning and regeneration of selective catalytic reduction denitration catalysts: Recent advances and future perspectives. *Energy Fuels* **2022**, *36*, 5622–5646. [CrossRef]
75. Jouini, H.; Mejri, I.; Martinez-Ortigosa, J.; Cerillo, J.L.; Petitto, C.; Mhamdi, M.; Blasco, T.; Delahay, G. Alkali poisoning of Fe-Cu-ZSM-5 catalyst for the selective catalytic reduction of NO with NH_3. *Res. Chem. Intermed.* **2022**, *48*, 3415–3428. [CrossRef]
76. Xu, G.; Guo, X.; Cheng, X.; Yu, J.; Fang, B. A review of Mn-based catalysts for low-temperature NH_3-SCR: NO_x removal and H_2O/SO_2 resistance. *Nanoscale* **2021**, *13*, 7052–7080. [CrossRef]
77. Peng, H.; Li, G.; An, T. Core-shell confinement $MnCeO_x$@ZSM-5 catalyst for NO_x removal with enhanced performances to water and SO_2 resistance. In *Core-Shell and Yolk-Shell Nanocatalysts*; Springer: Singapore, 2021; pp. 165–179.
78. Wang, B.; Zhang, Y.; Fan, X. Deactivation of Cu SCR catalysts based on small-pore SSZ-13 zeolites: A review. *Chem. Phys. Impact* **2023**, *6*, 100207. [CrossRef]
79. Sun, G.; Yu, R.; Xu, L.; Wang, B.; Zhang, W. Enhanced hydrothermal stability and SO_2-tolerance of Cu-Fe modified AEI zeolite catalysts in NH_3-SCR of NO_x. *Catal. Sci. Technol.* **2022**, *12*, 3898–3911. [CrossRef]
80. Kwak, J.H.; Tran, D.; Burton, S.D.; Szanyi, J.; Lee, J.H.; Charles, H.; Peden, F. Effects of hydrothermal aging on NH_3-SCR reaction over Cu/zeolites. *J. Catal.* **2012**, *287*, 203–209. [CrossRef]
81. Shi, X.; He, H.; Xie, L. The effect of Fe species distribution and acidity of Fe-ZSM-5 on the hydrothermal stability and SO_2 and hydrocarbons durability in NH_3-SCR reaction. *Chin. J. Catal.* **2015**, *36*, 649–656. [CrossRef]
82. Zhao, S.; Peng, J.; Ge, R.; Yang, K.; Wu, S.; Qian, Y.; Xu, T.; Gao, J.; Chen, Y.; Sun, Z. Poisoning and regeneration of commercial V_2O_5-WO_3/TiO_2 selective catalytic reduction (SCR) catalyst in coal-fired power plants. *Process Saf. Environ. Prot.* **2022**, *168*, 971–992. [CrossRef]

Disclaimer/Publisher's Note: The statements, opinions and data contained in all publications are solely those of the individual author(s) and contributor(s) and not of MDPI and/or the editor(s). MDPI and/or the editor(s) disclaim responsibility for any injury to people or property resulting from any ideas, methods, instructions or products referred to in the content.

Article

Selective Alkylation of Benzene with Methanol to Toluene and Xylene over H-ZSM-5 Zeolites: Impact of Framework Al Spatial Distribution

Shu Ren [1,2], Fan Yang [2], Chao Tian [1], Yinghong Yue [1], Wei Zou [2], Weiming Hua [1,*] and Zi Gao [1]

[1] Shanghai Key Laboratory of Molecular Catalysis and Innovative Materials, Department of Chemistry, Fudan University, Shanghai 200438, China; 19110220067@fudan.edu.cn (S.R.); 21110220105@m.fudan.edu.cn (C.T.); yhyue@fudan.edu.cn (Y.Y.); zigao@fudan.edu.cn (Z.G.)

[2] State Key Laboratory of Green Chemical Engineering and Industrial Catalysis, SINOPEC Shanghai Research Institute of Petrochemical Technology Co., Ltd., Shanghai 201208, China; yangfan.sshy@sinopec.com (F.Y.); zouw.sshy@sinopec.com (W.Z.)

* Correspondence: wmhua@fudan.edu.cn; Tel.: +86-21-31249121

Abstract: The alkylation of benzene with methanol can effectively generate high-value-added toluene and xylene out of surplus benzene, which is now achieved primarily using solid acids like H-ZSM-5 zeolites as catalysts. In this work, two H-ZSM-5 samples with distinct framework aluminum (Al_F) distributions, but otherwise quite similar textural and acidic properties, have been prepared by employing tetrapropylammonium hydroxide (TPAOH) and *n*-butylamine (NBA) as organic structure-directing agents (OSDAs). Systematic investigations demonstrate that Al_F is preferentially located at the intersections in MFI topology when TPAOH is adopted. In contrast, less Al_F is positioned therein as NBA is utilized. Density functional theory (DFT) calculations reveal that the transition-state complexes cannot be formed in the straight and sinusoidal channels due to their much smaller sizes than the dynamic diameters of transition states, whereas there are adequate spaces for the formation of transition states at the intersections. Benefitting from abundant Al_F at the intersections, which provides more acid sites therein, H-ZSM-5 synthesized from TPAOH is more active relative to the counterpart obtained from NBA. At a WHSV of 4 h^{-1} and 400 °C, the former catalyst gives a 52.8% conversion, while the latter one affords a 45.9% conversion. Both catalysts display close total selectivity towards toluene and xylene (ca. 84%). This study provides an efficient way to regulate the distribution of acid sites, thereby enhancing the catalytic performance of H-ZSM-5 zeolite in the titled reaction.

Keywords: benzene alkylation; H-ZSM-5 zeolite; Al spatial distribution; DFT calculation

1. Introduction

Benzene, toluene and xylene are very important chemical raw materials, which are extensively utilized in the production of organic solvents, gasoline octane number adjusting additives, vitamins and drugs [1–3]. In recent years, with the upgrading of gasoline quality, benzene, which was previously allowed to add into gasoline, has been greatly limited, resulting in a severe surplus, and concurrently, its price has witnessed a steep drop. However, the added values of toluene and xylene, as homologues of benzene, have kept growing steadily because of their reduced toxicity and wide usages in the manufacture of chemical intermediates, fine chemicals and polymers. Hence, the alkylation of benzene with methanol is a promising way to make better use of the surplus benzene, which has advantages such as mild operation conditions and an abundant source of methanol. This process can also seek a new way out for the current aromatic market [4–12].

It is generally accepted that benzene alkylation with methanol is catalyzed by Brønsted acid sites (BASs) from solid acid catalysts [13,14]; particularly, H-ZSM-5 zeolites are considered an excellent catalyst, owing to its unique acidity and pore structure. Previous research

has confirmed that the reaction can proceed via two mechanisms, namely the stepwise mechanism and the concert mechanism, according to both experimental and theoretical work [15,16]. In the stepwise mechanism (Figure S1), methanol is firstly adsorbed on a BAS to be protonated into $CH_3OH_2^+$, which is further dehydrated to obtain a methyl moiety bounced on the framework oxygen of the zeolite, i.e., the formation of surface-bonded methoxy group from methanol adsorption [17–20]. A benzene molecule then enters into the zeolite channel, which is adsorbed neighboring to the methoxy group. Thereafter, a methyl-benzene complex regarded as the transition state is encountered for the addition of the methoxy group to benzene [14,18]. A protonated toluene cation is subsequently attained when the energy barrier is overcome, whose back-donating a proton to the zeolite framework brings about the generation of a toluene molecule. For the concerted pathway (Figure S1), the dehydration of $CH_3OH_2^+$ and the attack from the methyl group on the benzene molecule simultaneously happen, leading to a transition state in the form of H_2O-CH_3-C_6H_6, while other elementary reactions are close to the stepwise ones [14,18]. It is rather difficult to quantitatively distinguish the relative contributions of two mechanisms; consequently, the real reaction is possibly a combination of the above-mentioned two pathways. Apart from the main reaction, side reactions, including methanol to olefins (MTO), isomerization and methyl-transfer reaction, will also exist in the system, particularly for MTO, which gives rise to the generation of C_1–C_5 hydrocarbons and whose extent can be enhanced by higher acid density and sluggish diffusion [21].

It is generally considered that low acid density, hierarchical pores and small particle size of H-ZSM-5 zeolites are conducive to avoiding MTO reactions, reducing the by-product ethylbenzene (as a product via the alkylation of benzene with ethylene) and improving catalytic activity [8,22–24]. By far, properties including but not limited to morphology, Si/Al ratio and structure (core-shell, etc.) of the H-ZSM-5 zeolite have been deliberately explored to establish their relationships with the catalytic performances [8,25,26], yet there are other parameters of the H-ZSM-5 zeolite (surface area, crystal size, Al spatial distribution) that could exert significant influences on the catalytic behavior of H-ZSM-5 in the alkylation of benzene with methanol.

Numerous studies have shown that the distribution of Al atoms in the zeolite framework is not random but determined by the interaction between organic structural-directing agents (OSDAs) and zeolites [27–30]. Al sitting variations can be achieved by changing OSDAs, altering the gel composition, and adjusting the feeding sequence, etc. [30–36], which then determine the reaction path or carbon deposition rate, finally affecting the catalytic performance. Accordingly, Brønsted acid sites derived from framework aluminum (Al_F) of H-ZSM-5 can be divided into three types according to their exact locations, i.e., in the straight channels, sinusoidal channels and intersections. Although these three types of BASs can all be deemed as active sites, the adsorption enthalpy and entropy may be rather distinct due to the steric hindrances considering the different sizes between channels and intersections (5.1 × 5.5 Å for sinusoidal channels, 5.3 × 5.6 Å for straight channels and ca. 9 Å for intersection cavity). In particular, the steric constraints of bulky transition states can be more prominent for those acid sites located in the straight and sinusoidal channels. In consequence, previous reports have demonstrated significant differences in the catalytic performances of ethane and ethylene aromatization, MTO and 1-octene cracking reactions catalyzed by H-ZSM-5 with Al_F concentrated in intersections or channels [37–39]. Wang et al. found that the location of Al_F in H-ZSM-5 could influence catalytic performance in the alkylation of benzene with methanol, and they attributed this phenomenon to the co-adsorption effect of benzene and methanol when Al_F was located at intersections [40]. However, the chemical environment of Al_F at different locations is also believed to be capable of altering the host–guest interaction between the transition state and the H-ZSM-5 zeolite framework, thus determining catalytic properties. On the one hand, the implementation of the above catalyst preparation will often introduce other variables aside from the aluminum location, thereby affecting the accuracy of the results. On the other hand, the function of Al sittings on the catalytic performance of H-ZSM-5 in the alkylation of benzene with methanol is still far from sufficient understanding.

In this work, to study the influence of Al_F distribution on benzene alkylation with methanol from a mechanistic perspective, two H-ZSM-5 samples possessing very similar textural and acidic properties were synthesized utilizing tetrapropylammonium hydroxide and *n*-butylamine as OSDAs (i.e., templates). This work unveils that the acid sites (derived from Al_F) located in the straight and sinusoidal channels are not as active as those at the intersections due to a confined effect, i.e., the smaller spaces of the channels on the transition states as demonstrated by DFT calculations, which provides a new insight into the structure–activity relationship on the titled reaction.

2. Results and Discussion

2.1. Structural and Textural Properties

As shown in Figure 1, XRD patterns of both Z5-NBA and Z5-TPA exhibit typical diffraction peaks at 8.0°, 8.9°, 9.1°, 23.1°, 23.3°, 23.7°, 24.0° and 24.4° out of the (101), (020), (111), (332), (051), (151), (303) and (133) crystal planes in MFI topology (PDF #44-0003), respectively, indicating the successful formation of the H-ZSM-5 zeolite without detectable impurities [41,42]. The relative crystallinity was calculated by integrating the areas of diffraction peaks within 22.5−25°, and the sum of the areas for Z5-NBA was set as 100% for reference. As listed in Table 1, the XRD crystallinity of Z5-TPA and Z5-NBA was 105% and 100%, respectively. The close crystallinity suggests that both TPA$^+$ and NBA OSDAs are capable of generating H-ZSM-5 zeolites with high crystallinity.

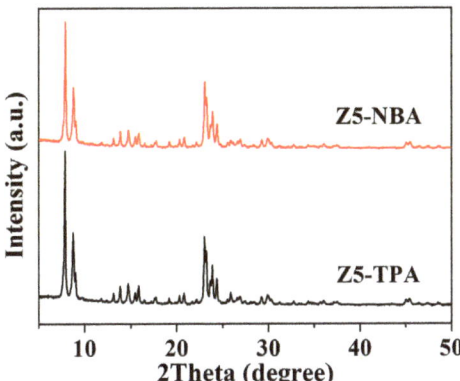

Figure 1. XRD patterns of Z5-TPA and Z5-NBA.

Table 1. Textural properties of Z5-TPA and Z5-NBA.

Sample	Si/Al Ratio [a]	Crystallinity (%)	Surface Area (m^2/g)		Pore Volume (cm^3/g)	
			Total	Micro [b]	Micro [b]	Meso [c]
Z5-TPA	151	105	398	361	0.17	0.10
Z5-NBA	149	100	393	358	0.17	0.10

[a] Determined by XRF; [b] Micropore volume calculated by the *t*-plot method; [c] Mesopore volume.

The SEM images for Z5-TPA and Z5-NBA are displayed in Figure 2. Both Z5-TPA and Z5-NBA zeolites are composed of spherical particles with diameters of around 3 μm for Z5-TPA and 2.6 μm for Z5-NBA (Figure 2a,c). The high-resolution micrographs (Figure 2b,d) evidence that the two samples are constructed by strip-like crystals of ~350 nm in length and ~100 nm in width for Z5-TPA and ~460 nm in length and ~120 nm in width for Z5-NBA with smooth surfaces. The SEM result demonstrates that both Z5-TPA and Z5-NBA zeolites have the same morphology and similar particle and crystal sizes.

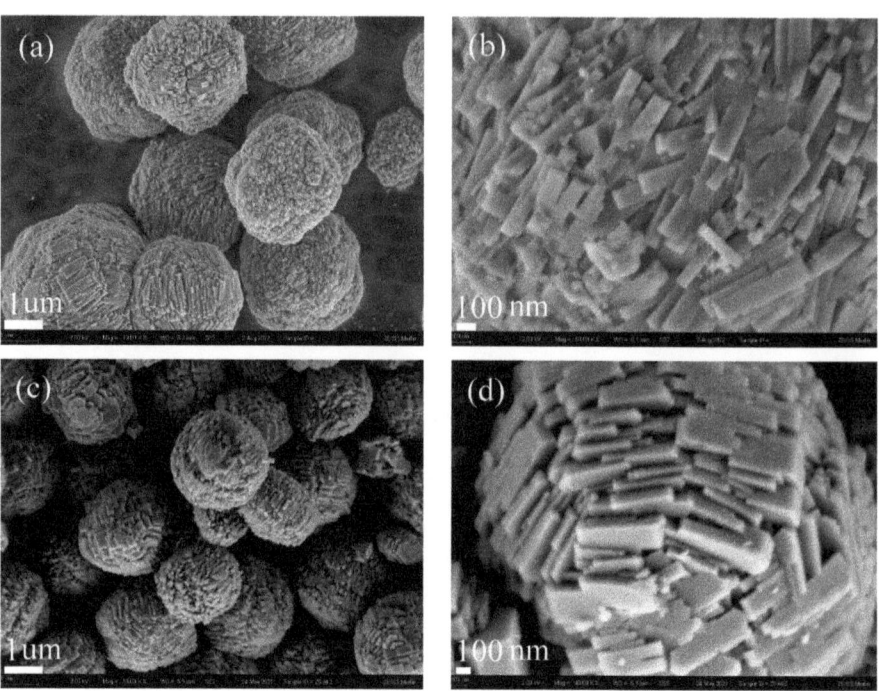

Figure 2. SEM graphs of Z5-TPA (**a,b**) and Z5-NBA (**c,d**).

As shown in Table 1, both Z5-TPA and Z5-NBA samples possess similar BET surface areas (ca. 395 m^2/g) and micropore surface areas (ca. 360 m^2/g). The total pore volumes (0.27 cm^3/g) and micropore volumes (0.17 cm^3/g) are the same for both samples. The Si/Al molar ratios measured by XRF are close for both zeolites (151 vs. 149), which are equivalent to those of the initial gels (150). These observations confirm that both Z5-TPA and Z5-NBA display very similar textural properties and chemical compositions.

2.2. Acidic Properties

Acidity is an important property of zeolites, which is directly related to the Si/Al ratio. The surface acidity of H-ZSM-5 samples was determined by NH$_3$-TPD (Figure 3). As shown on the NH$_3$-TPD curves, both Z5-TPA and Z5-NBA have two distinctive desorption peaks at 185 °C and 376 °C, which are attributed to the desorption of NH$_3$ molecules that interact with the weak acids and strong acids, respectively [43,44]. Judging from the peak temperature, the acid strength of Z5-TPA is the same as that of Z5-NBA. The quantitative results listed in Table 2 indicate that both Z5-TPA and Z5-NBA possess quite similar amounts of weak acid sites (59 vs. 60 µmol/g), strong acid sites (70 vs. 70 µmol/g) and total acid sites (129 vs. 130 µmol/g). This finding is a consequence of the close Si/Al ratio for two samples.

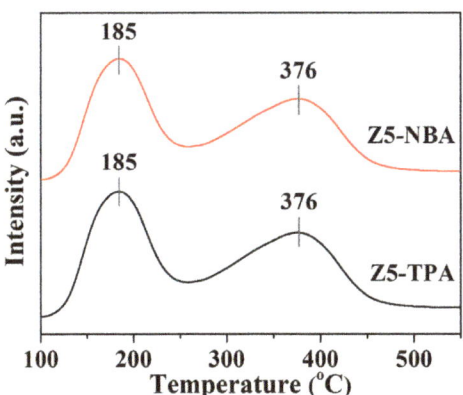

Figure 3. NH$_3$-TPD profiles of Z5-TPA and Z5-NBA.

Table 2. Acidic properties of Z5-TPA and Z5-NBA.

Sample	Acidity by NH$_3$-TPD (μmol/g)			Acidity by Py-IR (μmol/g)			BAS Distribution (%) [a]			Conv. of Cumene Cracking (%)
	Weak	Strong	Total	Brønsted	Lewis	Total	Weak	Medium	Strong	
Z5-TPA	59	70	129	78	17	95	22	31	47	33.9
Z5-NBA	60	70	130	78	17	95	22	31	47	33.6

[a] The acid strength distribution of Brønsted acid sites (BASs) was estimated from Py-IR after the evacuation at different temperatures; the difference in BAS between 200 and 300 °C, the difference in BAS between 300 and 400 °C and the remaining BAS at 400 °C correspond to weak, medium and strong acid sites, respectively.

Considering that NH$_3$-TPD cannot distinguish the type of acid sites, Py-IR experiments were also carried out to determine the amounts of Brønsted acid sites (BASs) and Lewis acid sites (LASs), as well as the distribution of Brønsted acid strength. On the Py-IR spectra (Figure 4), the peak at 1455 cm^{-1} is attributed to pyridine adsorbed on LASs, whereas the 1541 cm^{-1} band is assigned to pyridine adsorbed on BASs [45,46]. Furthermore, the peak at 1489 cm^{-1} is caused by pyridine adsorbed on both BASs and LASs [47]. The quantitative results are summarized in Table 2, in which the total acidity is measured at 200 °C. For the two samples, the amounts of BAS and LAS are the same. Brønsted acid sites account for the majority of the overall acid sites with a ratio of Brønsted acidity to Lewis acidity of 4.6. Moreover, the proportions of weak, medium and strong BAS amounts among the total Brønsted acidity are the same for both Z5-TPA and Z5-NBA samples, indicating that they have the same distribution across the Brønsted acid strength.

Cumene cracking is a typical reaction catalyzed by Brønsted acid sites [48,49]. As shown in Table 2, the cumene conversion over the two samples is very close, which is 33.9% for Z5-TPA and 33.6% for Z5-NBA. This observation demonstrates very similar Brønsted acidity of the two samples, which is consistent with the result of Py-IR. The above results demonstrate that both Z5-TPA and Z5-NBA display very close acidic properties, including the acid amount, acid strength, acid type and the strength distribution of BAS.

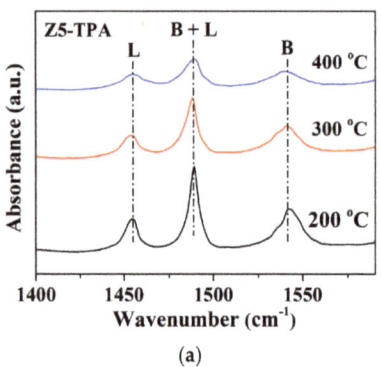
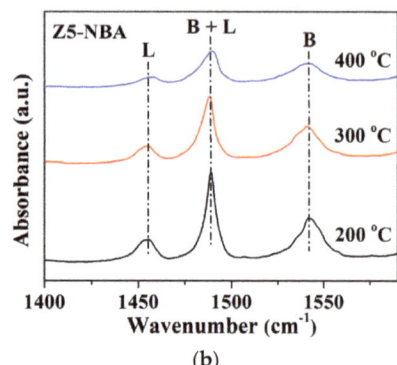

Figure 4. Py-IR spectra of (**a**) Z5-TPA and (**b**) Z5-NBA after desorption at 200 °C, 300 °C and 400 °C.

2.3. ^{29}Si and ^{27}Al MAS NMR, Constraint Index

As the textural properties, composition, morphology and acidic properties of Z5-NBA and Z5-TPA are almost identical, solid-state MAS NMR measurements were carried out to investigate the chemical environments of Si and Al. The ^{29}Si MAS NMR spectra illustrated in Figure 5 show closely similar features, and four sub-peaks are fitted from the origin curve at approximately −103, −107, −113 and −116 ppm, respectively. Two peaks of −113 and −116 ppm are attributed to symmetric and slightly asymmetric $Q^4(0Al) = Si(OSi)_4$-type silicon connecting to four Si tetrahedrons in the H-ZSM-5 framework [37,50]. The peak at −107 ppm in the fitting curve is assigned to $Q^4(1Al) = Si(OSi)_3(OAl)$-type silicon linking to one Al tetrahedron and three Si tetrahedrons in the H-ZSM-5 framework [51–53]. The peak centered at −103 ppm is associated with $Q^3(0Al) = Si(OSi)_3(OH)$-type silicon connecting to three Si tetrahedrons in the zeolite framework and one hydroxyl, which is located on the surface of H-ZSM-5 [27,37,54]. The quantitative results given in Table 3 show close relative proportions of various Si species for Z5-TPA and Z5-NBA, revealing that both zeolites have quite similar SiO_4 environments.

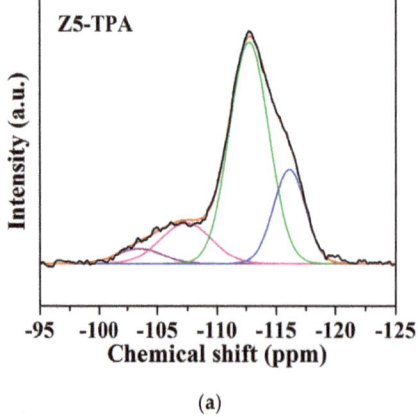
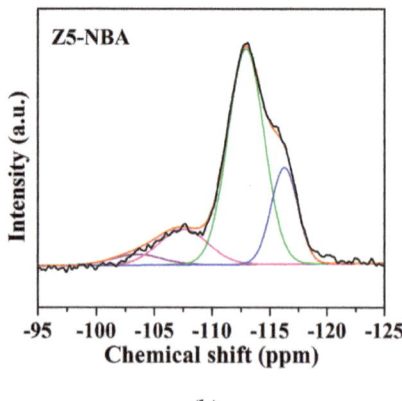

Figure 5. ^{29}Si MAS NMR spectra of (**a**) Z5-TPA and (**b**) Z5-NBA.

Table 3. Deconvolution results of the ^{29}Si MAS NMR spectra for Z5-TPA and Z5-NBA based on the normalized peak areas of different Si species.

Sample	Proportion of Various Peaks (%)			
	Q^3(0Al), −103 ppm	Q^4(1Al), −107 ppm	Q^4(0Al), −113 ppm	Q^4(0Al), −116 ppm
Z5-TPA	4.7	14.4	60.1	20.8
Z5-NBA	4.6	14.5	60.4	20.5

In the ^{27}Al MAS NMR spectra, as shown in Figure 6a, both Z5-TPA and Z5-NBA samples exhibit one intense resonance peak at 55 ppm corresponding to tetrahedrally coordinated Al species, while the resonance signal at 0 ppm attributed to octahedrally coordinated extra-framework Al species is almost invisible, corroborating that the great majority of Al atoms have been incorporated into the H-ZSM-5 zeolite framework [37,55–58]. The ^{27}Al MAS NMR spectrum at 55 ppm is deconvolved into five peaks located at 48, 52, 55, 57 and 60 ppm, respectively, as illustrated in Figure 6b,c. The peaks centered at 48, 52 and 55 ppm can be assigned to the framework Al atoms at the intersections, whereas the peaks at 57 and 60 ppm correspond to framework Al atoms in the sinusoidal and straight channels of the H-ZSM-5 zeolite, respectively [38]. The framework aluminum distribution was then calculated based on the proportion of various peak areas, and the results are presented in Table 4. The relative content of Al at the intersections is obviously higher for Z5-TPA (86.4%) than Z5-NBA (69.0%), while the former sample has noticeably lower content of Al in the sinusoidal channels than the latter one (9.9% vs. 25.7%). This result suggests that Al_F is more favored to be positioned at the intersections when TPA$^+$ was employed as OSDA.

Table 4. Aluminum distribution obtained from the curve fitting of ^{27}Al MAS NMR spectra and CI values for C_6 paraffins cracking over Z5-TPA and Z5-NBA.

Sample	Aluminum Distribution (%)			CI [a]
	Straight	Sinusoidal	Intersection	
Z5-TPA	3.7	9.9	86.4	3.4
Z5-NBA	5.3	25.7	69.0	5.0

[a] Constraint index.

Notably, during the synthesis of aluminosilicate zeolites, the isomorphic substitution of framework Si^{4+} by Al^{3+} leads to the formation of negative charges, which must be balanced with positively charged species; these species could be OSDAs (e.g., TPA$^+$, hydrolyzed and dissociated amines), extra-framework Al species (e.g., AlO$^+$, Al(OH)$^{2+}$) and inorganic cations (e.g., Na$^+$, K$^+$) [28]. To guarantee a neutral framework and the continuation of crystallization, the framework Al atoms will be located near positions that are more suitable for the accommodation of the above-mentioned cations. In the classical crystallization mechanism of the ZSM-5 zeolite, when the large-sized TPA$^+$ was used as OSDA, it can only be located at the intersections with larger void spaces [37,59], which leads to the enrichment of Al_F at the intersections. Z5-NBA is obtained with the assistance of NBA and Na$^+$, both of which are small-sized and will be randomly distributed within the MFI framework during the crystallization process, resulting in less framework Al atoms distributed at the intersections than Z5-TPA. The difference in Al_F location is further demonstrated by the following constraint index.

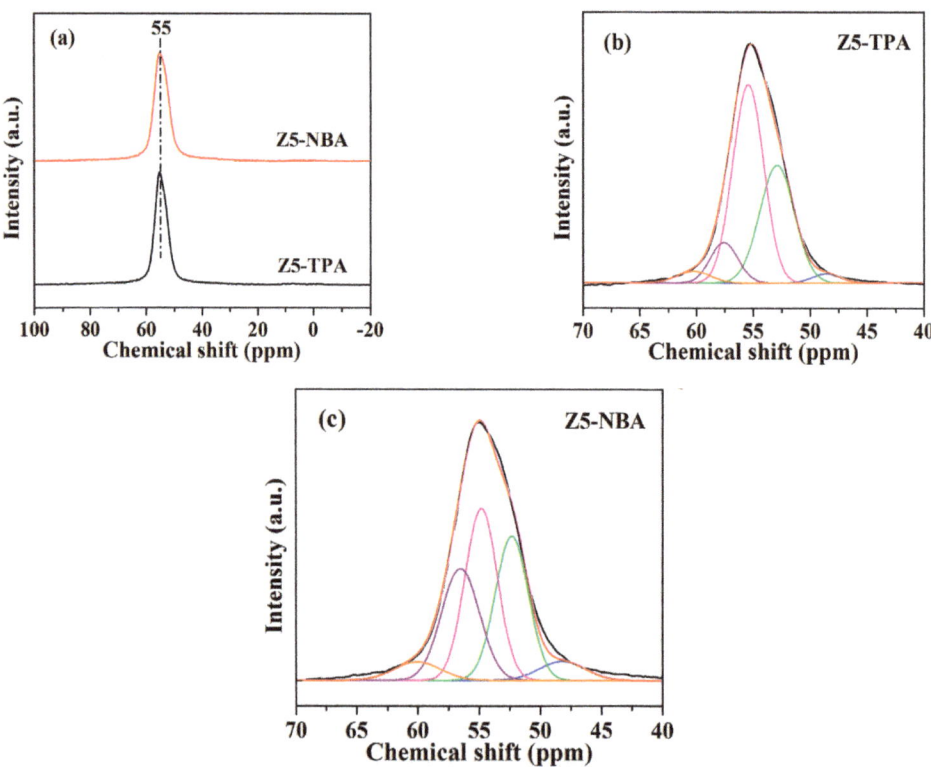

Figure 6. ^{27}Al MAS NMR spectra of Z5-TPA and Z5-NBA (a) and their corresponding curve fittings of Z5-TPA (b) and Z5-NBA (c).

The constraint index (CI) is used to estimate the distribution of acid sites derived from the framework Al atoms. The CI value is determined from the cleavage rate ratio of *n*-hexane to 3-methylpentane, which was measured at a conversion rate below 15%. Considering that both Z5-TPA and Z5-NBA zeolites have an identical MFI structure and quite similar textural and acidic properties, the variation in the CI value can be attributed to the difference in the distribution of framework Al atoms. The CI values of H-ZSM-5 zeolites synthesized with different OSDAs are shown in Table 4. The CI value of Z5-NBA is higher than that of Z5-TPA (5.0 vs. 3.4). This result suggests that the amount of acid sites at the intersections is greater for Z5-TPA than Z5-NBA, since the cracking reaction of 3-methylpentane proceeds faster at the intersections than in the straight and sinusoidal channels due to the larger spaces of intersections that can accommodate the bulky bimolecular transition state [28]. Combined with the ^{27}Al NMR result, it is clear that the preferential positioning of TPA$^+$ at the intersections causes more framework Al atoms to sit therein, leading to more acid sites concentrated at the intersections. Compared to Z5-TPA, the Z5-NBA sample attained from NBA and Na$^+$ displays less framework Al atoms and acid sites at the intersections, i.e., more framework Al atoms and acid sites in the straight and sinusoidal channels.

2.4. Catalytic Performance

The selective alkylation of benzene with methanol to toluene and xylene is catalyzed by Brønsted acid sites of the zeolites derived from Al$_F$ [13,14]. We compared catalytic behavior of Z5-TPA and Z5-NBA to investigate the effect of Al$_F$ distribution. The reaction data after 4 h on stream given in Table 5 shows that the total selectivity towards toluene and xylene is very close for the two samples, which is 84.4% and 84.0% for Z5-TPA and Z5-NBA, respectively. However, the benzene conversion of Z5-TPA (52.8%) is noticeably higher than that of Z5-NBA (45.9%), and a higher total yield of toluene and xylene is also observed over Z5-TPA (44.6%) compared to Z5-NBA (38.6%). In terms of reaction rate, Z5-TPA displays a much higher rate than Z5-NBA (2445 vs. 1510 mmol/(g h)). The catalytic activity strongly depends on Brønsted acid properties of the zeolites, i.e., amount, strength and distribution [60]. Larger surface area provides more accessible acid sites in the zeolites, resulting in higher conversion of benzene. Hu et al. found that introducing mesopores in H-ZSM-5 could improve benzene conversion, due to enhanced diffusion of reactants and products, as well as easier access to the active sites in micropores [23]. The aforementioned characterization results reveal that both Z5-TPA and Z5-NBA zeolites have the same mesopores, very similar surface areas and very close acidic properties, including amount, strength, and strength distribution of the BAS. Therefore, the difference in catalytic activity should be attributed to the distribution of acid sites, i.e., the Al_F distribution.

Table 5. Reaction data of the Z5-TPA and Z5-NBA catalysts at a WHSV of 4 h^{-1} and 400 °C.

Catalyst	Conversion (%)	Selectivity (%) [a]						S_{T+X} [b] (%)	Y_{T+X} [c] (%)	Rate [d] (mmol/(g h))
		T	PX	MX	OX	EB	C_{9+}			
Z5-TPA	52.8	52.4	8.7	16.5	6.8	10.5	5.1	84.4	44.6	2445
Z5-NBA	45.9	48.5	9.2	18.1	8.2	7.3	8.7	84.0	38.6	1510

[a] T: toluene; PX: *p*-xylene; MX: *m*-xylene; OX: *o*-xylene; EB: ethylbenzene; C_{9+}: trimethylbenzene and higher alkyl aromatics. [b] S_{T+X} refers to the total selectivity of toluene and xylene. [c] Y_{T+X} refers to the total yield of toluene and xylene. [d] Millimoles of benzene converted per gram of catalyst per hour at 400 °C obtained at a benzene conversion below 10%.

The apparent activation energies in the temperature range of 375–450 °C were measured for Z5-TPA and Z5-NBA. As illustrated in Figure 7, both zeolites have equivalent activation energies (90 vs. 89 kJ/mol), suggesting that the reaction mechanism on these catalysts is the same.

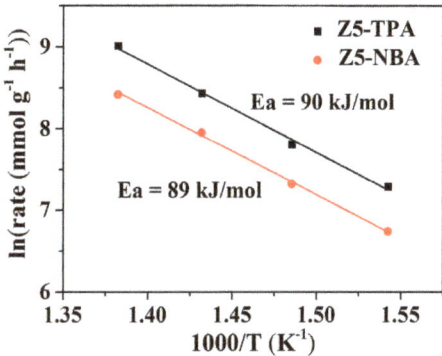

Figure 7. Arrhenius plot of the reaction rate on Z5-TPA and Z5-NBA.

Spectroscopic studies have verified that methoxy groups may indeed be formed from methanol, which is in accordance with a stepwise mechanism [15,60]. On the other hand, some reports favor the concerted mechanism [61,62]. Quantum chemical calculations have demonstrated the possibility of both mechanisms [16]. We think that both routes may exist during the reaction.

2.5. Periodic Density Functional Study

According to the above analysis, the only difference between Z5-NBA and Z5-TPA lies in the location of Al_F, which determines the sittings of active centers, i.e., the acid sites. To understand the link between the Al_F location and reaction mechanism, a periodic density functional study was conducted using the Vienna Ab-initio Simulation Package (VASP 4.6).

Firstly, the small size of NBA makes straight channels, sinusoidal channels and channel intersections all feasible locations for its existence, and the adsorption energies of NBA at the three locations are all close to −1.47 eV. Both the configurations and energies for NBA in ZSM-5 concrete the regular dispersion of NBA in ZSM-5; meanwhile, given that NBA can be positively charged after hydrolyzation and dissociation, a more even distribution of Al_F at the three locations can be anticipated. More importantly, the addition of Na^+ cations, which are randomly distributed in the MFI framework, also contributes significantly to the uniform dispersion of Al_F in Z5-NBA [37]. Different from NBA, it has been found that TPA^+ could only be resided at the intersections due to its large size, with an adsorption energy of −4.18 eV. Considering the fact that no other cations can be found in the synthesis of Z5-TPA, Al_F will be directed into the channel intersections of ZSM-5 zeolite via the charge interaction between TPA^+ and AlO_4^-. The simulations are consistent with the ^{27}Al MAS NMR and CI results, which further corroborates that the Al_F is more concentrated at the intersections for Z5-TPA than Z5-NBA.

Since the acid sites come from the compensating protons adjacent to Al_F, various Al_F locations bring about simultaneous changes in the coordinates of acid sites. Specifically, the acid sites in the straight/sinusoidal channels face confined spaces, yet the acid sites in the cross-sections (i.e., channel intersections) hold more spacious room. As mentioned in the introduction, two possible mechanisms are involved in the alkylation of benzene with methanol, namely the concerted pathway and the stepwise pathway, and the largest intermediates in the two processes are believed to be transition-state complexes before the generation of protonated toluene cations. Theoretically, the configurations of transition-state complexes could exist in the channels or intersections; however, unluckily, only those at the intersections can be found after searching for a series of potential models no matter how these precursors are placed, and to the best of our knowledge, no evidence has been found to support the formation of transition states in the straight and sinusoidal channels. In the determined configuration of the concerted pathway, an oxygen atom in the water, a carbon atom in the methyl and a carbon atom in the benzene are aligned almost linearly, with the angle of 171.8°, C-O length of 2.2 Å and C-C length of 2.1 Å (Figure 8). Similarly, in the determined configuration of the stepwise pathway, a framework oxygen atom, a carbon atom in the methyl and a carbon atom in the benzene are also linearly arranged, with C-O distance of 2.2 Å, C-C distance of 2.1 Å and O-C-C angle of 178.0° (Figure 9). Accordingly, the dynamic diameters of the two transition-state complexes are calculated using the method put forward by Mehio et al. [63], which is 7.3 Å for the stepwise pathway and 9.1 Å for concerted pathway. Obviously, the kinetic diameters are much larger than the sizes of ZSM-5 apertures (5.1 × 5.5 Å for sinusoidal channels and 5.3 × 5.6 Å for straight channels). Consequently, there are insufficient spaces for the formation of transition states in the straight and sinusoidal channels, which hinders the proceeding of the alkylation reaction.

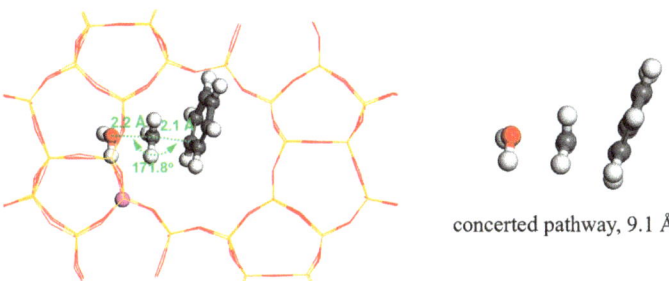

Figure 8. The transition-state complex in the optimized geometry by the concerted mechanism (red, oxygen; purple, aluminum; white, hydrogen; gray, carbon; similarly hereinafter).

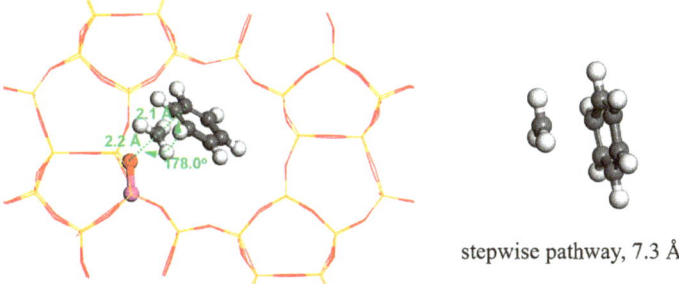

Figure 9. The transition-state complex in the optimized geometry by the stepwise mechanism.

Liu et al. [37] reported that Pt-modified ZSM-5 with acid sites mainly located in the intersections tended to produce more aromatic compounds for ethane aromatization than a catalyst with acid sites located in the straight and/or sinusoidal channels. The reason for this is that acid sites located in the intersections with more space void were conducive for the geometry transformation of intermediates. Zhu et al. [40] studied the effect of Al_F distribution on the catalytic performance in the alkylation of benzene with methanol. The results demonstrated that H-ZSM-5 with more Al_F in the intersections can improve benzene conversion due to the larger space and co-adsorption effect. Taken everything into consideration, in the synthesis of ZSM-5, TPA^+ will be only located at the intersections, which causes the preferential enrichment of Al_F as well as acid sites therein. As there are larger spaces near to the acid sites at the intersections, the alkylation of benzene with methanol could progress smoothly, pertaining to the minimized steric hindrances for the large transition-state complexes. Compared to TPA^+, the engagement of NBA and Na^+ induces more acid sites present in the straight and sinusoidal channels. However, these channel acid sites are incapable of catalyzing the alkylation reaction because of the narrow spaces which could not hold the large transition-state complexes. Resultantly, Z5-TPA displays higher activity than Z5-NBA in the alkylation of benzene with methanol attributing to the rational Al_F locations.

3. Materials and Methods

3.1. Synthesis of H-ZSM-5 Zeolites

Two ZSM-5 zeolites were synthesized using the procedures described by Liu et al. with minor modifications [37]. Typically, silica sol (SiO_2, 40 wt%) was added slowly to $Al(NO_3)_3$ aqueous solution, and stirred for 1 h. Then, tetrapropylammonium hydroxide (TPAOH, 40 wt%) was added, followed by stirring for 4 h to form a homogeneous suspension with a molar composition of 1.0 SiO_2:0.00333 Al_2O_3:0.5 TPAOH:50 H_2O. The suspension was transferred into a Teflon-lined stainless-steel autoclave and statically crystallized at 170 °C for 120 h, followed by centrifuging, rinsing, drying overnight at 120 °C and calcining at 550 °C in air for 5 h.

Another ZSM-5 zeolite was prepared as follows: silica sol was added slowly to the mixed aqueous solution containing Al(NO$_3$)$_3$ and NaOH. Then, n-butylamine (NBA) was added, followed by stirring for 4 h to obtain a gel with a molar composition of 1.0 SiO$_2$:0.00333 Al$_2$O$_3$:0.98 NBA:0.08 Na$_2$O:10 H$_2$O. The prepared mother gel was transferred into a Teflon-lined stainless-steel autoclave and crystallized at 170 °C for 120 h, followed by centrifuging, rinsing, drying overnight at 120 °C and calcining at 550 °C in air for 5 h.

The NH$_4$-type zeolites were obtained by ion-exchanging the above ZSM-5 zeolites with 1 M NH$_4$NO$_3$ solution three times with a solution/zeolite ratio of 10 mL/g at 90 °C for a total of 9 h. The H-type ZSM-5 samples were acquired by calcination of the NH$_4$-type zeolites at 450 °C for 5 h. The resulting H-ZSM-5 catalysts were denoted as Z5-TPA and Z5-NBA, respectively.

3.2. Characterization of H-ZSM-5 Zeolites

Zeolite topologies were determined from powder X-ray diffraction (XRD) patterns on a D8 Advance X-ray diffractometer (Brucker, Madison, WI, USA) with a Cu Kα radiation source at 40 kV and 50 mA. Scanning electron microscopy (SEM) images of the samples were observed on a Zeiss Merlin scanning electron microscope (Merlin, Carl Zeiss AG, Oberkochen, Germany). The Si/Al molar ratios of H-ZSM-5 zeolites were determined by means of an S4 Pioneer X-ray fluorescence (XRF) spectrometer (Bruker, Rheinstetten, Germany). The surface areas and pore volumes of H-ZSM-5 zeolites were measured by a Micromeritics TriStar 3000 automatic absorption instrument (Micromeritics, Atlanta, GA, USA). Magic-angle-spinning nuclear-magnetic-resonance (MAS NMR) spectra of ^{27}Al and ^{29}Si were collected on a Bruker Avance III 600 MHz Wide Bore spectrometer (Bruker, Rheinstetten, Germany). The single-pulse sequence was adopted with a 10° pulse and a delay time of 0.3 s. The chemical shifts for ^{27}Al and ^{29}Si were calibrated by referring to AlCl$_3$ and tetramethylsilane, respectively. The ^{27}Al MAS NMR and ^{29}Si MAS NMR spectra were deconvoluted by using the mixed Gaussian–Lorentzian equation.

The strength and quantity of acid sites present on H-ZSM-5 zeolites were analyzed by temperature-programmed desorption of NH$_3$ (NH$_3$-TPD) on a Micrometritics AutoChem AMI-3300 apparatus (Micromeritics, Atlanta, GA, USA). A total of 0.1 g of sample (40–60 mesh) was loaded into a U-type tube and pretreated at 550 °C for 1 h in a N$_2$ flow. Then, the sample was cooled to 100 °C and exposed to a 10 vol.% NH$_3$/N$_2$ mixture (30 mL/min) for 0.5 h. After purging with helium (30 mL/min) for an additional 2 h to remove the physically adsorbed NH$_3$, the temperature was ramped from 100 °C to 550 °C at a rate of 10 °C/min using helium (30 mL/min) as the carrier gas. The desorbed ammonia was monitored by a thermal conductivity detector. The amounts of Brønsted and Lewis acid sites were determined by Fourier-transform infrared spectroscopy of pyridine adsorption (Py-IR) on a Nicolet iS50 spectrometer (Nicolet, Madison, WI, USA). Prior to each measurement, all samples were pressed into self-support wafers with a diameter of 13 mm and a weight of ca. 15 mg. Then, the sample was degassed at 400 °C for 2 h under vacuum (<10^{-2} Pa) to remove the impurities and adsorbed water, before being cooled to an ambient temperature. Thereafter, pyridine was introduced into the testing cell as saturated vapor for 10 min to allow sufficient adsorption, after which the cell was evacuated again for 20 min at 200 °C, 300 °C and 400 °C. The spectra were recorded using the background taken at the same temperature. Brønsted and Lewis acidities were quantified from the integrated areas of Py-IR bands at ca. 1540 and 1450 cm^{-1}, employing the molar extinction coefficients of 1.67 and 2.22 cm/µmol, respectively [64,65].

3.3. Computational Method

Density functional theory (DFT) calculations were carried out using Vienna Ab Initio Simulation Package 5.4. Project augmented wave method and Perdew–Burke–Ernzerhof exchange–correlation function were adopted, with a plane wave basis set kinetic energy cut-off of 400 eV. The Brillouin zone sampling was limited to the Γ-point, and the convergence criterion that forces on each atom was smaller than 0.03 eV/Å. The MFI structure containing pure SiO_2 was downloaded from International Zeolite Association, whose lattice parameters were then optimized to acquire a unit cell of a = 20.32 Å, b = 20.16 Å and c = 13.46 Å. A Brønsted acid site was created by replacing a Si (T12) atom with an Al atom, and the neighboring O between Al (T12) and Si (T3) was protonated. The dimer method was utilized to determine the transition states. The obtained configurations were confirmed by the existence of one and the only imaginary frequency. The adsorption energy was calculated by the following equation:

$$\Delta E_{ads} = E_{zeo+temp} - E_{zeo} - E_{temp} \qquad (1)$$

wherein ΔE_{ads} is the adsorption energy, $E_{zeo+temp}$ is the total energy of zeolite and template complexes, and E_{zeo} and E_{temp} are the energies of the zeolite framework and the template, respectively.

3.4. Catalytic Performace Evaluation

3.4.1. Estimation of Constraint Index

A total of 0.2 g of the catalyst (40–60 mesh) was loaded into the constant temperature zone of the fixed bed reactor. Prior to each test, the H-ZSM-5 zeolite was firstly activated at 450 °C for 3 h in a N_2 flow, and the temperature was cooled to 400 °C. Next, a mixed feedstock of n-hexane and 3-methylpentane with a molar ratio of 1:1 was pumped into the reactor using N_2 as the carrier gas to ensure a C_6 paraffin conversion below 15%. The reactor effluent was analyzed online by a gas chromatograph equipped with a flame ionization detector and a HP-PLOT Q capillary column (30 m × 0.32 mm × 20 μm). The constraint index (CI) value was calculated by the following equation [66]:

$$\text{constraint index} = \frac{\log(\text{fraction of n-hexane remaining})}{\log(\text{fraction of 3 - methylpentane remaining})} \qquad (2)$$

3.4.2. Benzene Alkylation with Methanol

The alkylation of benzene with methanol was carried out in a fixed bed reactor. The catalyst (3 g, 20–40 mesh) was loaded into the thermostatic region of the reactor and pretreated at 450 °C for 3 h in a N_2 flow. Then, the temperature was cooled to 400 °C. Next, a mixture of benzene/methanol as the reactant (molar ratio of 1:1) was pumped in the reactor with a co-feed N_2 flow (50 mL/min). The weight hourly space velocity (WHSV) was 4.0 h^{-1}. The products were analyzed offline by a gas chromatograph equipped with a flame ionization detector and a HP-INNOWAX capillary column (50 m × 0.32 mm × 0.5 μm).

3.4.3. Cumene Cracking

The cracking reaction of cumene was used to evaluate Brønsted acidity of H-ZSM-5 zeolites, which was carried at 300 °C in a pulsed microreactor loaded with 0.03 g of the catalyst (40–60 mesh). The catalyst was activated at 450 °C for 3 h in a He flow before reaction. The carrier gas was He (30 mL/min), and the amount of injected cumene was 1 μL.

4. Conclusions

In this work, two micron-sized spherical H-ZSM-5 zeolites with very similar surface areas, SiO_4 environments, acidic properties comprising the acid amount, acid strength, acid type and the strength distribution of BAS, but different Al_F distributions, were synthesized by using TPAOH and NBA as OSDAs, respectively. When the large-sized TPA^+ cations that could only be located at the intersections with larger void spaces are used, Al_F is preferentially concentrated at the intersections. In contrast, the random existence of the small-sized NBA and Na^+ in the MFI framework during the crystallization process leads to less Al_F

distributed at the intersections. DFT calculations unveil that the Al_F at the intersections exerts less steric hindrances for the transition states, whereas there are insufficient spaces for the formation of transition states in the straight and sinusoidal channels, which hinders the proceeding of the alkylation reaction. The Z5-TPA zeolite synthesized from TPAOH gives a 52.8% conversion in benzene alkylation with methanol at a WHSV of 4 h^{-1} and 400 °C, which is more active than the Z5-NBA zeolite synthesized from NBA (45.9% conversion). The comparable total selectivity towards toluene and xylene (ca. 84%) was achieved over both H-ZSM-5 zeolites. The higher activity observed for Z5-TPA than Z5-NBA is attributed to more Al_F distributed at the intersections of the former catalyst, i.e., more Brønsted acid sites located therein. This work not only provides a feasible method to control the Al distribution in the H-ZSM-5 zeolite framework, thus improving the catalytic performance in the titled reaction, but is also helpful in understanding the structure–performance relationship.

Supplementary Materials: The following supporting information can be downloaded at: https://www.mdpi.com/article/10.3390/catal13091295/s1, Figure S1: Schematic representation of the stepwise (middle and bottom row) and concerted mechanism (top row) for benzene alkylation with methanol on H-ZSM-5 zeolite.

Author Contributions: S.R.: investigation, data curation, writing—original draft. F.Y.: methodology, writing—original draft. C.T.: investigation, data curation. Y.Y.: methodology, formal analysis. W.Z.: methodology, formal analysis. W.H.: conceptualization, supervision, writing—review and editing, project administration. Z.G.: validation, formal analysis. All authors have read and agreed to the published version of the manuscript.

Funding: This work was financially supported by the National Natural Science Foundation of China, grant number 22072027, the Science and Technology Commission of Shanghai Municipality, grant number 19DZ2270100 and SINOPEC Shanghai Research Institute of Petrochemical Technology Co., Ltd., grant number 33750000-19-ZC0607-0005.

Data Availability Statement: Not applicable.

Conflicts of Interest: The authors declare no conflict of interest.

References

1. Ahn, J.H.; Kolvenbach, R.; Al-Khattaf, S.S.; Jentys, A.; Lercher, J.A. Methanol usage in toluene methylation with medium and large pore zeolites. *ACS Catal.* **2013**, *3*, 817–825. [CrossRef]
2. Gao, K.; Li, S.; Wang, L.; Wang, W. Study on the alkylation of benzene with methanol to selective formation of toluene and xylene over Co_3O_4-La_2O_3/ZSM-5. *RSC Adv.* **2015**, *5*, 45098–45105. [CrossRef]
3. Liu, C.; Su, J.; Liu, S.; Zhou, H.; Yuan, X.; Ye, Y.; Wang, Y.; Jiao, W.; Zhang, L.; Lu, Y.; et al. Insights into the key factor of zeolite morphology on the selective conversion of syngas to light aromatics over a Cr_2O_3/ZSM-5 catalyst. *ACS Catal.* **2020**, *10*, 15227–15237. [CrossRef]
4. Qian, J.; Xiong, G.; Liu, J.; Liu, C.; Guo, H. A preliminary study on the role of the internal and external surfaces of nano-ZSM-5 zeolite in the alkylation of benzene with methanol. *Ind. Eng. Chem. Res.* **2019**, *58*, 9006–9016. [CrossRef]
5. Wang, X.; Xu, J.; Qi, G.; Li, B.; Wang, C.; Deng, F. Alkylation of benzene with methane over ZnZSM-5 zeolites studied with solid-state NMR spectroscopy. *J. Phys. Chem. C* **2013**, *117*, 4018–4023. [CrossRef]
6. Rakoczy, J.; Romotowski, T. Alkylation of benzene with methanol on zeolites: Infrared spectroscopy studies. *Zeolites* **1993**, *13*, 256–260. [CrossRef]
7. Zhu, Z.; Chen, Q.; Xie, Z.; Yang, W.; Li, C. The roles of acidity and structure of zeolite for catalyzing toluene alkylation with methanol to xylene. *Micropor. Mesopor. Mater.* **2006**, *88*, 16–21. [CrossRef]
8. Hu, H.; Lyu, J.; Rui, J.; Cen, J.; Zhang, Q.; Wang, Q.; Han, W.; Li, X. The effect of Si/Al ratio on the catalytic performance of hierarchical porous ZSM-5 for catalyzing benzene alkylation with methanol. *Catal. Sci. Technol.* **2016**, *6*, 2647–2652. [CrossRef]
9. Hu, H.; Lyu, J.; Cen, J.; Zhang, Q.; Wang, Q.; Han, W.; Rui, J.; Li, X. Promoting effects of MgO and Pd modification on the catalytic performance of hierarchical porous ZSM-5 for catalyzing benzene alkylation with methanol. *RSC Adv.* **2015**, *5*, 63044–63049. [CrossRef]
10. Niziolek, A.M.; Onel, O.; Guzman, Y.A.; Floudas, C.A. Biomass-based production of benzene, toluene, and xylenes via methanol: Process synthesis and deterministic global optimization. *Energy Fuels* **2016**, *30*, 4970–4998. [CrossRef]
11. Zhang, J.; Zhou, A.; Gawande, K.; Li, G.; Shang, S.; Dai, C.; Fan, W.; Han, Y.; Song, C.; Ren, L.; et al. b-Axis-oriented ZSM-5 nanosheets for efficient alkylation of benzene with methanol: Synergy of acid sites and diffusion. *ACS Catal.* **2023**, *13*, 3794–3805. [CrossRef]

12. Ren, S.; Tian, C.; Yue, Y.; Zou, W.; Hua, W.; Gao, Z. Selective alkylation of benzene with methanol to toluene and xylene over sheet-like ZSM-5 with controllable b-oriented length. *Catal. Lett.* **2023**. [CrossRef]
13. Anderson, J.R.; Mole, T.; Christov, V. Mechanism of some conversions over ZSM-5 catalyst. *J. Catal.* **1980**, *61*, 477–484. [CrossRef]
14. Svelle, S.; Visur, M.; Olsbye, U.; Saepurahman; Bjørgen, M. Mechanistic aspects of the zeolite catalyzed methylation of alkenes and aromatics with methanol: A review. *Top. Catal.* **2011**, *54*, 897–906. [CrossRef]
15. Ivanova, I.I.; Corma, A. Surface species formed and their reactivity during the alkylation of toluene by methanol and dimethyl ether on zeolites as determined by in situ ^{13}C MAS NMR. *J. Phys. Chem. B* **1997**, *101*, 547–551. [CrossRef]
16. Vos, A.M.; Nulens, K.H.L.; De Proft, F.; Schoonheydt, R.A.; Geerlings, P. Reactivity descriptors and rate constants for electrophilic aromatic substitution: Acid zeolite catalyzed methylation of benzene and toluene. *J. Phys. Chem. B* **2002**, *106*, 2026–2034. [CrossRef]
17. Wen, Z.; Zhu, H.; Zhu, X. Density functional theory study of the zeolite-catalyzed methylation of benzene with methanol. *Catal. Lett.* **2019**, *150*, 21–30. [CrossRef]
18. Wen, Z.; Yang, D.; He, X.; Li, Y.; Zhu, X. Methylation of benzene with methanol over HZSM-11 and HZSM-5: A density functional theory study. *J. Mol. Catal. A* **2016**, *424*, 351–357. [CrossRef]
19. Maihom, T.; Boekfa, B.; Sirijaraensre, J.; Nanok, T.; Probst, M.; Limtrakul, J. Reaction mechanisms of the methylation of ethene with methanol and dimethyl ether over H-ZSM-5: An ONIOM study. *J. Phys. Chem. C* **2009**, *113*, 6654–6662. [CrossRef]
20. Van der Mynsbrugge, J.; Visur, M.; Olsbye, U.; Beato, P.; Bjørgen, M.; Van Speybroeck, V.; Svelle, S. Methylation of benzene by methanol: Single-site kinetics over H-ZSM-5 and H-beta zeolite catalysts. *J. Catal.* **2012**, *292*, 201–212. [CrossRef]
21. Kaeding, W.W. Conversion of methanol to hydrocarbons III. Methylation, ethylation, and propylation of benzene with methanol. *J. Catal.* **1988**, *114*, 271–276. [CrossRef]
22. Hu, H.; Lyu, J.; Wang, Q.; Zhang, Q.; Cen, J.; Li, X. Alkylation of benzene with methanol over hierarchical porous ZSM-5: Synergy effects of hydrogen atmosphere and zinc modification. *RSC Adv.* **2015**, *5*, 32679–32684. [CrossRef]
23. Hu, H.; Zhang, Q.; Cen, J.; Li, X. Catalytic activity of Pt modified hierarchical ZSM-5 catalysts in benzene alkylation with methanol. *Catal. Lett.* **2015**, *145*, 715–722. [CrossRef]
24. Khare, R.; Bhan, A. Mechanistic studies of methanol-to-hydrocarbons conversion on diffusion-free MFI samples. *J. Catal.* **2015**, *329*, 218–228. [CrossRef]
25. Liu, S.; Zhang, H.; Chen, H.; Chen, Z.; Zhang, L.; Ren, J.; Wen, X.; Yang, Y.; Li, Y. Fabrication of a core–shell MFI@TON material and its enhanced catalytic performance for toluene alkylation. *Catal. Sci. Technol.* **2020**, *10*, 1281–1291. [CrossRef]
26. Jalil, A.A.; Zolkifli, A.S.; Triwahyono, S.; Rahman, A.F.A.; Ghani, N.N.M.; Hamid, M.Y.S.; Mustapha, F.H.; Izan, S.M.; Nabgan, B.; Ripin, A. Altering dendrimer structure of fibrous-silica-HZSM5 for enhanced product selectivity of benzene methylation. *Ind. Eng. Chem. Res.* **2019**, *58*, 553–562. [CrossRef]
27. Magusin, P.C.M.M.; Zorin, V.E.; Aerts, A.; Houssin, C.J.Y.; Yakovlev, A.L.; Martens, J.A.; van Santen, R.A. Template-aluminosilicate structures at the early stages of zeolite ZSM-5 formation. A combined preparative, solid-state NMR, and computational study. *J. Phys. Chem. B* **2005**, *109*, 22767–22774. [CrossRef]
28. Yokoi, T.; Mochizuki, H.; Namba, S.; Kondo, J.N.; Tatsumi, T. Control of the Al distribution in the framework of ZSM-5 zeolite and its evaluation by solid-state NMR technique and catalytic properties. *J. Phys. Chem. C* **2015**, *119*, 15303–15315. [CrossRef]
29. Pashkova, V.; Sklenak, S.; Klein, P.; Urbanova, M.; Dedecek, J. Location of framework Al atoms in the channels of ZSM-5: Effect of the (hydrothermal) synthesis. *Chem. Eur. J.* **2016**, *22*, 3937–3941. [CrossRef]
30. Dedecek, J.; Balgová, V.; Pashkova, V.; Klein, P.; Wichterlová, B. Synthesis of ZSM-5 zeolites with defined distribution of Al atoms in the framework and multinuclear MAS NMR analysis of the control of Al distribution. *Chem. Mater.* **2012**, *24*, 3231–3239. [CrossRef]
31. Gábová, V.; Dědeček, J.; Čejka, J. Control of Al distribution in ZSM-5 by conditions of zeolite synthesis. *Chem. Commun.* **2003**, *39*, 1196–1197. [CrossRef] [PubMed]
32. Liang, T.; Chen, J.; Qin, Z.; Li, J.; Wang, P.; Wang, S.; Wang, G.; Dong, M.; Fan, W.; Wang, J. Conversion of methanol to olefins over H-ZSM-5 zeolite: Reaction pathway is related to the framework aluminum siting. *ACS Catal.* **2016**, *6*, 7311–7325. [CrossRef]
33. Hur, Y.G.; Kester, P.M.; Nimlos, C.T.; Cho, Y.; Miller, J.T.; Gounder, R. Influence of tetrapropylammonium and ethylenediamine structure-directing agents on the framework Al distribution in B-Al-MFI zeolites. *Ind. Eng. Chem. Res.* **2019**, *58*, 11849–11860. [CrossRef]
34. Biligetu, T.; Wang, Y.; Nishitoba, T.; Otomo, R.; Park, S.; Mochizuki, H.; Kondo, J.N.; Tatsumi, T.; Yokoi, T. Al distribution and catalytic performance of ZSM-5 zeolites synthesized with various alcohols. *J. Catal.* **2017**, *353*, 1–10. [CrossRef]
35. Kim, S.; Park, G.; Woo, M.H.; Kwak, G.; Kim, S.K. Control of hierarchical structure and framework-Al distribution of ZSM-5 via adjusting crystallization temperature and their effects on methanol conversion. *ACS Catal.* **2019**, *9*, 2880–2892. [CrossRef]
36. Al-Nahari, S.; Dib, E.; Cammarano, C.; Saint-Germes, E.; Massiot, D.; Sarou-Kanian, V.; Alonso, B. Impact of mineralizing agents on aluminum distribution and acidity of ZSM-5 zeolites. *Angew. Chem. Int. Ed.* **2023**, *62*, e202217992. [CrossRef] [PubMed]
37. Liu, H.; Wang, H.; Xing, A.; Cheng, J. Effect of Al distribution in MFI framework channels on the catalytic performance of ethane and ethylene aromatization. *J. Phys. Chem. C* **2019**, *123*, 15637–15647. [CrossRef]
38. Wang, S.; Wang, P.; Qin, Z.; Chen, Y.; Dong, M.; Li, J.; Zhang, K.; Liu, P.; Wang, J.; Fan, W. Relation of catalytic performance to the aluminum siting of acidic zeolites in the conversion of methanol to olefins, viewed via a comparison between ZSM-5 and ZSM-11. *ACS Catal.* **2018**, *8*, 5485–5505. [CrossRef]

39. Park, S.; Biligetu, T.; Wang, Y.; Nishitoba, T.; Kondo, J.N.; Yokoi, T. Acidic and catalytic properties of ZSM-5 zeolites with different Al distributions. *Catal. Today* **2018**, *303*, 64–70. [CrossRef]
40. Wang, Y.; He, X.; Yang, F.; Su, Z.; Zhu, X. Control of framework aluminum distribution in MFI channels on the catalytic performance in alkylation of benzene with methanol. *Ind. Eng. Chem. Res.* **2020**, *59*, 13420–13427. [CrossRef]
41. Yue, Y.; Gu, L.; Zhou, Y.; Liu, H.; Yuan, P.; Zhu, H.; Bai, Z.; Bao, X. Template-free synthesis and catalytic applications of microporous and hierarchical ZSM-5 zeolites from natural aluminosilicate minerals. *Ind. Eng. Chem. Res.* **2017**, *56*, 10069–10077. [CrossRef]
42. Sun, Y.; Ma, T.; Cao, S.; Wang, J.; Meng, X.; Gong, Y.; Zhang, Z.; Ma, A.; Liu, P. Defective sites in ZSM-5 zeolite synthesized by n-butylamine template facilitating uniform meso-microporosity by alkali-treatment. *Micropor. Mesopor. Mater.* **2021**, *326*, 11360–11368. [CrossRef]
43. Yang, J.; Gong, K.; Miao, D.; Jiao, F.; Pan, X.; Meng, X.; Xiao, F.; Bao, X. Enhanced aromatic selectivity by the sheet-like ZSM-5 in syngas conversion. *J. Energy Chem.* **2019**, *35*, 44–48. [CrossRef]
44. Wu, D.; Yu, X.; Chen, X.; Yu, G.; Zhang, K.; Qiu, M.; Xue, W.; Yang, C.; Liu, Z.; Sun, Y. Morphology-controlled synthesis of H-type MFI zeolites with unique stacked structures through a one-pot solvent-free strategy. *ChemSusChem* **2019**, *12*, 3871–3877. [CrossRef]
45. Soghrati, E.; Ong, T.K.C.; Poh, C.K.; Kawi, S.; Borgna, A. Zeolite-supported nickel phyllosilicate catalyst for C-O hydrogenolysis of cyclic ethers and polyols. *Appl. Catal. B* **2018**, *235*, 130–142. [CrossRef]
46. Shang, S.; Li, W.; Zhou, A.; Zhang, J.; Yang, H.; Zhang, A.; Guo, X. Fe-Substituted Pt/HZSM-48 for superior selectivity of i-C_{12} in n-dodecane hydroisomerization. *Ind. Eng. Chem. Res.* **2022**, *61*, 1056–1065. [CrossRef]
47. Parry, E.P. An infrared study of pyridine adsorbed on acidic solids. Characterization of surface acidity. *J. Catal.* **1963**, *2*, 371–379. [CrossRef]
48. Shishido, T.; Hattori, H. Hydrogen effects on cumene cracking over zirconium oxide promoted by sulfate ion and platinum. *J. Catal.* **1996**, *161*, 194–197. [CrossRef]
49. Nie, Y.; Shang, S.; Xu, X.; Hua, W.; Yue, Y.; Gao, Z. In_2O_3-doped $Pt/WO_3/ZrO_2$ as a novel efficient catalyst for hydroisomerization of n-heptane. *Appl. Catal. A* **2012**, *433–434*, 69–74. [CrossRef]
50. Meng, L.; Zhu, X.; Mezari, B.; Pestman, R.; Wannapakdee, W.; Hensen, E.J.M. On the role of acidity in bulk and nanosheet [T]MFI (T = Al^{3+}, Ga^{3+}, Fe^{3+}, B^{3+}) zeolites in the methanol-to-hydrocarbons reaction. *ChemCatChem* **2017**, *9*, 3942–3954. [CrossRef] [PubMed]
51. Fyfe, C.A.; Gobbi, G.C.; Kennedy, G.J. Investigation of the conversion (dealumination) of ZSM-5 into silicalite by high-resolution solid-state silicon-29 and aluminum-27 MAS NMR spectroscopy. *J. Phys. Chem.* **1984**, *88*, 3248–3253. [CrossRef]
52. Wu, Q.; Liu, X.; Zhu, L.; Ding, L.; Gao, P.; Wang, X.; Pan, S.; Bian, C.; Meng, X.; Xu, J.; et al. Solvent-free synthesis of zeolites from anhydrous starting raw solids. *J. Am. Chem. Soc.* **2015**, *137*, 1052–1055. [CrossRef]
53. Zhu, X.; Wu, L.; Magusin, P.; Mezari, B.; Hensen, E. On the synthesis of highly acidic nanolayered ZSM-5. *J. Catal.* **2015**, *327*, 10–21. [CrossRef]
54. Petushkov, A.; Yoon, S.; Larsen, S.C. Synthesis of hierarchical nanocrystalline ZSM-5 with controlled particle size and mesoporosity. *Micropor. Mesopor. Mater.* **2011**, *137*, 92–100. [CrossRef]
55. Nagy, J.B.; Gabelica, Z.; Debras, G.; Derouane, E.G.; Gilson, J.P.; Jacobs, P.A. ^{27}Al-n.m.r. characterization of natural and synthetic zeolites. *Zeolites* **1984**, *4*, 133–139. [CrossRef]
56. Deng, F.; Du, Y.; Ye, C.; Wang, J.; Ding, T.; Li, H. Acid sites and hydration behavior of dealuminated zeolite HZSM-5: A high-resolution solid state NMR study. *J. Phys. Chem.* **1995**, *99*, 15208–15214. [CrossRef]
57. Blasco, T.; Corma, A.; Martínez-Triguero, J. Hydrothermal stabilization of ZSM-5 catalytic cracking additives by phosphorus addition. *J. Catal.* **2006**, *237*, 267–277. [CrossRef]
58. Saenluang, K.; Imyen, T.; Wannapakdee, W.; Suttipat, D.; Dugkhuntod, P.; Ketkaew, M.; Thivasasith, A.; Wattanakit, C. Hierarchical nanospherical ZSM-5 nanosheets with uniform Al distribution for alkylation of benzene with ethanol. *ACS Appl. Nano Mater.* **2020**, *3*, 3252–3263. [CrossRef]
59. Li, R.; Chawla, A.; Linares, N.; Sutjianto, J.G.; Chapman, K.W.; Martínez, J.G.; Rimer, J.D. Diverse physical states of amorphous precursors in zeolite synthesis. *Ind. Eng. Chem. Res.* **2018**, *57*, 8460–8471. [CrossRef]
60. Jiang, Y.; Hunger, M.; Wang, W. On the reactivity of surface methoxy species in acidic zeolites. *J. Am. Chem. Soc.* **2006**, *128*, 11679–11692. [CrossRef]
61. Svelle, S.; Rønning, P.O.; Kolboe, S. Kinetic studies of zeolite-catalyzed methylation reactions: 1. Coreaction of [^{12}C]ethene and [^{13}C]methanol. *J. Catal.* **2004**, *224*, 115–123. [CrossRef]
62. Svelle, S.; Kolboe, S.; Swang, O.; Olsbye, U. Methylation of alkenes and methylbenzenes by dimethyl ether or methanol on acidic zeolites. *J. Phys. Chem. B* **2005**, *109*, 12874–12878. [CrossRef] [PubMed]
63. Mehio, N.; Dai, S.; Jiang, D.E. Quantum mechanical basis for kinetic diameters of small gaseous molecules. *J. Phys. Chem. A* **2014**, *118*, 1150–1154. [CrossRef] [PubMed]
64. Emeis, C.A. Determination of integrated molar extinction coefficients for infrared absorption bands of pyridine adsorbed on solid acid catalysts. *J. Catal.* **1993**, *141*, 347–354. [CrossRef]

65. Yang, F.; Zhong, J.; Liu, X.; Zhu, X. A novel catalytic alkylation process of syngas with benzene over the cerium modified platinum supported on HZSM-5 zeolite. *Appl. Energy* **2018**, *226*, 22–30. [CrossRef]
66. Frillette, V.J.; Haag, W.O.; Lago, R.M. Catalysis by crystalline aluminosilicates: Characterization of intermediate pore-size zeolites by the "Constraint Index". *J. Catal.* **1981**, *67*, 218–222. [CrossRef]

Disclaimer/Publisher's Note: The statements, opinions and data contained in all publications are solely those of the individual author(s) and contributor(s) and not of MDPI and/or the editor(s). MDPI and/or the editor(s) disclaim responsibility for any injury to people or property resulting from any ideas, methods, instructions or products referred to in the content.

Article

Effect of Different Zinc Species on Mn-Ce/CuX Catalyst for Low-Temperature NH$_3$-SCR Reaction: Comparison of ZnCl$_2$, Zn(NO$_3$)$_2$, ZnSO$_4$ and ZnCO$_3$

Lin Chen [1,2], Shan Ren [1,*], Tao Chen [1], Xiaodi Li [1], Zhichao Chen [1], Mingming Wang [1], Qingcai Liu [1] and Jie Yang [3,*]

[1] College of Materials Science and Engineering, Chongqing University, Chongqing 400044, China; lin.chen@cqu.edu.cn (L.C.); 202209131182@stu.cqu.edu.cn (T.C.); lxd199611@126.com (X.L.); 20162749@cqu.edu.cn (Z.C.); 202009131184@cqu.edu.cn (M.W.); liu_qingcai@163.com (Q.L.)
[2] Bioenergy and Catalysis Laboratory, Paul Scherrer Institute (PSI), CH-5232 Villigen, Switzerland
[3] College of Chemistry and Chemical Engineering, Wuhan Textile University, Wuhan 430200, China
* Correspondence: shan.ren@cqu.edu.cn (S.R.); yangjie@wtu.edu.cn (J.Y.)

Abstract: The effects of four distinct zinc species (ZnCl$_2$, Zn(NO$_3$)$_2$, ZnSO$_4$, and ZnCO$_3$) on a Mn-Ce co-doped CuX (MCCX)catalyst were investigated and contrasted in the low-temperature NH$_3$-SCR process. Aqueous solutions of ZnCl$_2$, Zn(NO$_3$)$_2$, ZnSO$_4$, and ZnCO$_3$ were used to poison the catalysts. The catalytic activity of all catalysts was assessed, and their physicochemical properties were studied. There was a notable drop trend in catalytic activity in the low temperature range (200 °C) after zinc species poisoning on MCCX catalyst. Interestingly, ZnSO$_4$ and ZnCO$_3$ on MCCX catalyst had more serious effect on catalytic activity than Zn(NO$_3$)$_2$ and ZnCl$_2$ from 150 °C to 225 °C, in which NO conversion of the MCCX-Zn-S and MCCX-Zn-C catalysts dropped about 20–30% below 200 °C compared with the fresh MCCX catalyst. The zeolite X structure was impacted by Zn species doping on the MCCX catalyst, and the Zn-poisoned catalysts had less acidic and lower redox ability than fresh Mn-Ce/CuX catalysts. Through the results of in situ DRIFTS spectroscopy experiments, all catalysts were governed by both Langmuir–Hinshelwood (L–H) and Eley–Rideal (E–R) mechanisms, and the possible mechanism for poisoning the Mn-Ce/CuX catalyst using various zinc species was revealed.

Keywords: Mn-Ce/CuX catalyst; zinc species poisoning; low-temperature NH$_3$-SCR

1. Introduction

The worldwide deterioration in air quality has led to the establishment of stronger rules on the emission of nitrogen oxides (NO$_x$) [1,2]. The selective catalytic reduction (SCR) of NO$_x$ with NH$_3$ is a well-established and widely utilized technology [3]. According to a number of studies [4–6], despite its popularity, the commercial V$_2$O$_5$-WO$_3$(MoO$_3$)/TiO$_2$ catalyst has a few limitations that prevent it from being truly helpful. It is hazardous to living things because of vanadium species, does not work well at lower temperature, and can only be used effectively between 350 and 450 °C. This shows how important it is to look into low-temperature SCR catalysts that do not use vanadium to control NO$_x$ emissions.

Compared to vanadium-based catalysts, copper-based catalysts have better redox properties and thermodynamic stabilities [7,8], making them a viable alternative. It was discovered by Wang et al. [9] that the maximum NO conversion of the Cu-ZSM-5 catalyst occurred between 200 and 400 °C, with practically full conversion, because of the highly preferential production of active copper species and the quick flip between Cu^{2+} and Cu$^+$ species during the NH$_3$-SCR process. Cu-exchanged zeolite catalysts have recently been the subject of extensive study in NH$_3$-SCR reaction [10,11]. Tarach et al. [12] compared the activity and stability of Cu-exchanged zeolites with different framework topologies in NH$_3$-SCR reaction and found that the Cu-ZSM-5 catalyst displayed an extraordinary activity in the NH$_3$-SCR process, achieving nearly 100% NO conversion at 175 °C with almost

100% N_2 selectivity. Our prior research [13,14] found that zeolite X could be generated cheaply from the blast furnace slag, making ion-exchange techniques a viable option for producing CuX zeolite catalyst. Doping Mn and Ce oxides on CuX zeolite could improve its SO_2 + H_2O resistance and enhance its low-temperature NH_3-SCR activity. Generally, the catalysts nevertheless often experienced severe poisoning by alkali, alkaline earth, and heavy metals in the actual fuel gas state. The deactivation rate of various poisonous zinc species was shown to be in the order of $ZnCl_2$ > $ZnSO_4$ > ZnO by Wang et al. [15], who studied the effects of three distinct zinc species (ZnO, $ZnSO_4$, and $ZnCl_2$) on Sb-CeZr_2O_x catalysts. According to our earlier research [16,17], $ZnCl_2$ had a more severe poisoning effect on the Mn-Ce/AC catalyst than $ZnSO_4$, and Zn^{2+} would occupy the acidic site of the catalyst after loading zinc salts. Nonetheless, the poisoning impact of various zinc species on Cu-exchanged zeolite catalysts has received insufficient attention till now.

The current research looked into how different zinc ions affected the catalytic performance of the Mn-Ce/CuX catalyst in low-temperature denitration reactions. To poison the Mn-Ce/CuX catalyst, we used an impregnation technique with $ZnCl_2$, $Zn(NO_3)_2$, $ZnSO_4$, and $ZnCO_3$ solutions. The causes of zinc salt poisoning and deactivation of the Mn-Ce/CuX catalyst have been identified, and the possible mechanism for poisoning the Mn-Ce/CuX catalyst using various zinc salts has been proposed, by comparing the catalytic activity, dispersion degree of active component, surface morphology and pore structure, surface acidity, redox performance, element concentration of active component, and surface reaction path of the catalyst before and after poisoning.

2. Results and Discussions

2.1. Denitration Performance

The catalytic activity for fresh MCCX and catalysts poisoned with different Zn species (MCCX-Zn-Cl, MCCX-Zn-N, MCCX-Zn-S, and MCCX-Zn-C catalysts) can be found in Figure 1. NO conversion of Zn-poisoned catalysts is much lower than that of the fresh MCCX catalyst at temperatures below 200 °C, as seen in Figure 1a. In detail, NO conversion of the MCCX-Zn-C catalyst surpassed the other three poisoned catalysts at temperatures below 125 °C, whereas NO conversion of the MCCX-Zn-Cl catalyst outperformed the other three Zn-poisoned catalysts at temperatures over 150 °C, reaching nearly 100% at 275 °C. The NO conversion rate for the MCCX-Zn-S and MCCX-Zn-C catalysts dropped by about 20–30% below 200 °C, making it the worst performing catalyst over the whole reaction temperature range. It was discovered that the conversion of the catalysts is most affected by the different Zn species in the lower temperature range, whereas they have almost no effect at higher temperatures. According to Figure 1b, comparing the N_2 selectivity of the MCCX-Zn-Cl, MCCX-Zn-N, MCCX-Zn-S, and MCCX-Zn-C catalysts to that of the fresh MCCX catalyst, there was an approximate 20% improvement. Figure 1c displays the results of a measurement of the concentration of NH_3 and N_2O in both Zn-poisoned and new MCCX catalysts. Zn-poisoned catalysts had lower concentrations of N_2O and higher concentrations of NH_3 compared to the fresh MCCX catalyst, among which the MCCX-Zn-S sample also had more NH_3 than N_2O.

2.2. Structural Properties

X-ray diffraction (XRD) patterns of MCCX, MCCX-Zn-Cl, MCCX-Zn-N, MCCX-Zn-S, and MCCX-Zn-C catalysts are shown in Figure 2. The zeolite X structure was described by peaks at 2theta = 6.8°, 15.5°, 18.5°, 20.2°, 23.4°, 26.8°, and 30.5°, while the CuO phases were characterized by peaks at 2theta = 35.5°, 38.9°, 48.8°, and 61.5°. It was discovered that the peak intensity of zeolite X on Zn-poisoned catalysts was lesser than on the fresh catalyst, whereas the peak intensity of CuO phase on Zn-poisoned catalysts was stronger than on the fresh one. It was noted that the peak intensity of zeolite X on the MCCX-Zn-Cl catalyst was much stronger than the other three catalysts, while the peak intensity of zeolite X on MCCX-Zn-S and MCCX-Zn-C catalysts were weaker. This result suggested that the zeolite X structure was impacted by Zn species doping on MCCX catalyst, and the zinc species contributed to the migration of CuO species. The absence of the Zn species peak indicates that they are uniformly dispersed on the catalyst surface. Micrographs of the catalysts (Figure 3) show that Zn-poisoned catalysts had a zeolite X structure that deviated from the usual octahedron, and some of the zeolite X particles agglomerated. In particular, the zeolite X structure of the MCCX-Zn-C catalyst had more serious damage than the other three catalysts. Furthermore, EDS mapping revealed that Cu aggregated on catalysts due to Zn poisoning, leading to an insufficiency of Cu active sites.

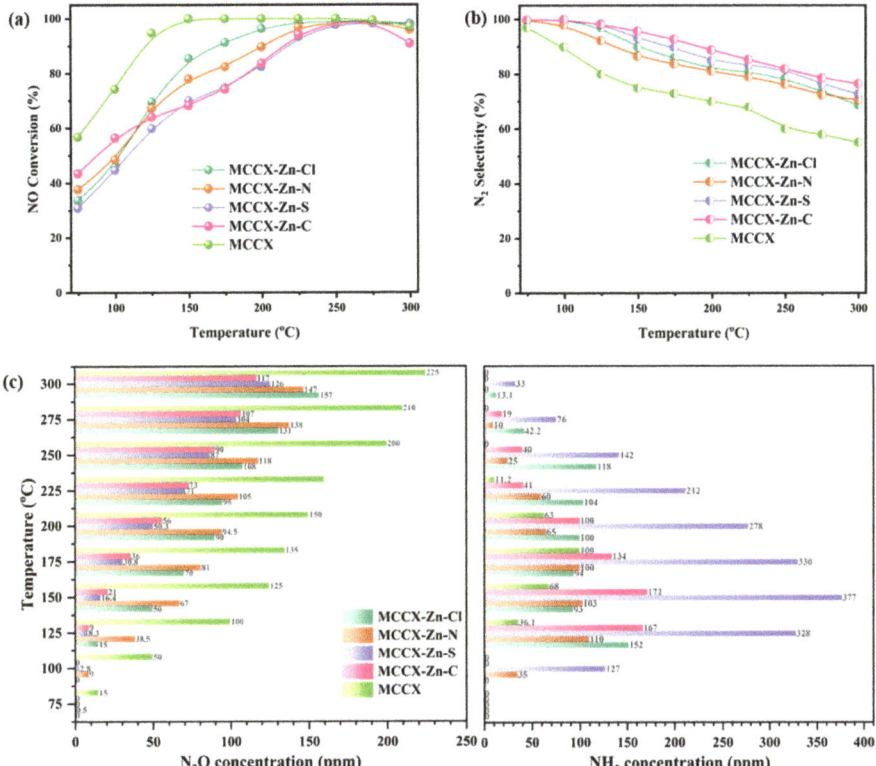

Figure 1. NH_3-SCR performance with temperature for MCCX, MCCX-Zn-Cl, MCCX-Zn-N, MCCX-Zn-S, and MCCX-Zn-C catalysts: (**a**) NO conversion; (**b**) N_2 selectivity; (**c**) N_2O and NH_3 concentration. Reaction conditions: 500 ppm NO, 500 ppm NH_3, 11% O_2, N_2 as balance, and GHSV of 36,000 h^{-1}.

2.3. XPS Analysis

To find out more about the elemental compositions of the MCCX-Zn-Cl, MCCX-Zn-N, MCCX-Zn-S, and MCCX-Zn-C catalysts, XPS spectra of Cu 2p, O 1s, Mn 2p, and Zn 2p were measured, as displayed in Figure 4. Peaks at the 948–966 eV, 940–945 eV, and 930–948 eV regions in the Cu 2p spectra (Figure 4a) were attributed to Cu 2p$_{1/2}$, satellite peaks, and Cu 2p$_{3/2}$, respectively [11,12]. Additionally, the spectra of Cu 2p$_{3/2}$ could be divided into two peaks; one at 935 eV was attributed to Cu^{2+}, and another at 933 eV was attributed to Cu^+ ions [9,10,18,19]. By calculating the area of the corresponding peaks, we were able to determine the relative abundance of the different Cu species and establish the following ranking for the concentration of Cu^{2+} species: the MCCX-Zn-S catalyst (45.1%) was more effective than the MCCX-Zn-N catalyst (43.5%), MCCX-Zn-Cl catalyst (39.7%), and MCCX-Zn-C catalyst (37.1%). In general, the isolated state of Cu^{2+} species in the zeolite lattice was an active site during the catalytic process [8,20,21].

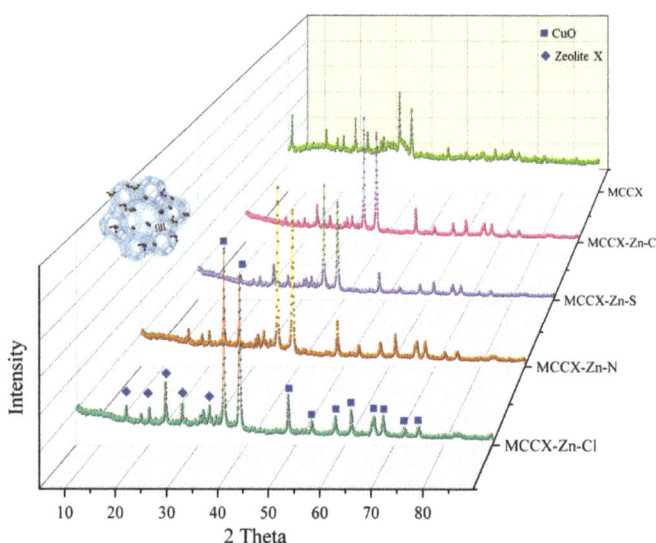

Figure 2. XRD patterns of MCCX, MCCX-Zn-Cl, MCCX-Zn-N, MCCX-Zn-S, and MCCX-Zn-C catalysts.

Figure 4b shows the XPS spectra of Mn 2p for MCCX-Zn-Cl, MCCX-Zn-N, MCCX-Zn-S, and MCCX-Zn-C catalysts. Mn $2p_{3/2}$ and Mn $2p_{1/2}$ were responsible for the two large peaks seen at about 642 and 650 eV, respectively [22]. In addition, three fitting peaks at about 641.5 eV, 640.5 eV, and 639.2 eV were identified in the spectra of Mn $2p_{3/2}$, which corresponded to Mn^{4+}, Mn^{3+}, and Mn^{2+} species, respectively [23–25]. The concentration order of species Mn4+ was as follows: MCCX-Zn-N > (32.5%) > MCCX-Zn-Cl (30.7%) > MCCX-Zn-C (26.8%) > MCCX-Zn-S (22.5%) catalysts. Previous studies indicated that high valve state Mn species (Mn^{4+} species) facilitated NO oxidation to NO_2 and accelerated the "fast SCR" process at low temperatures [26,27].

O 1s spectra of MCCX-Zn-Cl, MCCX-Zn-N, MCCX-Zn-S, and MCCX-Zn-C catalysts were depicted in Figure 4c. Three fitting peaks at around 531.6 eV, 529.5 eV, and 533.5 eV belonged to chemosorbed oxygen species (O_α), lattice oxygen species (O_β), and chemisorbed oxygen or weakly bound oxygen species (O_γ), respectively [28–30]. Generally, the catalysts exhibited the increased NH_3-SCR activity at higher concentrations of chemosorbed oxygen species due to their greater mobility in comparison to the other two oxygen species [13,23,31]. The concentration order of species O_α was as follows: MCCX-Zn-S (66.7%) > MCCX-Zn-C (57.1%) > MCCX-Zn-Cl (55.8%) > MCCX-Zn-N (50.5%) catalysts. The SO_4^{2-} and CO_3^{2-} might provide more chemosorbed oxygen species on catalysts, which was in line with the previous study [32]. In Figure 4d, there is one peak belonging to Zn $2p_{3/2}$ that could be found at Zn-poisoned catalysts [33]. The weakest peak intensity of the four catalysts was observed in the MCCX-Zn-S catalyst from the summary XPS spectra displayed in Figure 4e, attesting to the low amount of the surface's active components and, consequently, lesser SCR activity.

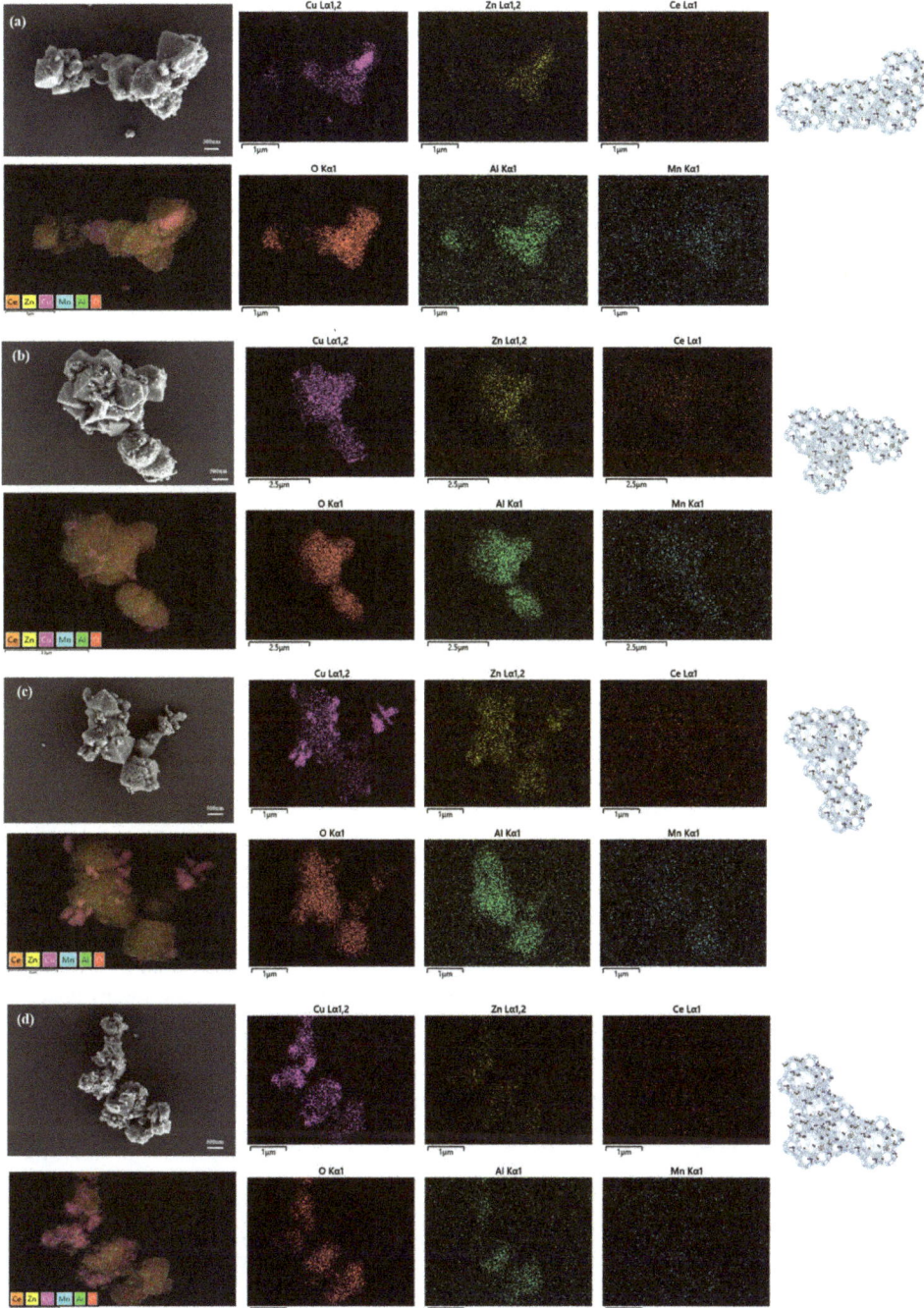

Figure 3. SEM images and elements mapping of (**a**) MCCX-Zn-Cl, (**b**) MCCX-Zn-N, (**c**) MCCX-Zn-S, and (**d**) MCCX-Zn-C catalysts.

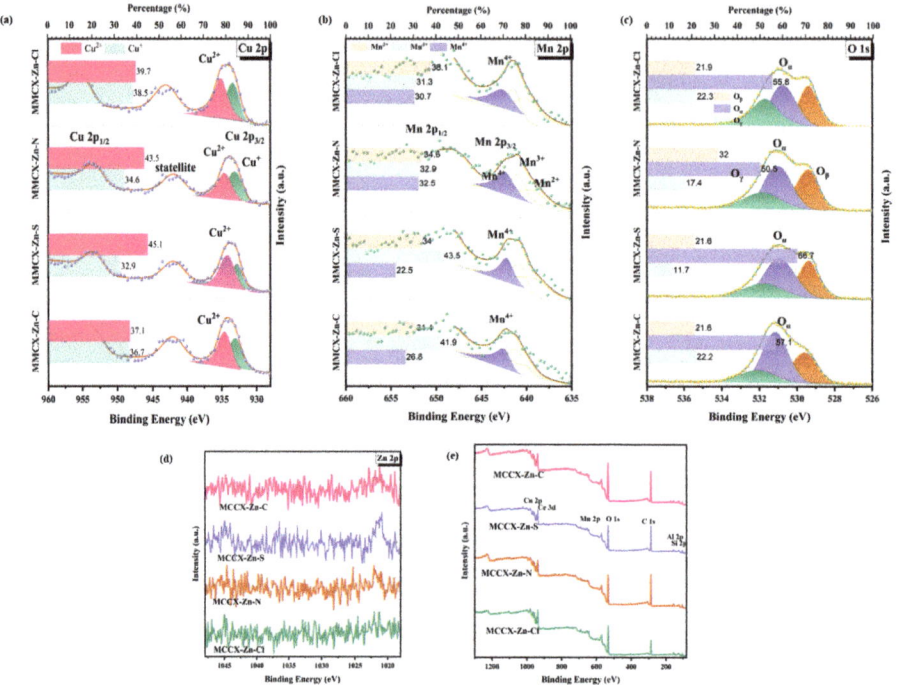

Figure 4. XPS spectra of MCCX-Zn-Cl, MCCX-Zn-N, MCCX-Zn-S, and MCCX-Zn-C catalysts: (**a**) Cu 2p (purple solid dots were assigned to the original data, red line was assigned to the fitting curve); (**b**) Mn 2p (green solid dots were assigned to the original data, deep yellow line was assigned to the fitting curve); (**c**) O 1 s (golden yellow solid dots were assigned to the original data, green line was assigned to the fitting curve); (**d**) Zn 2p; and (**e**) survey spectrum.

2.4. Acidity and Redox Property

The main factor in SCR activity is the acidity of the catalyst's surface. Figure 5a–c shows the NH_3-TPD results and in situ DRIFTS spectra of NH_3 desorption at temperatures between 35 and 300 °C for MCCX-Zn-Cl, MCCX-Zn-N, MCCX-Zn-S, and MCCX-Zn-C catalysts. Figure 5a shows that all Zn-poisoned catalysts exhibit several fitting peaks between 100 and 800 °C. The peak at 122 °C is associated with low acidity, those between ca. 200 and 400 °C are linked to medium acidity, and those beyond 400 °C are associated with high acidity [14,32,34]. In addition, as shown in Figure 5b, the area of fitting peaks was used to determine the amount of adsorbed NH_3. All the catalysts were found to have a greater middle acidity, and the following list shows what they ranked in terms of how much NH_3 they were able to absorb: concentrations of MCCX-Zn-Cl (478.1 μmoL/g), MCCX-Zn-C (435.9 μmoL/g), MCCX-Zn-N (299.1 μmoL/g), and MCCX-Zn-S (219.8 μmoL/g). Figure 5c shows the in situ DRIFTS spectra of NH_3 desorption, which shows multiple bands on each of the four catalysts between 3000 and 3800 cm^{-1}. These bands were assigned to Cu^{2+} absorbed NH_3 species (3182 and 3332 cm^{-1}) and NH_4^+ species on Brønsted acid sites (3272 cm^{-1}) [9,35]. Bands corresponding to coordinated adsorbed NH_3 species on Lewis acid sites (1610 and 1260 cm^{-1}) and NH_4^+ species on Brønsted acid sites (1390 cm^{-1}) were also observed [36,37]. The acidity reduced as the reaction temperature increased, which can be seen from the gradual dropping in peak intensity of the acid sites. Lewis acid sites on catalysts were also shown to be more stable than Brønsted acid sites, indicating that Lewis acid sites were the most important acid sites for the reactions.

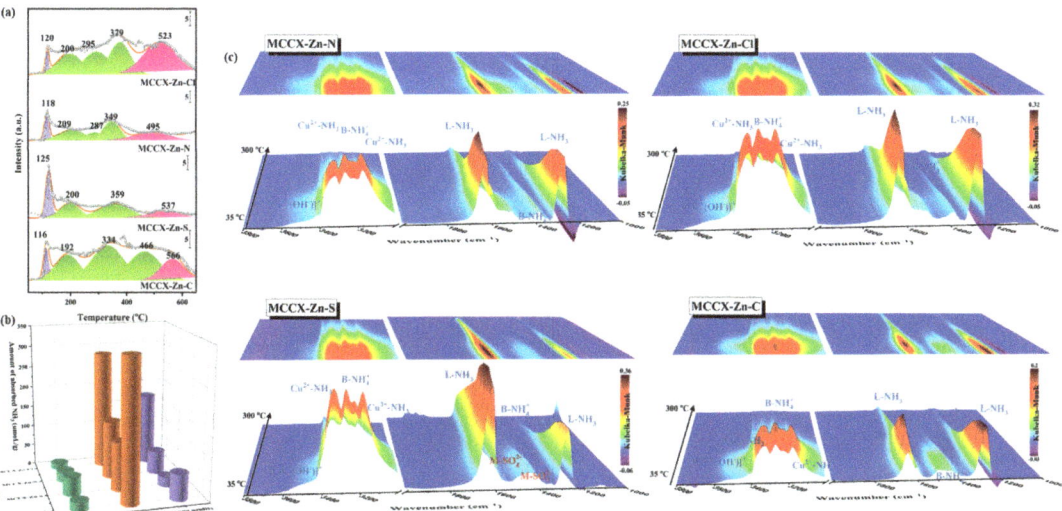

Figure 5. (a) NH$_3$-TPD profiles of all catalysts (purple fitting peaks belonged to weak acidity, green fitting peaks belonged to medium acidity and pink fitting peak belonged to strong acidity) and (b) integral area of peaks from NH$_3$-TPD profiles; (c) in situ DRIFTS spectra of NH$_3$ adsorption in 35–300 °C over MCCX-Zn-Cl, MCCX-Zn-N, MCCX-Zn-S, and MCCX-Zn-C catalysts. Conditions: 500 ppm NH$_3$ and N$_2$ as balance.

Profiles of H$_2$-TPR for MCCX-Zn-Cl, MCCX-Zn-N, MCCX-Zn-S, and MCCX-Zn-C catalysts are shown in Figure 6. The MCCX-Zn-Cl catalyst was found to have three fitting peaks at 200–400 °C, but the other three Zn-poisoned catalysts only had two fitting peaks at 200–400 °C. For the peak at ca. 200 °C, CuO was reduced to Cu^0; for the peak at 250 °C, Cu^{2+} species on the hydroxyl group of zeolite X were reduced; and for the peak at 300 °C, isolated Cu^{2+} ions located in lattice zeolite X were reduced. The peak over 400 °C was related to the reduction of Cu$^+$ ions, and no peak existed in the Zn-poisoned catalysts. Peaks at ca. 200–500 °C belonged to the reductions of MnO$_2$ to Mn$_2$O$_3$, Mn$_2$O$_3$ to Mn$_3$O$_4$, and Mn$_3$O$_4$ to MnO. The CeO$_2$ redox process of surface-capping oxygen occurs at temperatures of ca. 400 °C [38–40]. In addition, the consumption of H$_2$ for the Zn-poisoned catalysts was as follows: MCCX-Zn-C (6.6 mmoL/g) > MCCX-Zn-S (6.2 mmoL/g) > MCCX-Zn-Cl (5.5 mmoL/g) > MCCX-Zn-N (2.7 mmoL/g). It was discovered that the reduction of MCCX catalysts was affected by the presence of various Zn species, particularly the isolated Cu^{2+} ions in lattice zeolite X and Cu$^+$ ions. It could be found that the Zn poisoning produced more copper oxides on MCCX-Zn-S and MCCX-Zn-C catalysts, and reduced the Cu^{2+} active sites.

2.5. In Situ DRIFTS Studies

2.5.1. NO + O$_2$ Reacting with Pre-Adsorbed NH$_3$ Species

In order to study the impact of various Zn species on surface acidity types of the catalysts, the in situ DRIFTS spectra of MCCX-Zn-Cl, MCCX-Zn-N, MCCX-Zn-S, and MCCX-Zn-C catalysts for NH$_3$ adsorption at 200 °C were investigated. As shown in Figure 7, the bands of Cu^{2+}-absorbing NH$_3$ species at 3182 and 3332 cm^{-1}, NH$_4^+$ species on Brønsted acid sites at 3272 cm^{-1} and 1390 cm^{-1}, and coordinated adsorbed NH$_3$ species on Lewis acid sites at 1610 cm^{-1} and 1260 cm^{-1} [9,35] gradually increased in intensity after NH$_3$ was introduced for 5 min. It could be found that the bands belonging to Lewis acid sites were much stronger than that of Brønsted acid sites on Zn-poisoned catalysts.

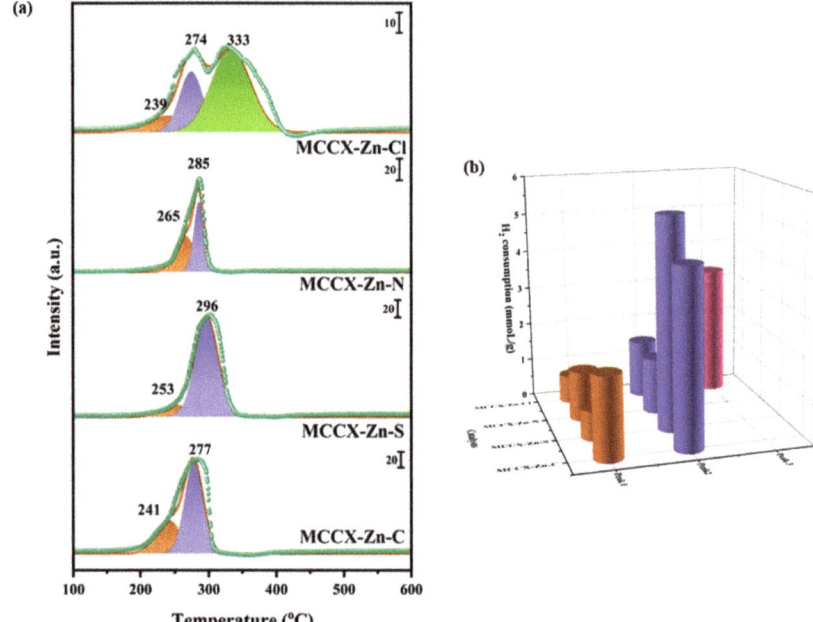

Figure 6. (a) H$_2$-TPR profiles of all catalyst (green solid dots were assigned to the original data, red line was assigned to the fitting curve, orange fitting peak related to peak 1, purple peak related to peak 2, green fitting peak related to peak 3) and (b) integral area of peaks from H$_2$-TPR profiles.

The intermediates of the reaction between NO + O$_2$ and pre-adsorbed NH$_3$ species at 200 °C were studied by performing in situ DRIFTS measurements over MCCX-Zn-Cl, MCCX-Zn-N, MCCX-Zn-S, and MCCX-Zn-C catalysts, as shown in Figure 8. When NH$_3$ was introduced, bands with intensities of Cu^{2+}-absorbing NH$_3$ species (3182 and 3332 cm^{-1}), NH$_4^+$ species on Brønsted acid sites (3272 cm^{-1} and 1390 cm^{-1}), and coordinated adsorbed NH$_3$ species on Lewis acid sites (1610 cm^{-1} and 1260 cm^{-1}) were observed on the four Zn-poisoned catalysts, and these bands disappeared after 20 min of NO + O$_2$ purging. After that, many striations, which could be attributed to various nitrate and nitrite species, began to show up. Nitrate species at the Cu^{2+} active site were attributed to the 1915 cm^{-1} band [41], while those at 1600–1650 cm^{-1} were characterized as bidentate bridging nitrate (M-O$_2$-NO), those at 1500–1570 cm^{-1} were described as bidentate chelating nitrate (M-O$_2$NO), those at 1480–1530 cm^{-1} as monodentate nitrate (M-O-NO$_2$), and those at 1317–1400 cm^{-1} as free-NO$_3^-$ species of antisymmetric N-O stretches [35,42–44]. Bidentate nitrite (M-O$_2$N) and bridging nitrite (M-O$_2$-N) contributed the bands at 1320 cm^{-1} and 1575 cm^{-1}, respectively [42,45]. It indicated that both the Langmuir–Hinshelwood (L–H) and Eley–Rideal (E–R) mechanisms were at play in all of the catalysts. ZnCl$_2$ species on MCCX catalysts had a slight influence on the Cu^{2+} active site, while the band intensity of nitrate species on the Cu^{2+} active site for MCCX-Zn-S was weaker than that of the other three catalysts.

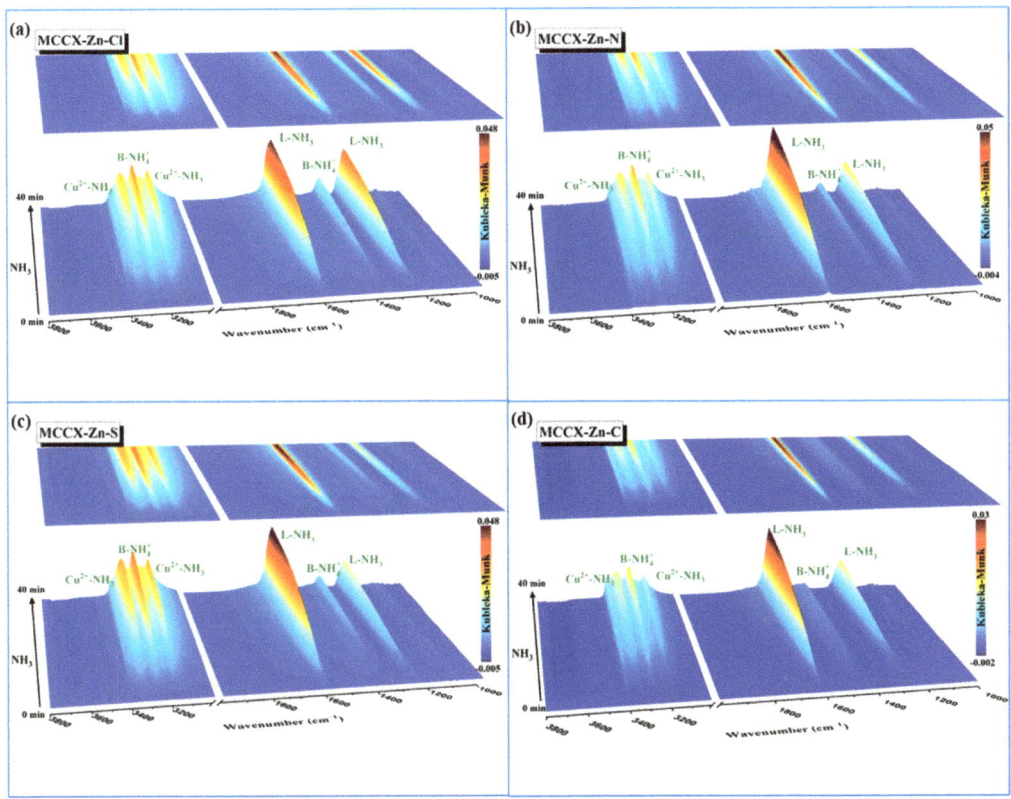

Figure 7. In situ DRIFTS spectra of (**a**) MCCX-Zn-Cl, (**b**) MCCX-Zn-N, (**c**) MCCX-Zn-S, and (**d**) MCCX-Zn-C catalysts for NH$_3$ adsorption at 200 °C. Conditions: 500 ppm NH$_3$ and N$_2$ as balance.

2.5.2. NH$_3$ Reacting with Pre-Adsorbed NO + O$_2$ Species

Figure 9 displays the in situ DRIFTS spectra of NO + O$_2$ co-adsorption at 200 °C over MCCX-Zn-Cl, MCCX-Zn-N, MCCX-Zn-S, and MCCX-Zn-C catalysts. The bands of bidentate bridging nitrate (M-O$_2$-NO) (at 1600–1650 cm^{-1}), bidentate chelating nitrate (M-O$_2$NO) (at 1500–1570 cm^{-1}), monodentate nitrate (M-O-NO$_2$) (at 1480–1530 cm^{-1}), and the free-NO$_3^-$ species (at 1300–1400 cm^{-1}) of antisymmetric N-O stretches appeared, and the bands of nitrate species on Cu^{2+} active site (at 1915 cm^{-1}) appeared over the MCCX-Zn-Cl, MCCX-Zn-N, MCCX-Zn-S, and MCCX-Zn-C catalysts [35,42–45]. The longer they were exposed to NO + O$_2$, the stronger their bands became. It was found that the nitrate species adsorbing on the Cu^{2+} active site over the MCCX-Zn-Cl, MCCX-Zn-N, MCCX-Zn-S, and MCCX-Zn-C catalysts had much stronger band intensity than the other bands of nitrites and nitrates, suggesting that the Cu^{2+} active site was the primary site involved in the NO + O$_2$ adsorption. Further, the nitrate and nitrite intermediate products were significantly affected by the poisoning impact of ZnCl$_2$, Zn(NO$_3$)$_2$, and ZnSO$_4$ on MCCX catalysts, as compared to ZnCO$_3$.

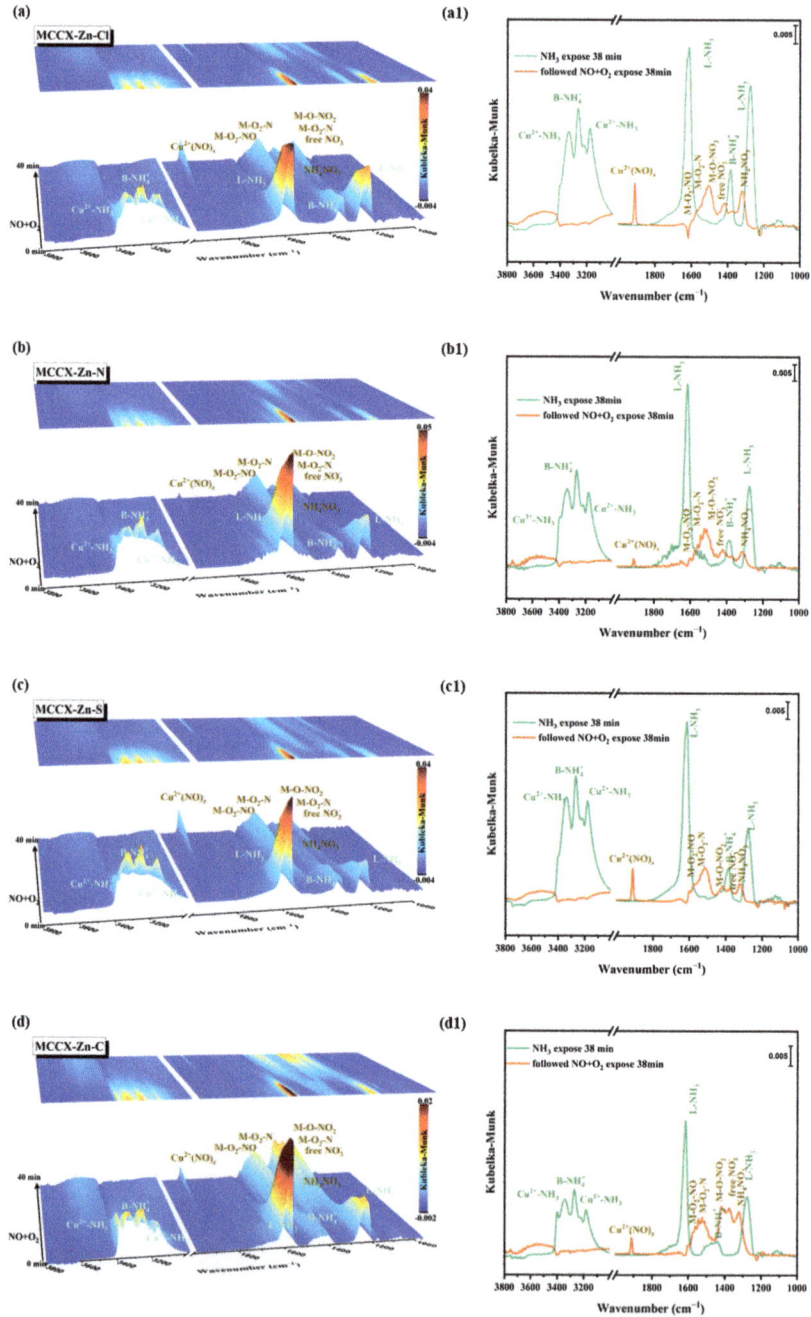

Figure 8. In situ DRIFTS spectra of (**a**) MCCX-Zn-Cl, (**b**) MCCX-Zn-N, (**c**) MCCX-Zn-S, and (**d**) MCCX-Zn-C catalysts for NO + O$_2$ reacting with pre-adsorbed NH$_3$ species at 200 °C; in situ DRIFTS spectra of (**a1**) MCCX-Zn-Cl, (**b1**) MCCX-Zn-N, (**c1**) MCCX-Zn-S, and (**d1**) MCCX-Zn-C catalysts for NH$_3$ adsorbed for 38 min and followed by NO + O$_2$ adsorbed for 38 min at 200 °C. Conditions: 500 ppm NH$_3$, 500 ppm NO, 11% O$_2$, and N$_2$ as balance.

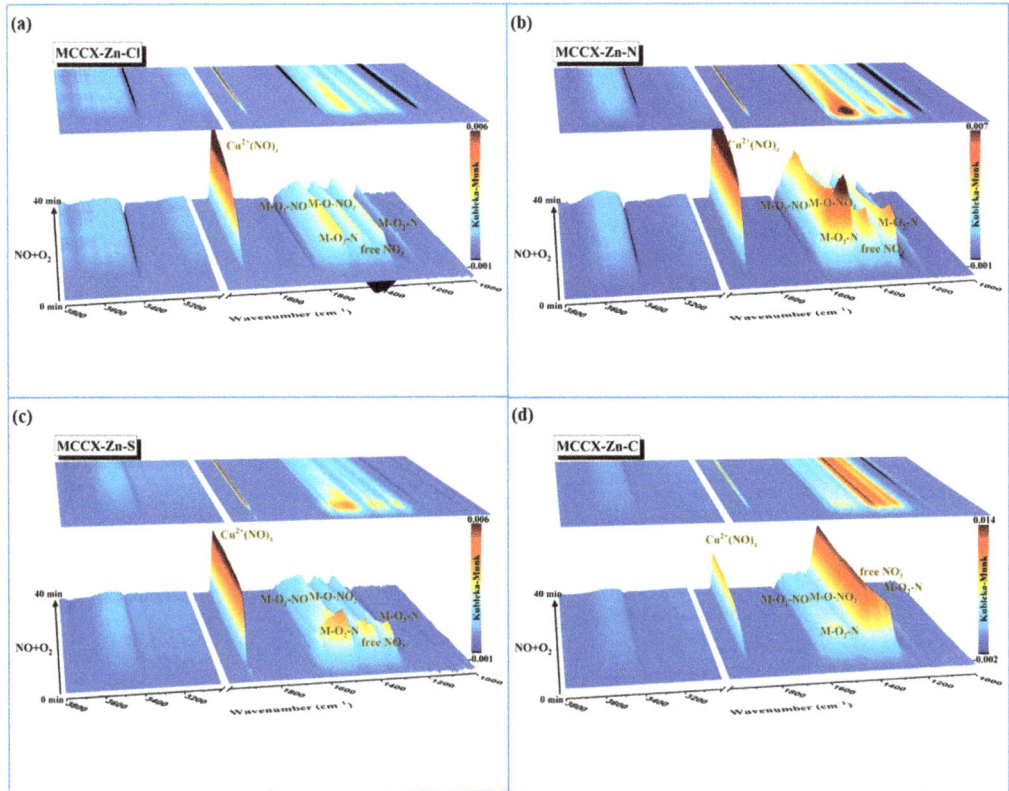

Figure 9. In situ DRIFTS spectra of (**a**) MCCX-Zn-Cl, (**b**) MCCX-Zn-N, (**c**) MCCX-Zn-S, and (**d**) MCCX-Zn-C catalysts for NO + O_2 co-adsorption at 200 °C. Conditions: 500 ppm NO, 11% O_2 and N_2 as balance.

Figure 10 displays the in situ DRIFTS measurements performed over the MCCX-Zn-Cl, MCCX-Zn-N, MCCX-Zn-S, and MCCX-Zn-C catalysts at 200 °C to investigate the reaction intermediates between NH_3 and pre-adsorbed NO + O_2 species. Nitrate species on the Cu^{2+} active site ($Cu(NO_3)_x$), bidentate bridging nitrate ($M-O_2-NO$), bidentate chelating nitrate ($M-O_2NO$), monodentate nitrate ($M-O-NO_2$), NH_4NO_3 species, and the free-NO_3^- species of antisymmetric N-O stretches all showed gradual decreasing trends in band intensity when NH_3 was introduced, which eventually disappeared after 15 min. Then, bands appeared at increasing intensities with increasing time of NH_3 exposure, which were attributed to Cu^{2+}-absorbing NH_3 species (3182 and 3332 cm^{-1}), NH_4^+ species on Brønsted acid sites (3272 cm^{-1}), and coordinated adsorbed NH_3 species on Lewis acid sites (1610 cm^{-1} and 1260 cm^{-1}).

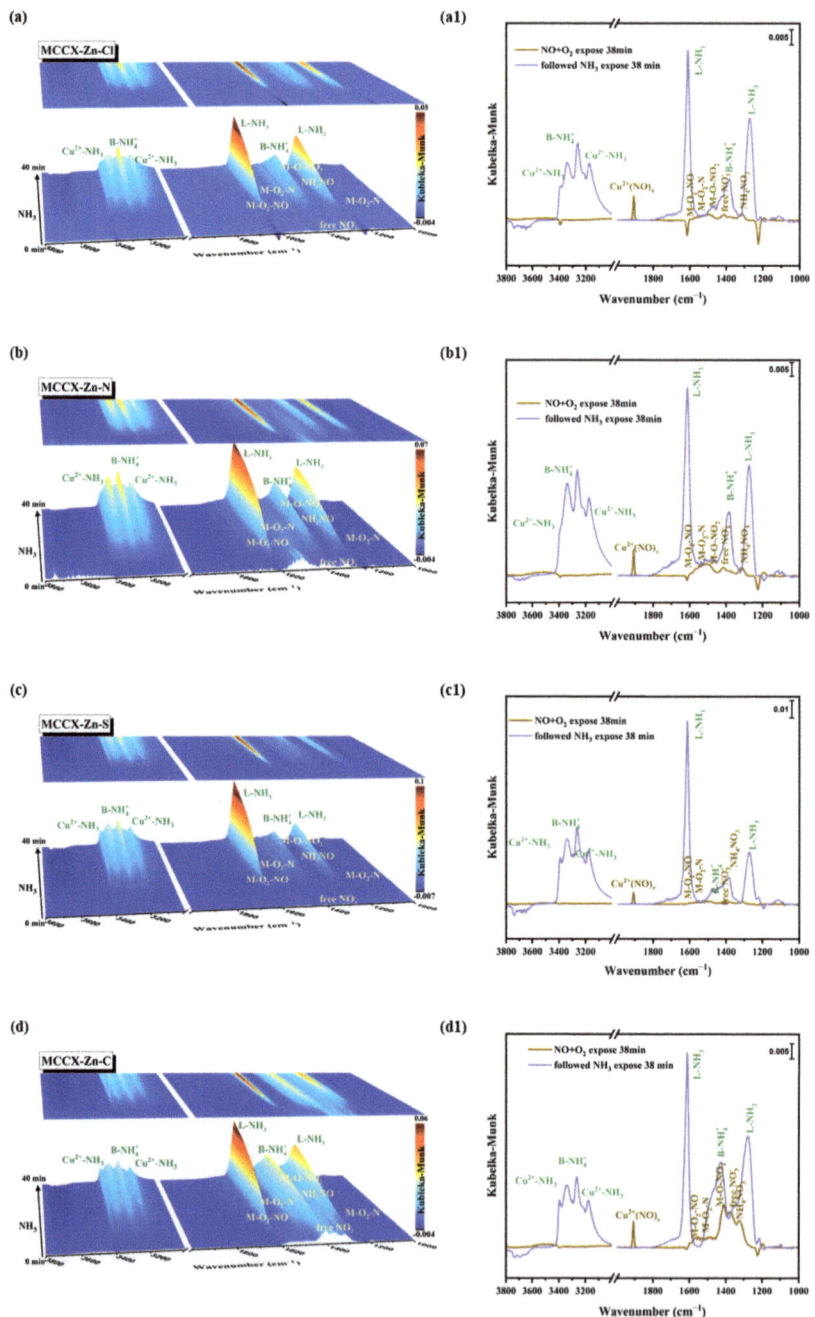

Figure 10. In situ DRIFTS spectra of (**a**) MCCX-Zn-Cl, (**b**) MCCX-Zn-N, (**c**) MCCX-Zn-S, and (**d**) MCCX-Zn-C catalysts for NH$_3$ reacting with pre-adsorbed NO + O$_2$ species at 200 °C; in situ DRIFTS spectra of (**a1**) MCCX-Zn-Cl, (**b1**) MCCX-Zn-N, (**c1**) MCCX-Zn-S, and (**d1**) MCCX-Zn-C catalysts for NO + O$_2$ adsorbed for 38 min and followed by NH$_3$ adsorbed for 38 min at 200 °C. Conditions: 500 ppm NH$_3$, 500 ppm NO, 11% O$_2$, and N$_2$ as balance.

3. Discussion

From the XRD results, it could be found that the zeolite X structure was impacted by Zn species doping on the MCCX catalyst, and the zinc species contributed to the migration of CuO species on the Zn-poisoned catalysts. Some of the zeolite X particles agglomerated, and Cu species aggregated on catalysts due to Zn poisoning, leading to insufficient Cu active sites from the SEM and EDS results. Furthermore, the zeolite X structure for the MCCX-Zn-S and MCCX-Zn-C catalysts had more serious damage. From the XPS analysis, it could be found that the MCCX-Zn-C catalyst had less Cu^{2+} species, while the MCCX-Zn-S catalyst had less Mn^{4+} species. The Cu^{2+} species and Mn^{4+} species were the main active sites for low-temperature SCR reaction, which might be the reason that the $ZnSO_4$ and $ZnCO_3$ on the MCCX catalyst had a more serious effect on catalytic activity. The MCCX-Zn-Cl catalyst had the strongest surface acidity, while the MCCX-Zn-S catalyst had the weakest surface acidity, which affected the adsorption of NH_3. From the H_2-TPR analysis, it could be found that Zn poisoning produced more copper oxides for MCCX-Zn-S and MCCX-Zn-C catalysts and reduced the Cu^{2+} active sites, thereby having an influence on NO and NH_3 adsorption. The bands belonging to Lewis acid sites were much stronger than that of Brønsted acid sites on Zn-poisoned catalysts from the in situ DRIFTS spectra of NH_3 adsorption. Based on the in situ DRIFTS spectra of NO + O_2 and NH_3 species, both the Langmuir–Hinshelwood (L–H) and Eley–Rideal (E–R) mechanisms were at play in all of the catalysts. $ZnCl_2$ species on MCCX catalysts had a slight influence on the Cu^{2+} active site, while the band intensity of nitrate species on the Cu^{2+} active site for MCCX-Zn-S was weaker than that of the other three catalysts. The scheme of mechanism effect for different zinc species on the Mn-Ce/CuX catalyst is summarized in Figure 11.

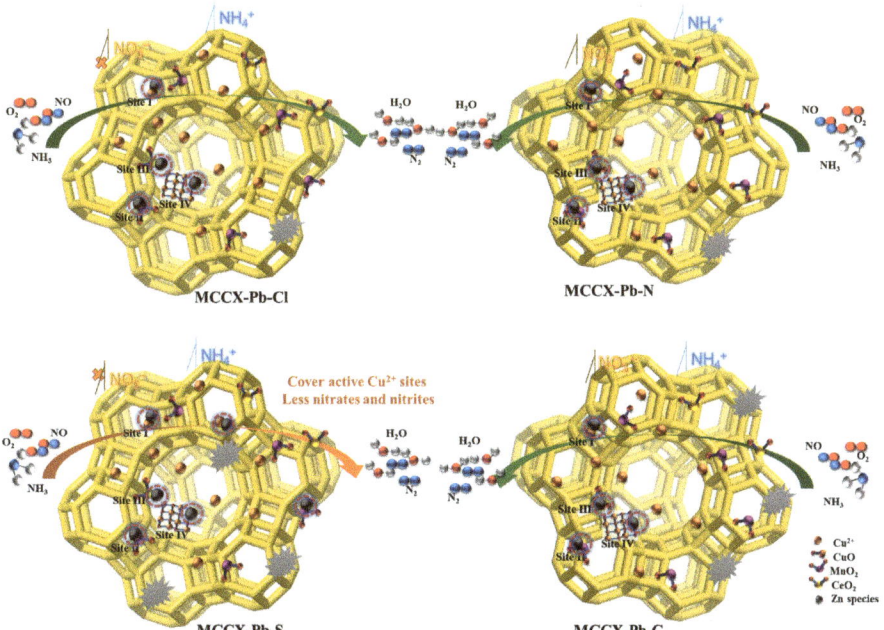

Figure 11. Scheme of mechanism effect for different zinc species on Mn-Ce/CuX catalyst.

4. Materials and Methods

4.1. Catalysts Preparation

The Mn-Ce/CuX catalysts utilized in this study were synthesized using the impregnation method, with the Mn-Ce mixed oxides loading of 5 wt.% (molar ratio of Mn/Ce was

7:3), and the details of the preparation process were described in our earlier study [19]. The first synthesis of CuX catalyst included both hydrothermal and ion exchange methods [13]. After obtaining CuX catalysts, they were added into a precursor solution including a predetermined amount of manganous nitrate and cerous nitrate, and the resulting slurry was dried out in a water bath at 80 °C. The Mn-Ce/CuX catalyst was produced by drying in an oven for 10 h at 80 °C and then calcining them in a muffle furnace for 3 h at 500 °C in air condition.

A total of 1 wt.% Zn species was used to poison the Mn-Ce/CuX catalysts, which were then composited using an impregnation approach in a variety of zinc compounds which were $ZnCl_2$, $Zn(NO_3)_2$, $ZnSO_4$, and $ZnCO_3$, respectively. After evaporating the solutions of $ZnCl_2$, $Zn(NO_3)_2$, $ZnSO_4$, and $ZnCO_3$ in an 80 °C water bath, the samples were dried in an 80 °C oven and then calcinated for 3 h at 500 °C in air. MCCX-Zn-Cl, MCCX-Zn-N, MCCX-Zn-S, and MCCX-Zn-C were the names assigned to the several Zn-poisoned catalysts for $ZnCl_2$, $Zn(NO_3)_2$, $ZnSO_4$, and $ZnCO_3$.

4.2. Catalytic Activity Tests

The NH_3-SCR activity of the MCCX-Zn-Cl, MCCX-Zn-N, MCCX-Zn-S, and MCCX-Zn-C catalysts was studied in a 10 mm diameter quartz reactor using 0.5 g (40–60 mesh) samples. The temperature went from 75 to 300 °C in increments of 25 °C, and each temperature point was maintained for 30 min. The simulant mixed gas contained 500 ppm NH_3, 500 ppm NO, 11 vol% O_2, and N_2 as balanced gas with a total flow gas of 100 mL/min and an hourly space velocity of 36,000 h^{-1}. The amount of NO, NH_3, NO_2, and N_2O coming out was measured by a gas analyzer (Thermo Scientific, Antaris IGS, Waltham, MA, USA). The NO conversion and N_2 selectivity of the MCCX-Zn-Cl, MCCX-Zn-N, MCCX-Zn-S, and MCCX-Zn-C catalysts were determined using the following equations (Equations (1) and (2)) [27,37].

$$\text{NO conversion (\%)} = \left(1 - \frac{[NO]_{out}}{[NO]_{in}}\right) \times 100\% \quad (1)$$

$$N_2 \text{ selectivity (\%)} = \left(1 - \frac{2[N_2O]_{out} - [NO_2]_{out}}{[NO]_{in} + [NH_3]_{in} - [NO]_{out} - [NH_3]_{out}}\right) \times 100\% \quad (2)$$

4.3. Catalyst Characterization

The X-ray diffraction (XRD) patterns were conducted to examine the composition and crystallinity of chemicals on the catalyst surface using the Rigaku D/max-2500 PC diffractometer which was recorded with a 2θ range of 5–90° at a scanning speed of 5°/min. Scanning electron microscope (SEM) analysis was used to investigate the surface microstructure of MCCX-Zn-Cl, MCCX-Zn-N, MCCX-Zn-S, and MCCX-Zn-C catalysts by the Thermo Scientific Quattro SEM. The content and distribution of surface elements were recorded using energy dispersive X-ray spectrometry (EDS) mapping.

X-ray photoelectron spectroscopy (XPS) experiments were used to measure the Cu 2p, O 1s, Mn 2p, and Zn 2p by the Thermo ESCALAB 250Xi with Al Kα radiation (1484.6 eV) at room temperature. The binding energy of the surface elements of the sample was calibrated using C 1s (284.6 eV), and the sample was degassed in a vacuum for 1 h before testing to minimize the influence of surface impurities.

A H_2-TPR test was used on an Auto Chem II TPR/TPD 2920 detector. The specific operation process was as follows: 150 mg of catalyst was placed in the U-shaped quartz reaction tube, and the sample was pretreated under Ar flow of 100 mL/min. Then, the sample was heated from room temperature to 300 °C at a heating rate of 5 °C/min and then cooled to 50 °C after purging for 1 h, after which the sample was heated to 600 °C (heating rate of 10 °C/min) under 5% H_2/Ar atmosphere (flow rate of 50 mL/min). At the same time, the detector recorded the TCD signal. NH_3-TPD tests were conducted by the Chembet Pulsar TPR/TPD detector (The American Konta Company), and the pretreated steps were the same as H_2-TPR. The sample was then exposed to a 5% NH_3/Ar atmosphere (flow rate of 100 mL/min) for 1 h and followed by purging in N_2 (100 mL/min). Then, the sample was heated to 800 °C at a constant heating rate of 5 °C/min under a flow of 100 mL/min Ar, and the TCD signal was detected.

In situ diffuse reflectance infrared Fourier transform (in situ DRIFTS) studies were investigated on the Thermo Fisher Nicolet iS50 spectrometer. Before the test, all the catalysts were purged at 300 °C for 30 min under N_2 and 11 vol% O_2 atmosphere (total flow rate of 100 mL/min); then, the background spectrum at a desired temperature was collected. Then, the catalysts were exposed under 500 ppm NH_3/N_2 or 500 ppm NO/N_2 and 11 vol% O_2 gas for 40 min (total flow rate of 100 mL/min), after which they were flushed under N_2 gas for 60 min.

5. Conclusions

This research investigated and contrasted the effects of four distinct zinc species ($ZnCl_2$, $Zn(NO_3)_2$, $ZnSO_4$, and $ZnCO_3$) on the Mn-Ce co-doped CuX catalyst in a low-temperature NH_3-SCR process. Poisoned catalysts were made by impregnating in aqueous solutions of zinc chloride, zinc nitrate, zinc sulfate, and zinc carbonate. All catalysts' NH_3-SCR activity were quantified, and the physicochemical characteristics were studied. There was a notable drop in catalytic activity for the catalyst in the low temperature range after zinc species poisoning, with the order of poisoning going as follows: $ZnSO_4 > ZnCO_3 > Zn(NO_3)_2 > ZnCl_2$. Zn species affected the reduction of the MCCX catalyst, especially isolated Cu^{2+} ions in lattice zeolite X, leading to a decrease in the SCR activity. $ZnSO_4$'s poisoning effect on MCCX catalysts was much more pronounced on the nitrate and nitrite intermediate products than that of $ZnCO_3$, $ZnCl_2$, and $Zn(NO_3)_2$, and all the catalysts were governed by both the Langmuir–Hinshelwood (L–H) and Eley–Rideal (E–R) mechanism as evidenced by the in situ DRIFTS spectra.

Author Contributions: L.C.: conceptualization, methodology, writing—original draft, investigation, formal analysis. S.R.: conceptualization, supervision, methodology, resources, funding acquisition, writing—review and editing, resources. T.C.: investigation, formal analysis. X.L.: investigation, formal analysis. Z.C.: investigation, visualization. M.W.: validation, writing—review and editing. Q.L.: conceptualization, project administration. J.Y.: investigation, formal analysis. All authors have read and agreed to the published version of the manuscript.

Funding: This research was funded by the National Natural Science Foundation of China, grant No. 52174298, and Shanxi Provincial Innovation Capacity Support Plan, grant number No. 2023-CX-TD-53.

Data Availability Statement: We are happy to share our research data.

Conflicts of Interest: The authors declare no conflict of interest.

References

1. Anenberg, S.C.; Miller, J.; Minjares, R.; Du, L.; Henze, D.K.; Lacey, F.; Malley, C.S.; Emberson, L.; Franco, V.; Klimont, Z.; et al. Impacts and mitigation of excess diesel-related NO_x emissions in 11 major vehicle markets. *Nature* **2017**, *545*, 467–471. [CrossRef]
2. Wuebbles, D.J. Atmosphere. Nitrous oxide: No laughing matter. *Science* **2009**, *326*, 56–57. [CrossRef] [PubMed]
3. Peng, H.; Dong, T.; Yang, S.; Chen, H.; Yang, Z.; Liu, W.; He, C.; Wu, P.; Tian, J.; Peng, Y.; et al. Intra-crystalline mesoporous zeolite encapsulation-derived thermally robust metal nanocatalyst in deep oxidation of light alkanes. *Nat. Commun.* **2022**, *13*, 295. [CrossRef] [PubMed]
4. Inomata, Y.; Kubota, H.; Hata, S.; Kiyonaga, E.; Morita, K.; Yoshida, K.; Sakaguchi, N.; Toyao, T.; Shimizu, K.I.; Ishikawa, S.; et al. Bulk tungsten-substituted vanadium oxide for low-temperature NO_x removal in the presence of water. *Nat. Commun.* **2021**, *12*, 557. [CrossRef] [PubMed]
5. Song, I.; Lee, H.; Jeon, S.W.; Ibrahim, I.A.M.; Kim, J.; Byun, Y.; Koh, D.J.; Han, J.W.; Kim, D.H. Simple physical mixing of zeolite prevents sulfur deactivation of vanadia catalysts for NO_x removal. *Nat. Commun.* **2021**, *12*, 901. [CrossRef]
6. Guo, Y.; Xu, X.; Gao, H.; Zheng, Y.; Luo, L.; Zhu, T. Ca-poisoning effect on V_2O_5-WO_3/TiO_2 and V_2O_5-WO_3-CeO_2/TiO_2 catalysts with different vanadium loading. *Catalysts* **2021**, *11*, 445. [CrossRef]
7. Negri, C.; Selleri, T.; Borfecchia, E.; Martini, A.; Lomachenko, K.A.; Janssens, T.V.W.; Cutini, M.; Bordiga, S.; Berlier, G. Structure and reactivity of oxygen-bridged diamino dicopper(II) complexes in Cu-ion-exchanged Chabazite catalyst for NH_3-mediated selective catalytic reduction. *J. Am. Chem. Soc.* **2020**, *142*, 15884–15896. [CrossRef]

8. Hu, W.; Gramigni, F.; Nasello, N.D.; Usberti, N.; Iacobone, U.; Liu, S.; Nova, I.; Gao, X.; Tronconi, E. Dynamic binuclear CuII sites in the reduction half-cycle of low-temperature NH$_3$–SCR over Cu-CHA catalysts. *ACS Catal.* **2022**, *12*, 5263–5274. [CrossRef]
9. Wang, H.; Jia, J.; Liu, S.; Chen, H.; Wei, Y.; Wang, Z.; Zheng, L.; Wang, Z.; Zhang, R. Highly efficient NO abatement over Cu-ZSM-5 with special nanosheet features. *Environ. Sci. Technol.* **2021**, *55*, 5422–5434. [CrossRef]
10. Hu, W.; Iacobone, U.; Gramigni, F.; Zhang, Y.; Wang, X.; Liu, S.; Zheng, C.; Nova, I.; Gao, X.; Tronconi, E. Unraveling the hydrolysis of Z_2Cu^{2+} to $ZCu^{2+}(OH)^-$ and its consequences for the low-temperature selective catalytic reduction of NO on Cu-CHA catalysts. *ACS Catal.* **2021**, *11*, 11616–11625. [CrossRef]
11. Abdul Nasir, J.; Guan, J.; Keal, T.W.; Desmoutier, A.W.; Lu, Y.; Beale, A.M.; Catlow, C.R.A.; Sokol, A.A. Influence of solvent on selective catalytic reduction of nitrogen oxides with ammonia over Cu-CHA zeolite. *J. Am. Chem. Soc.* **2023**, *145*, 247–259. [CrossRef] [PubMed]
12. Tarach, K.A.; Jabłońska, M.; Pyra, K.; Liebau, M.; Reiprich, B.; Gläser, R.; Góra-Marek, K. Effect of zeolite topology on NH$_3$-SCR activity and stability of Cu-exchanged zeolites. *Appl. Catal. B* **2021**, *284*, 119752. [CrossRef]
13. Chen, L.; Ren, S.; Xing, X.; Yang, J.; Yang, J.; Wang, M.; Chen, Z.; Liu, Q. Low-cost CuX catalyst from blast furnace slag waste for low-temperature NH$_3$-SCR: Nature of Cu active sites and influence of SO_2/H_2O. *ACS Sustain. Chem. Eng.* **2022**, *10*, 7739–7751. [CrossRef]
14. Chen, L.; Ren, S.; Jiang, Y.; Liu, L.; Wang, M.; Yang, J.; Chen, Z.; Liu, W.; Liu, Q. Effect of Mn and Ce oxides on low-temperature NH$_3$-SCR performance over blast furnace slag-derived zeolite X supported catalysts. *Fuel* **2022**, *320*, 123969. [CrossRef]
15. Wang, X.; Liu, Y.; Wu, Z. The poisoning mechanisms of different zinc species on a ceria-based NH$_3$-SCR catalyst and the co-effects of zinc and gas-phase sulfur/chlorine species. *J. Colloid Interface Sci.* **2020**, *566*, 153–162. [CrossRef]
16. Su, Z.; Ren, S.; Yang, J.; Yao, L.; Zhou, Y.; Chen, Z.; Zhang, T. Poisoning effect comparison of $ZnCl_2$ and $ZnSO_4$ on Mn-Ce/AC catalyst for low-temperature SCR of NO. *ChemistrySelect* **2020**, *5*, 9226–9234. [CrossRef]
17. Zhou, Y.; Su, B.; Ren, S.; Chen, Z.; Su, Z.; Yang, J.; Chen, L.; Wang, M. Nb_2O_5-modified Mn-Ce/AC catalyst with high $ZnCl_2$ and SO_2 tolerance for low-temperature NH$_3$-SCR of NO. *J. Environ. Chem. Eng.* **2021**, *9*, 106323. [CrossRef]
18. Chen, L.; Ren, S.; Peng, H.; Yang, J.; Wang, M.; Chen, Z.; Liu, Q. Low-cost Mn-Ce/CuX catalyst from blast furnace slag waste for efficient low-temperature NH$_3$-SCR. *Appl. Catal. A-Gen.* **2022**, *646*, 118868. [CrossRef]
19. Usberti, N.; Gramigni, F.; Nasello, N.D.; Iacobone, U.; Selleri, T.; Hu, W.; Liu, S.; Gao, X.; Nova, I.; Tronconi, E. An experimental and modelling study of the reactivity of adsorbed NH$_3$ in the low temperature NH$_3$-SCR reduction half-cycle over a Cu-CHA catalyst. *Appl. Catal. B* **2020**, *279*, 119397. [CrossRef]
20. Guo, A.; Xie, K.; Lei, H.; Rizzotto, V.; Chen, L.; Fu, M.; Chen, P.; Peng, Y.; Ye, D.; Simon, U. Inhibition effect of phosphorus poisoning on the dynamics and redox of Cu active sites in a Cu-SSZ-13 NH$_3$-SCR catalyst for NO$_x$ reduction. *Environ. Sci. Technol.* **2021**, *55*, 12619–12629. [CrossRef]
21. Chen, P.; Rauch, D.; Weide, P.; Schönebaum, S.; Simons, T.; Muhler, M.; Moos, R.; Simon, U. The effect of Cu and Fe cations on NH$_3$-supported proton transport in DeNO$_x$-SCR zeolite catalysts. *Catal. Sci. Technol.* **2016**, *6*, 3362–3366. [CrossRef]
22. Li, G.; Mao, D.; Chao, M.; Li, G.; Yu, J.; Guo, X. Significantly enhanced Pb resistance of a Co-modified Mn–Ce–O$_x$/TiO$_2$ catalyst for low-temperature NH$_3$-SCR of NO$_x$. *Catal. Sci. Technol.* **2020**, *10*, 6368–6377. [CrossRef]
23. Chen, L.; Ren, S.; Xing, X.; Yang, J.; Li, X.; Wang, M.; Chen, Z.; Liu, Q. Poisoning mechanism of KCl, K_2O and SO_2 on Mn-Ce/CuX catalyst for low-temperature SCR of NO with NH$_3$. *Process Saf. Environ. Prot.* **2022**, *167*, 609–619. [CrossRef]
24. Fang, X.; Liu, Y.; Cheng, Y.; Cen, W. Mechanism of Ce-modified birnessite-MnO$_2$ in promoting SO$_2$ poisoning resistance for low-temperature NH$_3$-SCR. *ACS Catal.* **2021**, *11*, 4125–4135. [CrossRef]
25. Yang, W.; Su, Z.a.; Xu, Z.; Yang, W.; Peng, Y.; Li, J. Comparative study of α-, β-, γ- and δ-MnO$_2$ on toluene oxidation: Oxygen vacancies and reaction intermediates. *Appl. Catal. B* **2020**, *260*, 118150. [CrossRef]
26. Fan, H.; Fan, J.; Chang, T.; Wang, X.; Wang, X.; Huang, Y.; Zhang, Y.; Shen, Z. Low-temperature Fe–MnO$_2$ nanotube catalysts for the selective catalytic reduction of NO$_x$ with NH$_3$. *Catal. Sci. Technol.* **2021**, *11*, 6553–6563. [CrossRef]
27. Yan, R.; Lin, S.; Li, Y.; Liu, W.; Mi, Y.; Tang, C.; Wang, L.; Wu, P.; Peng, H. Novel shielding and synergy effects of Mn-Ce oxides confined in mesoporous zeolite for low temperature selective catalytic reduction of NO$_x$ with enhanced SO_2/H_2O tolerance. *J. Hazard. Mater.* **2020**, *396*, 122592. [CrossRef]
28. Chen, J.; Zhao, R.; Zhou, R. A new insight into active Cu^{2+} species properties in one-pot synthesized Cu-SSZ-13 catalysts for NO$_x$ reduction by NH$_3$. *ChemCatChem* **2018**, *10*, 5182–5189. [CrossRef]
29. Chen, B.; Xu, R.; Zhang, R.; Liu, N. Economical way to synthesize SSZ-13 with abundant ion-exchanged Cu$^+$ for an extraordinary performance in selective catalytic reduction (SCR) of NO$_x$ by ammonia. *Environ. Sci. Technol.* **2014**, *48*, 13909–13916. [CrossRef]
30. Yao, X.; Zhang, L.; Li, L.; Liu, L.; Cao, Y.; Dong, X.; Gao, F.; Deng, Y.; Tang, C.; Chen, Z. Investigation of the structure, acidity, and catalytic performance of CuO/Ti$_{0.95}$Ce$_{0.05}$O$_2$ catalyst for the selective catalytic reduction of NO by NH$_3$ at low temperature. *Appl. Catal. B* **2014**, *150–151*, 315–329. [CrossRef]
31. Kang, K.; Yao, X.; Huang, Y.; Cao, J.; Rong, J.; Zhao, W.; Luo, W.; Chen, Y. Insights into the co-doping effect of Fe^{3+} and Zr^{4+} on the anti-K performance of CeTiO$_x$ catalyst for NH$_3$-SCR reaction. *J. Hazard. Mater.* **2021**, *416*, 125821. [CrossRef] [PubMed]
32. Wei, L.; Wang, Z.; Liu, Y.; Guo, G.; Dai, H.; Cui, S.; Deng, J. Support promotion effect on the SO$_2$ and K$^+$ co-poisoning resistance of MnO$_2$/TiO$_2$ for NH$_3$-SCR of NO. *J. Hazard. Mater.* **2021**, *416*, 126117. [CrossRef]
33. Yu, Y.; Geng, M.; Li, J.; Wang, J.; Wei, D.; He, C. The different effect of SO$_2$ on Zn-poisoned commercial V_2O_5-WO$_3$/TiO$_2$ catalysts with varied Zn loading. *Chem. Phys. Impact* **2023**, *6*, 100150. [CrossRef]

34. Xue, H.; Guo, X.; Meng, T.; Mao, D.; Ma, Z. Poisoning effect of K with respect to Cu/ZSM-5 used for NO reduction. *Colloid Interface Sci. Commun.* **2021**, *44*, 100465. [CrossRef]
35. Negri, C.; Hammershoi, P.S.; Janssens, T.V.W.; Beato, P.; Berlier, G.; Bordiga, S. Investigating the low temperature formation of Cu(II)-(N,O) species on Cu-CHA zeolites for the selective catalytic reduction of NO_x. *Chemistry* **2018**, *24*, 12044–12053. [CrossRef] [PubMed]
36. Wang, Z.; Guo, R.; Shi, X.; Liu, X.; Qin, H.; Liu, Y.; Duan, C.; Guo, D.; Pan, W. The superior performance of $CoMnO_x$ catalyst with ball-flowerlike structure for low-temperature selective catalytic reduction of NO_x by NH_3. *Chem. Eng. J.* **2020**, *381*, 122753. [CrossRef]
37. Zhang, Y.; Zhu, H.; Zhang, T.; Li, J.; Chen, J.; Peng, Y.; Li, J. Revealing the synergistic deactivation mechanism of hydrothermal aging and SO_2 poisoning on Cu/SSZ-13 under SCR condition. *Environ. Sci. Technol.* **2022**, *56*, 1917–1926. [CrossRef]
38. Wang, X.; Liu, Y.; Wu, Z. Highly active $NbOPO_4$ supported Cu-Ce catalyst for NH_3-SCR reaction with superior sulfur resistance. *Chem. Eng. J.* **2020**, *382*, 122941. [CrossRef]
39. Stanciulescu, M.; Bulsink, P.; Caravaggio, G.; Nossova, L.; Burich, R. NH_3-TPD-MS study of Ce effect on the surface of Mn- or Fe-exchanged zeolites for selective catalytic reduction of NO_x by ammonia. *Appl. Surf. Sci.* **2014**, *300*, 201–207. [CrossRef]
40. Zhou, X.; Huang, X.; Xie, A.; Luo, S.; Yao, C.; Li, X.; Zuo, S. V_2O_5-decorated Mn-Fe/attapulgite catalyst with high SO_2 tolerance for SCR of NO_x with NH_3 at low temperature. *Chem. Eng. J.* **2017**, *326*, 1074–1085. [CrossRef]
41. Li, Y.; Deng, J.; Song, W.; Liu, J.; Zhao, Z.; Gao, M.; Wei, Y.; Zhao, L. Nature of Cu species in Cu–SAPO-18 catalyst for NH_3–SCR: Combination of experiments and DFT calculations. *J. Phys. Chem. C* **2016**, *120*, 14669–14680. [CrossRef]
42. Yang, J.; Ren, S.; Zhou, Y.; Su, Z.; Yao, L.; Cao, J.; Jiang, L.; Hu, G.; Kong, M.; Yang, J.; et al. In situ IR comparative study on N_2O formation pathways over different valence states manganese oxides catalysts during NH_3–SCR of NO. *Chem. Eng. J.* **2020**, *397*, 125446. [CrossRef]
43. Li, Q.; Gu, H.C.; Li, P.; Zhou, Y.H.; Liu, Y.; Qi, Z.N.; Xin, Y.; Zhang, Z.L. In situ IR studies of selective catalytic reduction of NO with NH_3 on Ce-Ti amorphous oxides. *Chin. J. Catal.* **2014**, *35*, 1289. [CrossRef]
44. Yang, S.; Xiong, S.; Liao, Y.; Xiao, X.; Qi, F.; Peng, Y.; Fu, Y.; Shan, W.; Li, J. Mechanism of N2O formation during the low-temperature selective catalytic reduction of NO with NH_3 over Mn-Fe spinel. *Environ. Sci. Technol.* **2014**, *48*, 10354–10362. [CrossRef] [PubMed]
45. Yao, L.; Liu, Q.; Mossin, S.; Nielsen, D.; Kong, M.; Jiang, L.; Yang, J.; Ren, S.; Wen, J. Promotional effects of nitrogen doping on catalytic performance over manganese-containing semi-coke catalysts for the NH_3-SCR at low temperatures. *J. Hazard. Mater.* **2020**, *387*, 121704. [CrossRef] [PubMed]

Disclaimer/Publisher's Note: The statements, opinions and data contained in all publications are solely those of the individual author(s) and contributor(s) and not of MDPI and/or the editor(s). MDPI and/or the editor(s) disclaim responsibility for any injury to people or property resulting from any ideas, methods, instructions or products referred to in the content.

Article

Mechanistic Insight into the Propane Oxidation Dehydrogenation by N₂O over Cu-BEA Zeolite with Diverse Active Site Structures

Ruiqi Wu, Ning Liu *, Chengna Dai, Ruinian Xu, Gangqiang Yu, Ning Wang and Biaohua Chen

Faculty of Environment and Life, Beijing University of Technology, Beijing 100124, China; b202265055@emails.bjut.edu.cn (R.W.); daicn@bjut.edu.cn (C.D.); xuruinian@bjut.edu.cn (R.X.); yugq@bjut.edu.cn (G.Y.); ning.wang.1@bjut.edu.cn (N.W.); chenbh@bjut.edu.cn (B.C.)
* Correspondence: liuning@bjut.edu.cn

Abstract: The present work theoretically investigated propane oxidation dehydrogenation by utilizing N_2O as an oxidant (N_2O-ODHP) over Cu-BEA with three different types of active site, including monomeric Cu ($[Cu]^+$), dimeric Cu ($[Cu-Cu]^{2+}$), and distant monomeric Cu sites ($[Cu]^+$—$[Cu]^+$). Energetically, we calculated that the monomeric $[Cu]^+$ is favorable for the αH dehydrogenation step (ΔE = 0.05 eV), which, however, suffers from high barriers of N_2O dissociation and βH dehydrogenation steps of 1.40 and 1.94 eV, respectively. Although the dimeric $[Cu-Cu]^{2+}$ site with a Cu—Cu distance of 4.91 Å is much more favorable for N_2O dissociation (0.95 eV), it still needs to overcome an extremely high barrier (ΔE = 2.15 eV) for βH dehydrogenation. Interestingly, the distant $[Cu]^+$—$[Cu]^+$ site with the Cu—Cu distance of 5.82 Å exhibits low energy barriers for N_2O dissociation (0.89 eV) and ODHP steps (0.01 and 0.33 eV) due to the synergistic effect of distant $[Cu]^+$. The microkinetic analyses quantitatively verified the superior activity of the distant $[Cu]^+$—$[Cu]^+$ site with a reaction rate being eight to nine orders of magnitude higher than those of the monomeric and the dimeric Cu sites, and this is related to its ready charge-transfer ability, as shown by the partial Density of State (PDOS) analysis and the static charge differential density analysis in this study. Generally, the present work proposes that the distance between the $[Cu]^+$ sites plays a significant and important role in N_2O-ODHP over the Cu-based zeolite catalyst and modulates Cu—Cu distance, and this constitutes a promising strategy for highly-efficient Cu-zeolite catalyst design for N_2O-ODHP.

Keywords: propane oxidation dehydrogenation (ODHP); N_2O; Cu-BEA; density functional theory (DFT); microkinetic modelling

Citation: Wu, R.; Liu, N.; Dai, C.; Xu, R.; Yu, G.; Wang, N.; Chen, B. Mechanistic Insight into the Propane Oxidation Dehydrogenation by N₂O over Cu-BEA Zeolite with Diverse Active Site Structures. *Catalysts* **2023**, *13*, 1212. https://doi.org/10.3390/catal13081212

Academic Editors: De Fang and Yun Zheng

Received: 19 July 2023
Revised: 13 August 2023
Accepted: 14 August 2023
Published: 15 August 2023

Copyright: © 2023 by the authors. Licensee MDPI, Basel, Switzerland. This article is an open access article distributed under the terms and conditions of the Creative Commons Attribution (CC BY) license (https://creativecommons.org/licenses/by/4.0/).

1. Introduction

Propylene is one of the most important organic raw materials, and it can be used to synthesize petrochemical products, such as polyurethane, polypropylene, acetone, acrylonitrile, polyacrylonitrile, and propylene oxide. The traditional commercial production of propylene is mainly through fluid catalytic cracking (FCC) and steam cracking (SC) of petroleum by-products, such as naphtha and light diesel oil [1,2]. However, relatively low selectivity as well as limited resources cannot meet the growing demand of propylene, thereby making it highly desirable to develop some efficient and economical methods to produce propylene [3–5]. The massive exploitation and use of shale gas has increased the production of low-carbon alkanes, and propane has thus become a cheap chemical raw material. Converting abundant propane into propylene is not only an important topic in the field of the petrochemical industry but is also a research hotspot in the field of heterogeneous catalysis [4,5], and it is a promising means of meeting the huge demand of the propylene market [2,6].

The catalytic dehydrogenation of propane techniques include the direct dehydrogenation of propane (PDH) and the oxidative dehydrogenation of propane (ODHP). The direct PDH has been industrialized: one example is the Catofin process using chromium aluminum oxide as a catalyst by the Lummus Company [7,8], and another is the Oleflex process from UOP utilizing platinum-based catalysts [7,8]. However, the direct PDH is a reversible and strongly-endothermic reaction that is limited by thermodynamic equilibrium. Moreover, C−C bond breaking to produce methane/ethylene is much more likely to occur at high temperatures, which can further lower propylene selectivity [7,8]. Due to these shortcomings, researchers have conducted studies on the catalytic dehydrogenation of propane under aerobic conditions. Recently, it has been reported that utilizing N_2O as the oxidant is much more effective than O_2 for the ODHP [9–11]. For example, Bulanek et al. [9] have reported that the N_2O-ODHP is much more efficient relative to that of O_2-ODHP by displaying its higher propane conversion rate and its propylene selectivity. Katerinas et al. [10] have also reported that the selectivity of propylene increased from 33.6% to 68% when using N_2O as the oxidant, which can also improve the conversion rate of propane [12].

At the current stage, researchers are still trying to find the suitable catalyst for the N_2O-ODHP, and the zeolite catalyst is notable here due to its excellent N_2O dissociation activity, which generates αO [13]. The BEA zeolite, with a unique twelve-ring structure, possesses better N_2O dissociation activity than those of Y, MFI, FAU, and MOR [14,15]. Sobalikset al. [16] have reported that Cu (II) in BEA is the active site of N_2O decomposition. In our previous works, the catalytic dissociation of N_2O [17] as well as the N_2O oxidation of methane into methanol [18] were also investigated over the Cu-BEA, and it was found that both the monomeric and the dimeric Cu sites can function as the active sites for N_2O dissociation to generate αO. In the present work, the N_2O-ODHP was theoretically investigated over the Cu-BEA zeolite with diverse active site structures, including monomeric $[Cu]^+$, dimeric $[Cu-Cu]^{2+}$, and distant $[Cu]^+-[Cu]^+$ sites. The Mars van Krevelen mechanism is related to the reaction between the reactant and the lattice oxygen of the oxidation catalyst [19,20]. The first step is the oxidation of the reductant by the lattice oxygen of the catalyst that generates the product, with a simultaneous formation of oxygen vacancy. The second step is the regeneration of the catalytic active site through the dissociated oxygen in order to replenish oxygen vacancies. Being similar to such a mechanism, in the present work, the αO functions as the active site, and it is utilized to oxidize C_3H_8 into C_3H_6 and H_2O. The αO would thereby be further regenerated through the reoxidation by N_2O. The specific reaction mechanisms were well illustrated by DFT, and the microkinetic modeling was further conducted in order to quantitatively compare the reaction rates of the diverse active sites. Generally, the present work aims to shed a deeper mechanistic light on the active-site motif structural effect on N_2O-ODHP and, moreover, emphasize that modulating the Cu—Cu distance would constitute a promising method for a highly efficient Cu-based zeolite catalyst design for the N_2O-ODHP.

2. Result and Discussion

2.1. N_2O-ODHP Mechanism Simulation over Diverse ACTIVE Site

Three types of Cu-BEA models with different active centers (Figure S1a,c) were constructed for the N_2O-ODHP mechanism simulations, which comprised three steps: (i) N_2O dissociation to produce αO with the simultaneous release of N_2, and the dehydrogenation of (ii) the αH and (ii) the βH of C_3H_8. The derived energy diagrams along with a different reaction route (Routes A–C) over these active sites are depicted in Figure 1a–d, and based on this, we conducted an in-depth analysis of the specific reaction pathways and the evolution of the structures intermediately generated, and we then compared the derived energy barrier in order to determine the optimal active center for N_2O-ODHP.

2.1.1. N_2O-ODHP over Monomeric $[Cu]^+$ Site of Route A

(a) Dissociation of N_2O to form αO (Reaction Step A1). In this part, the N_2O-ODHP over the monomeric $[Cu]^+$ site of Cu-BEA (noted as Z^a−Cu) was simulated by DFT. Firstly, the N_2O molecule can be adsorbed over the Z^a−Cu site through the O end with a bond length of 2.00 Å. After overcoming a relatively high energy barrier of 1.40 eV (TSA1;

Figure 1a), the αO can be generated. As noted, compared with the structure of the N_2O-adsorption state, the bond of [CuO]−N was elongated from 1.22 to 1.89 Å, and the bond angle of Cu−O−N_2 bent from 121.4 to 124.0° (Figure 2a,b). Such structural distortion can be related to the pre-activation effects of the monomeric $[Cu]^+$. The αO of $[Cu-O]^+$ can be formed, as shown in Figure 2c, with a [CuO]−N bond of 3.76 Å.

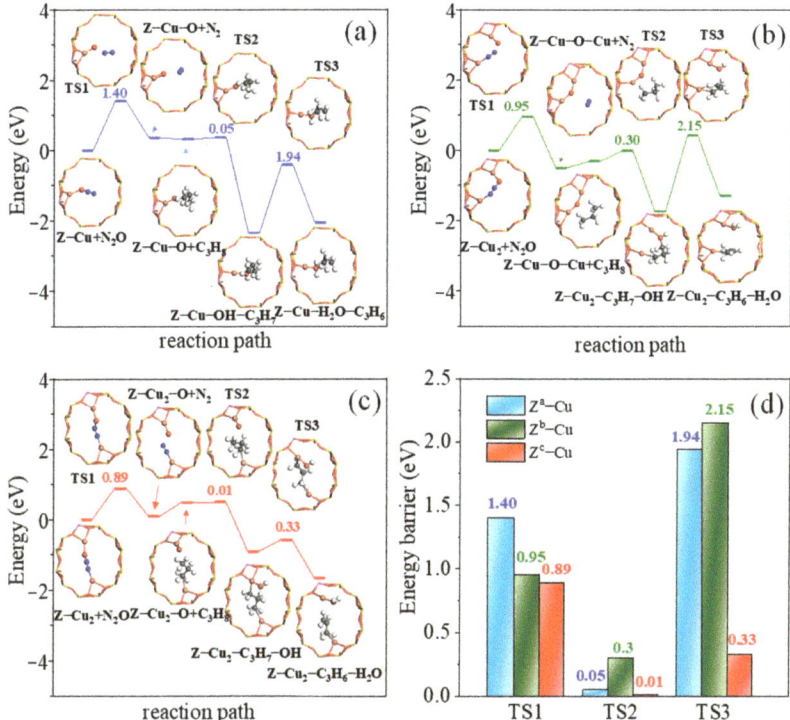

Figure 1. Energy diagram of N_2O-ODHP over Cu-BEA with different active sites of (**a**) Z^a−Cu (reaction Route A); (**b**) Z^b−Cu (reaction Route B); (**c**) Z^c−Cu (reaction Route C); and (**d**) energy barrier comparisons. Si (yellow), O (red), N (blue), Al (pink), Cu (orange).

(b) Propane dehydrogenation of αH (Reaction Step A2). Reaction Route A2 describes the propane dehydrogenation of αH over Z^a−Cu−O, wherein there exists one bond fracture of H−C_3H_7 and two bond formations of [Cu−O]−H and [Cu−O]−C_3H_7 (the αO being connected with the subtracted H and the radical of C_3H_7-). As shown in Figure 2d,e, the distance of O−αH slightly shrunk from 2.26 to 1.84 Å, and the bond of C−αH extended from 1.11 to 1.13 Å after the adsorption of C_3H_8 (ΔE = −0.06 eV, Figure 1a). The αH dehydrogenation occurred by crossing the energy barrier of 0.05 eV through the TSA2, which can be characterized by the Cu−O−C bond angle of 100.6° and the Cu−O−H bond angle of 97.5°. After that, the αH can be subtracted from C_3H_8, forming Z^a−Cu−OH with the OH bond of 0.97 Å (Figure 2e). In comparison to 1.49 eV of αH dehydrogenation over the CeO_2(111) [21], it would be easier for αH dehydrogenation to occur over the monomeric $[Cu]^+$ site of Cu-BEA.

(c) Propane dehydrogenation of βH to form propylene (Reaction Step A3). The dehydrogenation of βH occurred in Route A3 through another transition state of TSA3 with a greatly higher energy barrier of 1.94 eV (Figure 1a), wherein the βH can migrate from C_3H_7 to Z^a−Cu−OH, finally generating the C_3H_6 and the H_2O. The distance of the [CuO]−βH shrunk from 2.62 to 1.72 Å, and the bond of C−αH extended from 1.11 to 1.28 Å and formed the Z^a–Cu–H_2O–C_3H_8 that can be seen in Figure 2f,g. Therefore, according to the above DFT energy calculations, we can derive that the monomeric $[Cu]^+$ site possessing a relatively high barrier for N_2O-ODHP, especially for the βH dehydrogenation step, is not potentially active for the N_2O-ODHP.

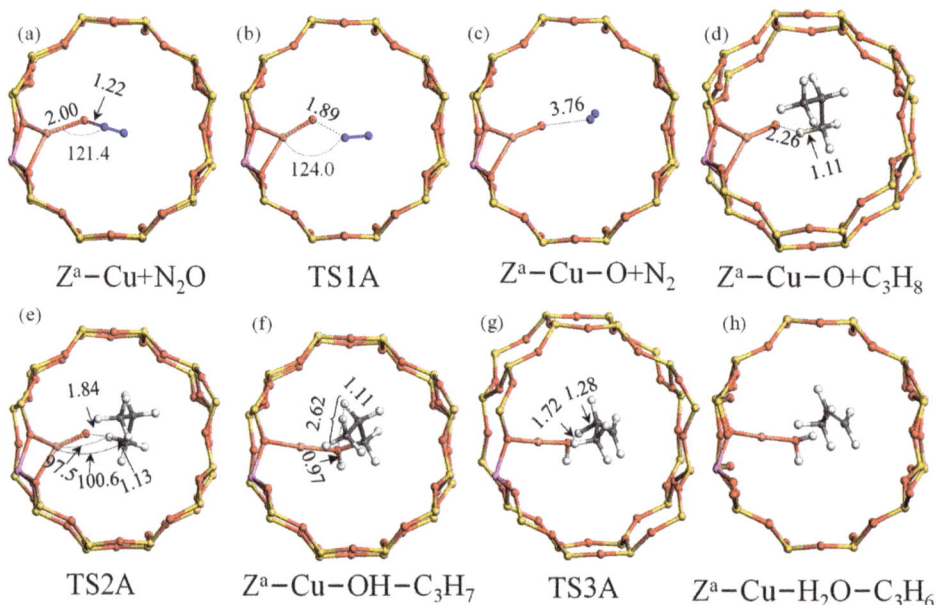

Figure 2. Optimized model of Cu-BEA with Z^a–Cu site for the N_2O-ODHP: (a) Z^a–Cu+N_2O (adsorption fo N_2O); (b) TS1A; (c) Z^a–Cu–O+N_2; (d) Z^a–Cu–O+C_3H_8 (adsorption of C_3H_8); (e) TS2A; (f) Z^a–Cu–OH–C_3H_7; (g) TS3A; (h) Z^a–Cu–H_2O–C_3H_6. Si (yellow), O (red), N (blue), Al (pink), Cu (orange).

2.1.2. N_2O-ODHP over Dimeric $[Cu-Cu]^{2+}$ Site of Route B

(a) Dissociation of N_2O to form αO (Reaction Step B1). In this part, the N_2O-ODHP was simulated over the dimeric $[Cu-Cu]^{2+}$ site (noted as Z^b–Cu). As shown in Figure 3a, the N_2O could be absorbed over Z^b–Cu through both its O and N end, with a Cu–ON_2 bond of 1.95 Å, a Cu–N_2O bond of 1.80 Å, and an N–N–O bond angle of 175.0°. Much stronger structural distortion can be observed for the adsorbed N_2O relative to that of the N_2O being adsorbed over the monomeric $[Cu]^+$ site, and this is closely related to the strong synergistic effect of the dimeric $[Cu-Cu]^{2+}$. Moreover, due to such a synergistic effect, the N_2O can be readily (ΔE = 0.95 eV, Figure 1b) dissociated to generate the αO ($[Cu-O-Cu]^{2+}$, Figure 3c) and the N_2 through the TS1B. Further comparing the structures of the adsorption state (Figure 3a) and the TS1B (Figure 3b), the bond length of Cu–ON_2 shrunk from 1.95 to 1.75 Å, while the N_2–O bond enlarged from 1.79 to 1.80 Å, and the bond angle of N–N–O decreased from 112.9 to 118.9°.

(b) Propane dehydrogenation of αH (Reaction step B2). The C_3H_8 can be initially adsorbed over Z^b–Cu–O–Cu (Figure 3d). Subsequently, the αH would migrate from C_3H_8 to the αO through the TS2B (Figure 3e), which is characterized by the Cu_a–O bond of 1.80 Å, the Cu_b–O bond of 1.79 Å, and the Cu–O–Cu bond angle of 143.2°. This crosses a low energy barrier of 0.30 eV (Figure 1b). After that, an intermediate structure of Z^b–Cu_2–C_3H_7–OH (Figure 3f) with a Cu_a–OH bond of 1.92 Å, a Cu_b–C bond of 2.11 Å, and a C–C–C bond angle of 115.9° can be formed, wherein the radical of the C_3H_7- can be inserted into the Z^b–Cu–O–Cu site.

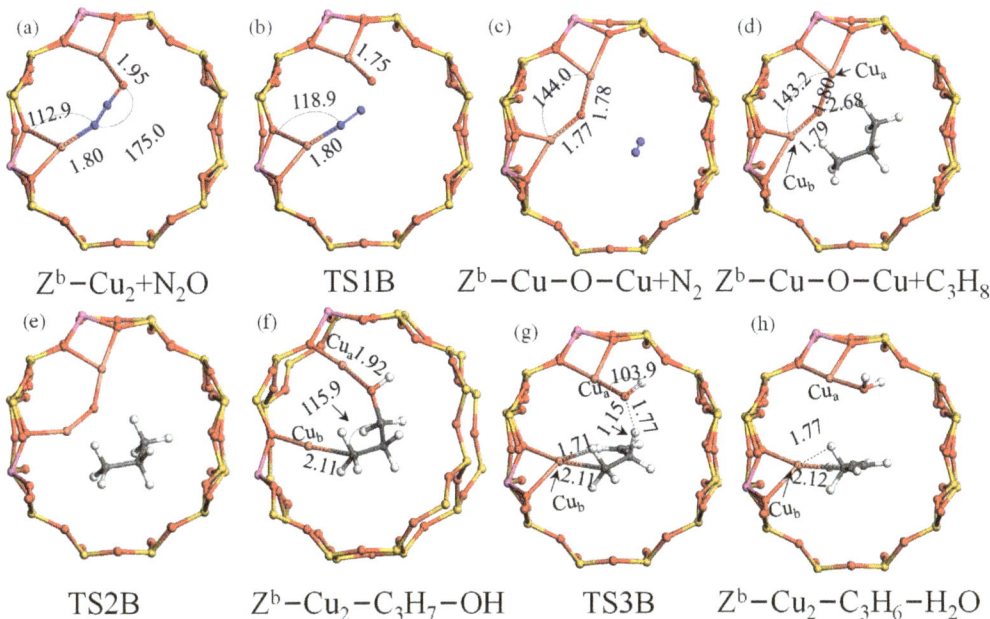

Figure 3. Optimized model of Cu-BEA with Z^b–Cu site for the N_2O-ODHP: (**a**) Z^b–Cu_2+N_2O (adsorption fo N_2O); (**b**) TS1B; (**c**) Z^b–Cu–O–Cu+N_2; (**d**) Z^b–Cu–O–Cu+C_3H_8 (adsorption of C_3H_8); (**e**) TS2B; (**f**) Z^b–Cu_2–OH–C_3H_7; (**g**) TS3B; (**h**) Z^b–Cu_2–C_3H_6–H_2O. Si (yellow), O (red), N (blue), Al (pink), Cu (orange).

(c) Propane dehydrogenation of βH to form propylene (Reaction Step B3). After Reaction Step B2, the βH further migrates to the αO of the Z^b–Cu_2–C_3H_7–OH site, producing C_3H_6 and H_2O (Figure 3h), which, however, needs to overcome a significantly high barrier of 2.15 eV (Figure 1b) that is comparable to the value of 1.94 eV (Figure 1b) for the scenario of the monomeric $[Cu]^+$ site. The TS3B (Figure 3g) can be characterized by a Cu_a–O bond of 1.82 Å, an O–βH bond of 1.77Å, a C–βH of 1.15 Å, and a Cu–O–αH bond angle of 103.9°. As noted, the Cu_b site forms a bridge with H and C in TS3B (Cu_b–H of 1.71 Å and Cu_b–C of 2.11 Å). Subsequently, the C–H bond would expand from 1.71 to 1.77 Å, eventually leading to the C–H bond being inviable in Figure 3h. Eventually, the C_3H_6 can be produced after the C–βH bond breaking and the Cu_b–C bond formation that is associated with the formation of H_2O. As noted, such a high barrier of 2.15 eV also indicates that the dimeric Cu site of Z^b–Cu is not suitable for the N_2O-ODHP.

2.1.3. N_2O-ODHP over Distant $[Cu]^+$—$[Cu]^+$ Site of Route C

(a) Dissociation of N_2O to form αO (Reaction step C1). The N_2O-ODHP was further simulated over the distant $[Cu]^+$—$[Cu]^+$ site noted as Z^c–Cu and with the Cu—Cu distance of 5.82 Å.

As shown in Figure 4a, given that it is similar to that of the dimeric $[Cu-Cu]^{2+}$ site, the N_2O molecule can also be adsorbed over Z^c–Cu through both its O and N end, with a bond length of 1.94 and 1.81 Å and a N–N–O bond angle of 173.8°. Relatively stronger structural distortion can be observed for the adsorbed N_2O over the Z^c–Cu site in comparison to that of the Z^b–Cu site, which indicates a stronger synergistic effect of the Z^c–Cu site compared to the Z^b–Cu for N_2O preactivation. A similar finding can also be observed for the N_2O dissociation step to generate αO, wherein the Z^c–Cu exhibits a relatively lower N_2O-dissociation energy barrier (0.89 eV, Figure 1c) than that of the $[Cu-Cu]^{2+}$ dimeric site (Z^b–Cu of 0.95 eV, Figure 1b), which is due to this type of synergistic effect. The TSC1

(Figure 4b) can be characterized by a Cu_a–O bond of the 1.81 Å, an N–N–O bond angle of 127.6°, and a Z^c–Cu-O bond angle of 147.9°. Most importantly, the Z^c–Cu site evolved into the motif structure of $[Cu-O]^+$—$[Cu]^+$ (Figure 4c, Cu–O bond of 1.71 Å), which contains two distant monomeric Cu sites that are greatly favorable to the further dehydrogenation of both the αH and the βH of the C_3H_8 molecule, as will be detailed below.

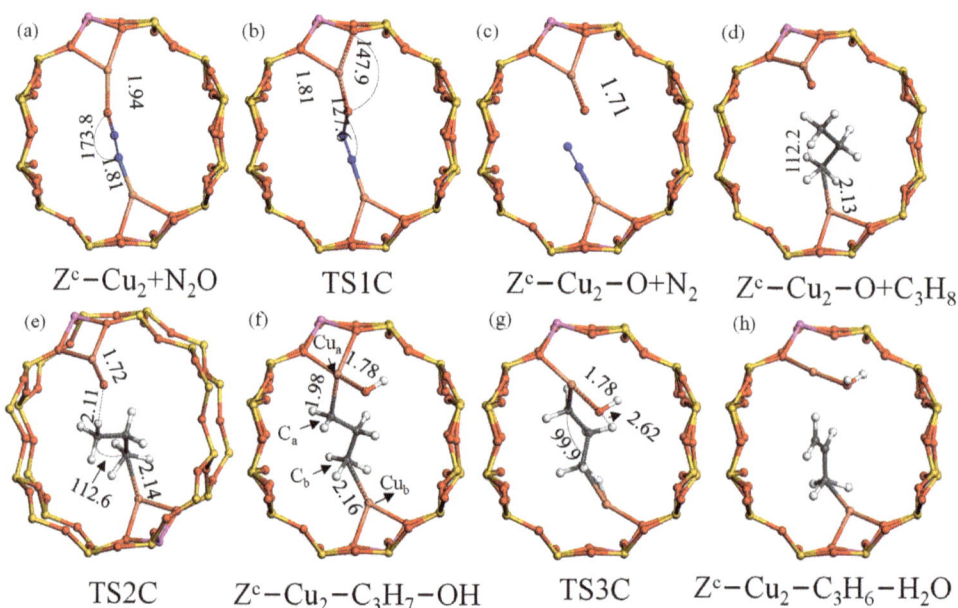

Figure 4. Optimized model of Cu-BEA with Z^c–Cu site for the N_2O-ODHP: (a) Z^c–Cu_2+N_2O (adsorption fo N_2O); (b) TS1C; (c) Z^c–Cu–O–Cu+N_2; (d) Z^c–Cu_2–O+C_3H_8 (adsorption of C_3H_8); (e) TS2C; (f) Z^c–Cu_2–C_3H_7–OH; (g) TS3C; (h) Z^c–Cu_2–C_3H_6–H_2O. Si (yellow), O (red), N (blue), Al (pink), Cu (orange).

(b) Propane dehydrogenation of αH (Reaction step C2). The C_3H_8 can be adsorbed over the $[Cu]^+$ site of $[Cu-O]^+$—$[Cu]^+$ through the C end, forming a C–Cu bond of 2.13 Å and a C_3H_8 (C–C–C) bond angle of 112.2°. As noted, much stronger structural distortion of C_3H_8 can also be observed over the Z^c–Cu site than that of the Z^b–Cu site, which eventually leads to an extremely low barrier of 0.01 eV (Figure 1c) during the αH dehydrogenation through a transition state of TS2C (Figure 4e), and this is characterized by a Cu–C bond of 2.14 Å, a Cu–O bond of 1.72 Å, a [Cu–O]–αH bond of 2.11 Å, and a C-C-C bond angle of 112.6°. Finally, an intermediate structure of Z^c–Cu_2–C_3H_7–OH (Figure 4f) can be formed with the Cu_a–OH bond of 1.78 Å. As noted, it is interesting to see that the generated C_3H_7- radical was well inserted between the distant $[Cu-OH]^+$—$[Cu]^+$ site, forming, respectively, a Cu_a–C bond of 1.98 Å and a Cu_b–C bond of 2.16 Å. This would be greatly favorable for βH dehydrogenation by taking advantage of the synergistic effect of the distant $[Cu-OH]^+$—$[Cu]^+$ site, as stated below in Reaction Step C3.

(c) Propane dehydrogenation of βH to form propylene (Reaction step C3). In this step, the βH would migrate from the C_3H_7- to the $[Cu_a$–OH$]^+$ site, generating the adsorbed H_2O and C_3H_6. Being totally different from the scenarios of both the monomeric and the dimeric Cu active sites (ΔE = 1.94 and 2.15 eV, respectively; Figure 1a,b), the βH can be readily dehydrogenated from C_3H_7- by crossing a significantly lower energy barrier of 0.33 eV (Figure 1c) over the Z^c–Cu site due to its strong synergistic effect. The TS3C can be characterized by a Cu-OH bond of 1.78 Å, a βH-O bond of 2.62 Å, and a C–C–C bond angle of 99.9° (Figure 4g). Carefully analyzing the motif structure of Z^b–Cu_2-C_3H_7-OH

(Figure 3f) and Z^c–Cu_2-C_3H_7-OH (Figure 4f), one can find that although the radical of C_3H_7- can be inserted between both the Z^b–Cu and the Z^c–Cu site, the specific adsorption mode was different for the two. The C_3H_7- was adsorbed over the Z^b–Cu site through the O–C_a and the Cu_b–C_c bond, and was adsorbed through the Cu_a–C_a and the Cu_b–C_b bond over the Z^c–Cu site. In this regard, the C_3H_7- has to break up the two bonds of O–C_a and βH–C_b in order to generate the C_3H_6 and H_2O during the transition state of the TS3B (as seen in Figure 3g), whereas, on the contrary, the C_3H_7- only needs to break up one βH–C_b bond during the transition state of the TS3C (seen in Figure 4g).

In the light of the above statements, we can therefore note that the Z^c–Cu site possessing the lowest energy barrier for N_2O dissociation (0.89 eV) as well as αH (0.01 eV) and βH (0.33 eV) dehydrogenation (Figure 1c,d), relative to those of the Z^b–Cu and the Z^a–Cu sites (especially for the βH dehydrogenation step), is the most active site for N_2O-ODHP. Moreover, this finding also shows that modulating the Cu—Cu distance may constitute a promising strategy for highly-efficient zeolite-based N_2O-ODHP catalyst design.

2.2. Microkinetic Modeling

Based on the above DFT simulations and transition state theory, microkinetic modeling was further conducted to explore the reaction dynamics (through the intermediate surface coverage variations) and to determine and compare the reaction rate of the rate of the determining step (RDS) over the three different Cu active sites. The N_2O-ODHP reaction can be described by five elementary steps, given that it is associated with the kinetic equations listed in Table 1. The calculated kinetic parameters, including reaction rate constant, pre-exponential factor, and specific forward and reverse reaction rate, were listed in Table S1.

Table 1. Elementary steps of micro-dynamics and the equations of reaction rate for the N_2O-ODHP over Cu-BEA.

Step	Elementary Steps	Reaction Rate Equations
R1	Z–Cu–$N_2O_{(g)}$ ↔ Z–Cu–N_2O	$r_1 = k_1 P_{N_2O}\theta_v - k_{-1}\theta_{N_2O}$
R2	Z–Cu–N_2O → Z–Cu–O+$N_{2(g)}$	$r_2 = k_2\theta_{N_2O}$
R3	Z–Cu–O+$C_3H_{8(g)}$ ↔ Z–Cu–O–C_3H_8	$r_3 = k_3 P_{C_3H_8}\theta_O - k_{-3}\theta_{O\text{-}C_3H_8}$
R4	Z–Cu–O–C_3H_8 ↔ Z–Cu–OH–C_3H_7	$r_4 = k_4\theta_{O\text{-}C_3H_8} - k_{-4}\theta_{C_3H_7\text{-}OH}$
R5	Z–Cu–OH–C_3H_7 ↔ Z–Cu–H_2O–C_3H_6	$r_5 = k_5\theta_{C_3H_7\text{-}OH} - k_{-5}\theta_{C_3H_6\text{-}H_2O}$

(a) Microkinetic modeling over Z^a–Cu. Figure 5a displays the intermediate coverage variations along with reaction time (t) over the Z^a–Cu (T = 823 K). Initially, the unoccupied active site coverage (θ_v) would decrease from 1 to 0.8 ML as it is associated with the increase of adsorbed N_2O (θ_{N_2O}) from 0 to 0.2 ML. This process corresponds with the N_2O dissociation step that generates αO, wherein the adsorbed N_2O (θ_{N_2O}) constitutes the major active-site covered species over the Z^a–Cu due to the relatively high energy barrier of R2 (1.40 eV, Figure 1a, N_2O dissociation to generate αO). Along with the reaction, both θ_v and θ_{N_2O} would quickly decrease to 0 ML as they are both accompanied with the rapid growth of the coverage of the intermediate of propanol ($\theta_{C3H7\text{-}OH}$), which finally reaches the equilibrium. This finding indicates that after the formation of αO, it would quickly participate in the ODHP reaction in order to generate the intermediate of propanol, and the intermediate of propanol ($\theta_{C3H7\text{-}OH}$) constitutes the major active-site covered species, which indicates that the R5 (ΔE = 1.94 eV, Figure 1a) constitutes the RDS during N_2O-ODHP over the Z^a–Cu site, leading to the accumulation of C_3H_7–OH over the active site.

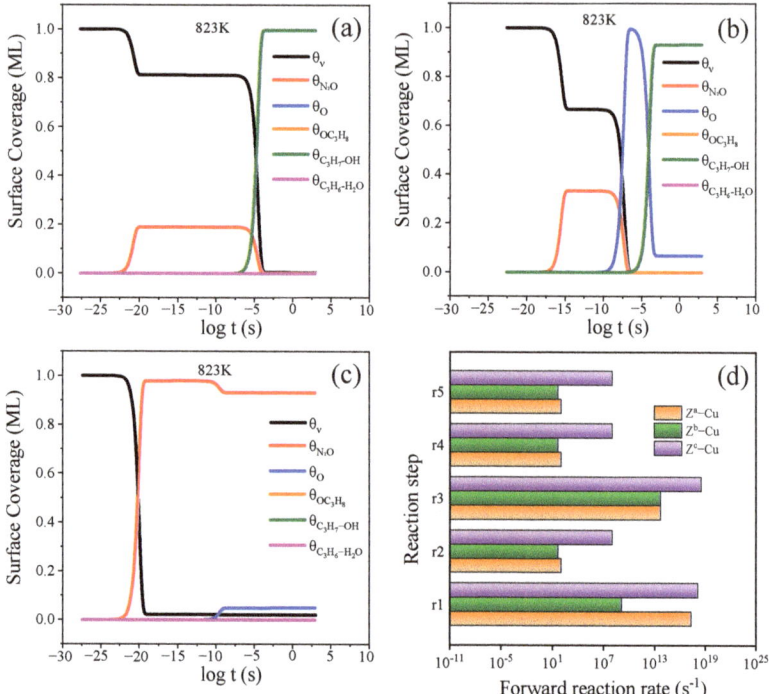

Figure 5. Microkinetic modeling results: the surface coverage variations along with reaction time (*t*) over the (**a**) Z^a–Cu, (**b**) Z^b–Cu, and (**c**) Z^c–Cu sites of Cu-BEA at 823 K, and (**d**) the forward reaction rate comparisons.

(b) Microkinetic modeling over Z^b–Cu. Figure 5b displays the variations of the intermediate coverages during the N_2O-ODHP over the Z^b–Cu site. Being similar to that of Z^a–Cu site, the adsorbed N_2O would initially cover the active site by displaying the θ_{N_2O} of 0.3 ML. However, the αO, given that it is in the form of $[Cu-O-Cu]^{2+}$, would shortly occupy the active site due to the relatively lower N_2O dissociation barrier (0.95 versus 1.40 eV of R2) and higher αH dehydrogenation barrier (0.3 versus 0.05 eV of R4) than that of Z^a–Cu site, which leads to the short accumulation of the αO over the active site. Along with the further reaction (reaching equilibrium), the active site would be eventually covered by the propanol ($\theta_{C3H7-OH}$), which is similar to the scenario of the Z^a–Cu site due to the extremely high barrier of R5 (2.15 eV). Thus, similar to that of Z^a–Cu, the R5 would constitute the RDS during the N_2O-ODHP over the Z^b–Cu site.

(c) Microkinetic modeling over Z^c–Cu. The surface coverage of the reactant as well as the generated intermediates during N_2O-ODHP over the Z^c–Cu site were both depicted in Figure 5c. Being totally different from the scenarios of the Z^a–Cu and the Z^b–Cu sites, the N_2O constitutes the major active-site covered species over the Z^c–Cu site at T = 823 K, wherein the θ_{N_2O} initially increases up to 1 ML and then decreases to a stable value of above 0.9 ML due to another intermediate αO (θ_o = 0.1 ML) after the reaction, which reaches the equilibrium. This finding indicates that the N_2O dissociation step (R2) would constitute the RDS during the N_2O-ODHP over the Z^c–Cu site. This finding correlates well with the highest energy barrier of 0.89 eV of R2 during N_2O-ODHP (see Figure 1c).

(d) Reaction rate comparisons. Figure 5d displays the forward reaction rate comparisons of each elementary step during N_2O-ODHP over the different active sites of Z^a–Cu, Z^b–Cu, and Z^c–Cu. Obviously, the Z^c–Cu site displays much higher reaction rates than those of the Z^a–Cu and the Z^b–Cu. Moreover, in the net reaction rate (NRR) comparisons, which were further depicted in Figure S1, the Z^c–Cu displays an NRR of 1.25+E8 mol·m^{-3}·s^{-1}, and it is five and six orders of magnitude higher, respectively, than those of the Z^a–Cu (120.57 mol·m^{-3}·s^{-1}) and the Z^b–Cu (49.21 mol·m^{-3}·s^{-1}) sites. These findings quantitatively verify the superior activity of the Z^c–Cu site relative to those of the Z^a–Cu and the Z^b–Cu sites.

2.3. Static Charge Difference PDOS and Analyses

To further illustrate the superior activity of the Z^c–Cu, the static charge difference and the partial density of state (PDOS) analyses were further conducted based on TS3 (corresponding to the βH dehydrogenation step). As shown in Figure 6a–d, large amounts of charge transfers occurred during the βH dehydrogenation, and the Cu of Z^c–Cu provided more charges relative to those of the Z^a–Cu and the Z^b–Cu (Figure 6d), which can be closely related to the smallest band gap between the Cu and C of C_3H_7-, as shown by the PDOS analyses of Figure 6e–g (4.97 versus 5.82 and 5.29 eV). This finding indicates that the Z^c–Cu would exhibit a stronger electric field effect on C_3H_7-, and that it is thereby greatly favorable for βH dehydrogenation.

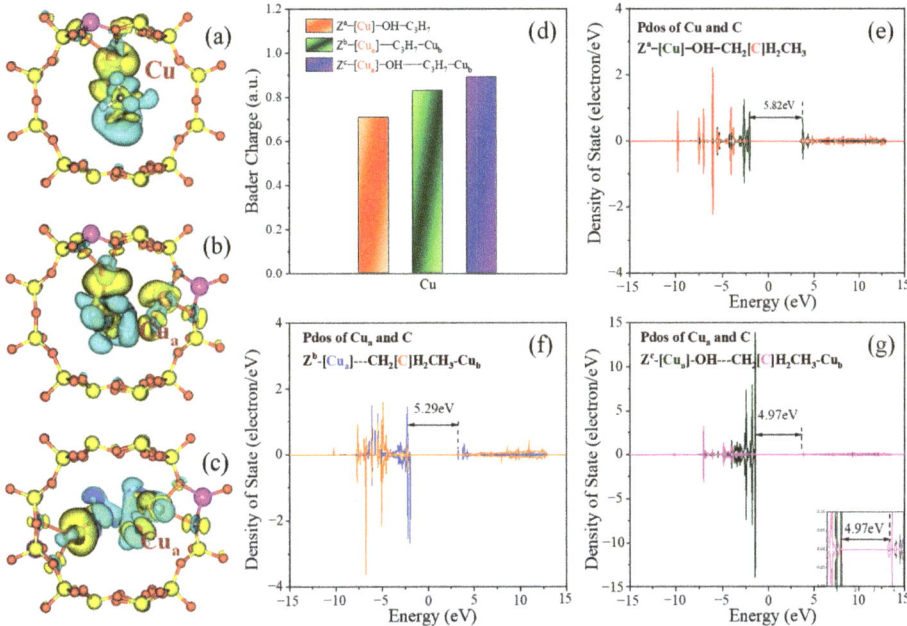

Figure 6. The static charge difference (a–c) and partial density of state (PDOS) analyses (e–g) of TS3 models for Z^a–, Z^b–, and Z^c–Cu-BEA; (d) bader charges analysis results; (e) partial density of state (PDOS) of atomic Cu and C over (e) Za–Cu; (f) Zb–Cu; and (g) Zc–Cu site. Yellow and blue colors represent the increase and decrease in electron density, respectively. Si (yellow), O (red), N (blue), Al (pink), Cu (orange).

3. N$_2$O-ODHP Activity Measurement

As is well known, in addition to the [Cu]$^+$ cations, the CuO$_x$ species can also exist over Cu-modified zeolite catalyst (Cu-Zeolite) and shed more light on the activity behaviors of these different Cu species. The 1%Cu-BEA and 1%CuO-SiO$_2$ were prepared by the impregnation method (the metal loading of 1wt.%), and they were further evaluated for the N$_2$O-ODHP. In addition, the Fe-modified zeolites (Fe-Zeolite) have also been reported to possess excellent N$_2$O dissociation activity in order to produce αO [17], and to make a comparison with the Cu modified zeolite, the 1%Fe-BEA, and 1%Fe-ZSM-5 were also prepared by the impregnation method (metal loading of 1 wt.%) and evaluated by N_2O-ODHP. The specific preparation method is stated in detail in the Supporting Information section of this paper. The activity measurement results, including the C_3H_8 and N_2O conversions as well as the product selectivity, were profiled in Figure 7a–c. As can be seen there, the 1%Cu-BEA displays a higher C_3H_8 conversion (31.5%) and a higher C_3H_6 selectivity (74.5%) than the other catalyst samples, especially in comparison with that of 1%CuO-SiO$_2$, displaying a C_3H_8 conversion of 3.5% and a C_3H_6 selectivity of 43.5%. This finding indicates that the CuO species would not constitute the major active species for the N_2O-ODHP.

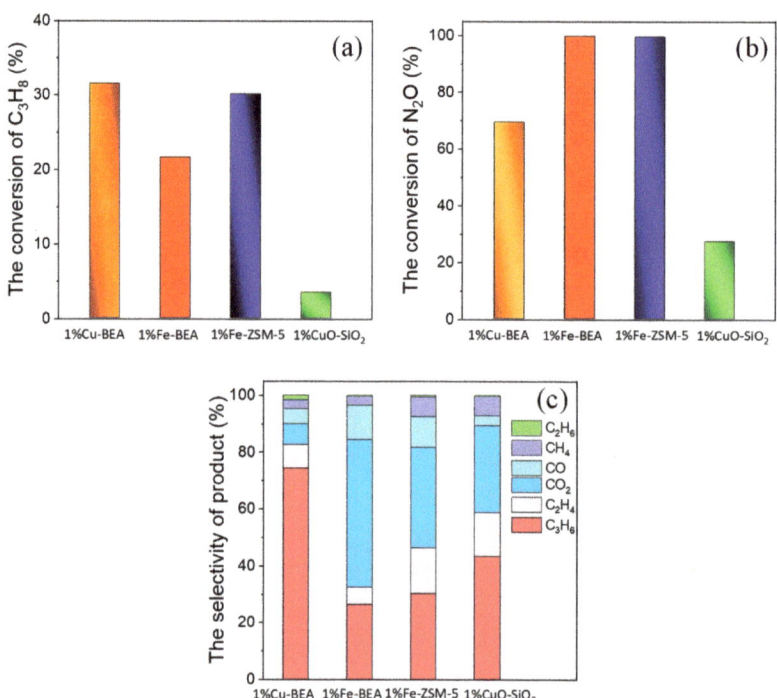

Figure 7. Activity measurement of N_2O-ODHP over 1%Cu-BEA (orange), 1%Fe-BEA (red), 1%Fe-ZSM-5 (dark blue), and 1%CuO-SiO$_2$ (green): (**a**) C_3H_8 conversion; (**b**) N_2O conversion; (**c**) product selectivity of C_2H_6 (light green), CH_4 (light purple), CO (light blue), CO_2 (blue), C_2H_4 (light pink), and C_3H_6 (light red). GHSV = 12,000 h^{-1}, C_3H_8:N_2O:He = 10:10:80, T = 550 °C.

As further shown by the N_2O conversion of Figure 7b, the 1%CuO-SiO$_2$ displays a much lower N_2O conversion (27.2%) than the 1%Cu-BEA (69.4%). This finding shows that the lower N_2O dissociation activity of CuO species probably constitutes one of the major reactions that leads to the extremely low N_2O-ODHP activity of the 1%CuO-SiO$_2$. Conversely, the 1%Cu-BEA possessing active Cu cations for N_2O dissociation that generate αO exhibits a much higher level of N_2O-ODHP activity relative to the 1%CuO-SiO$_2$. As has also been reported on a theoretical level [22], the CuO exhibits a much high energy barrier (2.71 eV) for N_2O dissociation, which indicates that it is very difficult to decompose N_2O and produce α-O over CuO.

As for the samples of 1%Fe-BEA and 1%Fe-ZSM-5, the nearly complete N_2O conversion (~100%) can be achieved due to the superior N_2O dissociation activity of Fe cations than those of the Cu cations [23], although this does lead to the ready overoxidation of C_3H_8 into CO$_x$ (CO and CO$_2$ of 64.1 and 46.0%, respectively; see Figure 7c). This finding shows that the Fe cations are active for N_2O dissociation, although they suffer from the overoxidation of C_3H_8. The Cu-based zeolites would therefore probably be much more suitable for the N_2O-ODHP relative to that of the Fe-based zeolite catalyst. However, we would also like to emphasize that such works still need further investigation. We would also like to note that the TPR and UV-vis correlate with the active-site-structure in the theoretical calculation. This will be further studied in our research in the future, which will focus on the influence of a diversely structured topologized zeolite (MFI, FER, MOR, and BEA) on a specific structure, as well as on the location of $[Cu]^+$ cations and how they are related to N_2O-ODHP catalytic behaviors, both experimentally and theoretically.

4. Computational Modeling and Methodology

Density functional theory (DFT) adopts the Vienna ab-initio simulation package (VASP). The Projection Enhanced Wave (PAW) method utilizes the interaction between electrons and the core, and it uses Generalized Gradient Approximation (GGA) and Perdew Burke Ernzerhof (PBE) functions to achieve electron exchange correlation [24]. In the process of geometric optimization, the convergence achieved at the energy difference is 10^{-5} eV, with an ion relaxation convergence standard of 0.05 eV/Å. The energy cutoff of the set plane wave is 400 eV. The K point of the Brillouin region is set to $2 \times 2 \times 1$ in the calculation of the structure, and it is set to $4 \times 4 \times 2$ in the partial wave density of states. In the International Zeolite Association (IZA) database, the cellular model data for BEA is A = 12, B = 12.632, and C = 9.421 Å [25]. The calculation of the transition state (TS) uses the climbing image light pushing elastic bond (CI-NEB) and the dimer method. Four points are inserted between the initial state and the final state, and the saddle points corresponding to the transition state are located in order to find the lowest energy path. The TS state is only identified on one imaginary frequency [25–29]. The method of micro-dynamic modeling is included in the Supporting Information section of this paper.

5. Conclusions

The present work theoretically investigates N_2O-ODHP over Cu-BEA with three types of Cu sites—-monomeric $[Cu]^+$ (Z^a–Cu), dimeric $[Cu-Cu]^{2+}$ (Z^b–Cu), and distant $[Cu]^+$—$[Cu]^+$ (Z^c–Cu). The Z^a–Cu is beneficial for αH dehydrogenation (0.05 eV), but it requires a high energy barrier in N_2O dissociation and βH dehydrogenation (1.40 and 1.94 eV) to be overcome. The Z^b–Cu, with a Cu—Cu distance of 4.91 Å, is suitable for the N_2O dissociation step (0.95 eV), but it is highly resistant to the βH hydrogenation step because it displays an extremely high barrier of 2.15 eV. Being contrary to the scenarios of the Z^a–Cu and the Z^b–Cu, the Z^c–Cu site with the Cu—Cu distance of 5.82 Å is not only favorable for N_2O dissociation (0.89 eV) but also greatly active for the ODHP steps of αH (0.01 eV) and βH (0.33 eV) dehydrogenation. The microkinetic modeling further showed that the Z^c–Cu exhibits a five to six orders of magnitude higher net reaction rate than those of the Z^a–Cu and the Z^b–Cu sites. This is closely correlated with the specific structure of the Z^c–Cu, which possesses a much stronger electric field effect on the C_3H_8 molecule due to the synergic effect of Cu, as this possesses the smallest band gaps and they are favorable to the charge transfers between the Cu active site and the C_3H_8 molecule. Generally, we ultimately propose that modulating the Cu active site distance (Cu—Cu) probably constitutes a promising strategy for highly-efficient Cu-zeolite design for the N_2O-ODHP. Additionally, we would like to mention that it is not possible for there to be only one type of Z—Cu site (Cu cation site)] experimentally, and other types of Cu site (such as $[Cu_3O_3]^{2+}$, which is proposed by Lercher et al. [30] for methane direct oxidation to methanol) may also be active for the N_2O-ODHP, which would be a good direction for further study of this subject.

Supplementary Materials: The following supporting information can be downloaded at: https://www.mdpi.com/article/10.3390/catal13081212/s1, Figure S1. Optimized Cu-BEA models with diverse Cu active site motifs (a) Z^a-Cu; (b) Z^b-Cu; (c) Z^c-Cu site; Si (yellow), O (red), N (blue), Al (pink), Cu (orange); Scheme S1. Schematic reaction for N_2O-ODHP over Z^a-Cu (reaction Route A), Z^b-Cu (reaction Route B) and Z^c-Cu (reaction Route C), respectively; Table S1. Micro-dynamics parameters of the reaction steps over Z^a-Cu, Z^b-Cu and Z^c-Cu site; Figure S2. The comparison of net reaction rate over Schematic reaction for N_2O-ODHP over Z^a-Cu, Z^b-Cu and Z^c-Cu site at 823 K [17,18,25,31–36].

Author Contributions: Methodology, N.W.; validation, C.D.; investigation, R.W.; resources, B.C.; data curation, R.X.; writing—original draft preparation, R.W.; writing—review and editing, N.L.; visualization, G.Y.; supervision, B.C. All authors have read and agreed to the published version of the manuscript.

Funding: This research was funded by National Natural Science Foundation of China (No. 22178011 and 22176006).

Data Availability Statement: Data is available upon request to the corresponding authors.

Acknowledgments: We acknowledge the final support from National Natural Science Foundation of China (No. 22178011 and 22176006).

Conflicts of Interest: The authors declare no conflict of interest.

References

1. Pérez-Ramírez, J.; Gallardo-Llamas, A. Framework composition effects on the performance of steam-activated FeMFI zeolites in the N_2O-mediated propane oxidative dehydrogenation to propylene. *J. Phys. Chem. B* **2005**, *109*, 20529–20538. [CrossRef] [PubMed]
2. Kondratenko, E.; Cherian, M.; Baerns, M.; Su, D.; Schlogl, R.; Wang, X.; Wachs, I. Oxidative dehydrogenation of propane over V/MCM-41 catalysts: Comparison of O_2 and N_2O as oxidants. *J. Catal.* **2005**, *234*, 131–142. [CrossRef]
3. Sanchezgalofre, O.; Segura, Y.; Perezramirez, J. Deactivation and regeneration of iron-containing MFI zeolites in propane oxidative dehydrogenation by N_2O. *J. Catal.* **2007**, *249*, 123–133. [CrossRef]
4. Wei, W.; Moulijn, J.A.; Mul, G. FAPO and Fe-TUD-1: Promising catalysts for N_2O mediated selective oxidation of propane. *J. Catal.* **2009**, *262*, 1–8. [CrossRef]
5. Kondratenko, E.V.; Brückner, A. On the nature and reactivity of active oxygen species formed from O_2 and N_2O on VO_x/MCM-41 used for oxidative dehydrogenation of propane. *J. Catal.* **2010**, *274*, 111–116. [CrossRef]
6. Orlyk, S.; Kyriienko, P.; Kapran, A.; Chedryk, V.; Balakin, D.; Gurgul, J.; Zimowska, M.; Millot, Y.; Dzwigaj, S. CO_2-assisted dehydrogenation of propane to propene over Zn-BEA zeolites: Impact of acid–base characteristics on catalytic performance. *Catalysts* **2023**, *13*, 681. [CrossRef]
7. Vogt, E.T.; Weckhuysen, B.M. Fluid catalytic cracking: Recent developments on the grand old lady of zeolite catalysis. *Chem. Soc. Rev.* **2015**, *44*, 7342–7370. [CrossRef] [PubMed]
8. Chen, J.Q.; Bozzano, A.; Glover, B.; Fuglerud, T.; Kvisle, S. Recent advancements in ethylene and propylene production using the UOP/Hydro MTO process. *Catal. Today* **2005**, *106*, 103–107. [CrossRef]
9. Bulánek, R.; Wichterlová, B.; Novoveská, K.; Kreibich, V. Oxidation of propane with oxygen and/or nitrous oxide over Fe-ZSM-5 with low iron concentrations. *Appl. Catal. A* **2004**, *264*, 13–22. [CrossRef]
10. Novoveská, K.; Bulánek, R.; Wichterlová, B. Oxidation of propane with oxygen, nitrous oxide and oxygen/nitrous oxide mixture over Co- and Fe-zeolites. *Catal. Today* **2005**, *100*, 315–319. [CrossRef]
11. Wu, G.; Hao, Y.; Zhang, N.; Guan, N.; Li, L.; Grünert, W. Oxidative dehydrogenation of propane with nitrous oxide over Fe–O–Al species occluded in ZSM-5: Reaction and deactivation mechanisms. *Microporous Mesoporous Mater.* **2014**, *198*, 82–91. [CrossRef]
12. Jiang, X.; Sharma, L.; Fung, V.; Park, S.J.; Jones, C.W.; Sumpter, B.G.; Baltrusaitis, J.; Wu, Z. Oxidative dehydrogenation of propane to propylene with soft oxidants via heterogeneous catalysis. *ACS Catal.* **2021**, *11*, 2182–2234. [CrossRef]
13. Patet, R.E.; Koehle, M.; Lobo, R.F.; Caratzoulas, S.; Vlachos, D.G. General acid-type catalysis in the dehydrative aromatization of furans to aromatics in H-[Al]-BEA, H-[Fe]-BEA, H-[Ga]-BEA, and H-[B]-BEA zeolites. *J. Phys. Chem. C* **2017**, *121*, 13666–13679. [CrossRef]
14. Chalupka, K.; Thomas, C.; Millot, Y.; Averseng, F.; Dzwigaj, S. Mononuclear pseudo-tetrahedral V species of VSiBEA zeolite as the active sites of the selective oxidative dehydrogenation of propane. *J. Catal.* **2013**, *305*, 46–55. [CrossRef]
15. Mauvezin, M.; Delahay, G.; Kißlich, F.; Coq, B.; Kieger, S. Catalytic reduction of N_2O by NH_3 in presence of oxygen using Fe-exchanged zeolites. *Catal. Lett.* **1999**, *62*, 41–44. [CrossRef]
16. Sobalik, Z.; Sazama, P.; Dedecek, J.; Wichterlová, B. Critical evaluation of the role of the distribution of Al atoms in the framework for the activity of metallo-zeolites in redox N_2O/NO_x reactions. *Appl. Catal. A* **2014**, *474*, 178–185. [CrossRef]

17. Liu, N.; Zhang, R.; Chen, B.; Li, Y.; Li, Y. Comparative study on the direct decomposition of nitrous oxide over M (Fe, Co, Cu)–BEA zeolites. *J. Catal.* **2012**, *294*, 99–112. [CrossRef]
18. Xu, R.; Liu, N.; Dai, C.; Li, Y.; Zhang, J.; Wu, B.; Yu, G.; Chen, B. H_2O-built proton transfer bridge enhances continuous methane oxidation to methanol over Cu-BEA zeolite. *Angew. Chem.* **2021**, *60*, 16634–16640. [CrossRef]
19. Widmann, D.; Behm, R.J. Dynamic surface composition in a Mars-van Krevelen type reaction: CO oxidation on Au/TiO_2. *J. Catal.* **2018**, *357*, 263–273. [CrossRef]
20. Wang, C.; Gu, X.-K.; Yan, H.; Lin, Y.; Li, J.; Liu, D.; Li, W.-X.; Lu, J. Water-mediated Mars–Van Krevelen mechanism for CO oxidation on ceria-supported single-atom Pt_1 catalyst. *ACS Catal.* **2016**, *7*, 887–891. [CrossRef]
21. Jan, F.; Lian, Z.; Zhi, S.; Yang, M.; Si, C.; Li, B. Revealing the role of HBr in propane dehydrogenation on CeO_2(111) via DFT-based microkinetic simulation. *Phys. Chem. Chem. Phys.* **2022**, *24*, 9718–9726. [CrossRef] [PubMed]
22. Suo, W.; Sun, S.; Liu, N.; Li, X.; Wang, Y. The adsorption and dissociation of N_2O on CuO(111) surface: The effect of surface structures. *Surf. Sci.* **2020**, *696*, 121596. [CrossRef]
23. Li, Y.; Liu, N.; Dai, C.; Xu, R.; Yu, G.; Wang, N.; Zhang, J.; Chen, B. Synergistic Effect of neighboring Fe and Cu cation sites boosts FenCum-BEA activity for the continuous direct oxidation of Methane to Methanol. *Catalysts* **2021**, *11*, 1444. [CrossRef]
24. Jianwen, Z.; Zhifeng, H.; Ziqian, Y.; Meijuan, L.; Fei, C.; Qiang, S. The influence of alkaline earth elements on electronic properties of α-Si_3N_4 via DFT calculation. *J. Wuhan Univ. Technol. Mater. Sci. Ed.* **2020**, *35*, 863–871.
25. Zhou, Z.; Qin, B.; Li, S.; Sun, Y. A DFT-based microkinetic study on methanol synthesis from CO_2 hydrogenation over the In_2O_3 catalyst. *Phys. Chem. Chem. Phys.* **2021**, *23*, 1888–1895. [CrossRef] [PubMed]
26. Montejo-Valencia, B.D.; Curet-Arana, M.C. Periodic DFT study of the opening of fructose and glucose rings and the further conversion of fructose to trioses catalyzed by M-BEA (M = Sn, Ti, Zr, or Hf). *J. Phys. Chem. C* **2019**, *123*, 3532–3540. [CrossRef]
27. Mahyuddin, M.H.; Staykov, A.; Shiota, Y.; Yoshizawa, K. Direct conversion of methane to methanol by metal-exchanged ZSM5 zeolite (Metal = Fe, Co, Ni, Cu). *ACS Catal.* **2016**, *6*, 8321–8331. [CrossRef]
28. Liu, N.; Zhang, R.; Li, Y.; Chen, B. Local electric field effect of TMI (Fe, Co, Cu)-BEA on N_2O direct dissociation. *J. Phys. Chem. C* **2014**, *118*, 10944–10956. [CrossRef]
29. Konsolakis, M. Recent advances on nitrous oxide (N_2O) decomposition over non-noble-metal oxide catalysts: Catalytic performance, mechanistic considerations, and surface chemistry aspects. *ACS Catal.* **2015**, *5*, 6397–6421. [CrossRef]
30. Ikuno, T.; Grundner, S.; Jentys, A.; Li, G.; Pidko, E.A.; Fulton, J.L.; Sanchez-Sanchez, M.; Lercher, J.A. Formation of Active Cu-oxo Clusters for Methane Oxidation in Cu-Exchanged Mordenite. *J. Phys. Chem. C* **2019**, *123*, 8759–8769. [CrossRef]
31. Liu, N.; Yuan, X.; Zhang, R.; Xu, R.; Li, Y. Mechanistic insight into selective catalytic combustion of acrylonitrile (C_2H_3CN): NCO formation and its further transformation towards N_2. *Phys. Chem. Chem. Phys.* **2017**, *19*, 7971–7979. [CrossRef] [PubMed]
32. Liu, N.; Yuan, X.; Zhang, R.; Li, Y.; Chen, B. Mechanistic insight into selective catalytic combustion of HCN over Cu-BEA: Influence of different active center structures. *Phys. Chem. Chem. Phys.* **2017**, *19*, 23960–23970. [CrossRef] [PubMed]
33. Dixit, M.; Baruah, R.; Parikh, D.; Sharma, S.; Bhargav, A. Autothermal reforming of methane on rhodium catalysts: Microkinetic analysis for model reduction. *Comput. Chem. Eng.* **2016**, *89*, 149–157. [CrossRef]
34. Ke, C.; Lin, Z. Density Functional Theory Based Micro- and Macro-Kinetic Studies of Ni-Catalyzed Methanol Steam Reforming. *Catalysts* **2020**, *10*, 349. [CrossRef]
35. Gokhale, A.A.; Kandoi, S.; Greeley, J.P.; Mavrikakis, M.; Dumesic, J.A. Molecular-level descriptions of surface chemistry in kinetic models using density functional theory. *Chem. Eng. Sci.* **2004**, *59*, 4679–4691. [CrossRef]
36. Cai, Q.-X.; Wang, J.-G.; Wang, Y.-G.; Mei, D. Mechanistic insights into the structure-dependent selectivity of catalytic furfural conversion on platinum catalysts. *AIChE J.* **2015**, *61*, 3812–3824. [CrossRef]

Disclaimer/Publisher's Note: The statements, opinions and data contained in all publications are solely those of the individual author(s) and contributor(s) and not of MDPI and/or the editor(s). MDPI and/or the editor(s) disclaim responsibility for any injury to people or property resulting from any ideas, methods, instructions or products referred to in the content.

Review

Research and Application Development of Catalytic Redox Technology for Zeolite-Based Catalysts

Wentao Zhang [1], De Fang [2,3,*], Guanlin Huang [2], Da Li [4] and Yun Zheng [5,*]

1. Department of Aircraft Maintenance Engineering, Guangzhou Civil Aviation College, Guangzhou 510403, China; wtzhang1991@163.com
2. State Key Laboratory of Silicate Materials for Architectures, Wuhan University of Technology, Wuhan 430070, China; 308192@whut.edu.cn
3. Center for Materials Research and Analysis, Wuhan University of Technology, Wuhan 430070, China
4. College of Physics and Optoelectronic Engineering, Shenzhen University, Shenzhen 256600, China; lida@szu.edu.cn
5. Key Laboratory of Optoelectronic Chemical Materials and Devices, Ministry of Education, Jianghan University, Wuhan 430056, China
* Correspondence: fangde@whut.edu.cn (D.F.); zhengyun@jhun.edu.cn (Y.Z.); Tel./Fax: +86-27-87651843 (D.F.); +86-27-84226806 (Y.Z.)

Abstract: Zeolites are porous materials with plentiful and adjustable pore structures, which are widely applied in various fields such as fossil fuel energy conversion, preparation of clean energy, chemical product conversion, CO_2 capture, VOC treatment, and so on. Zeolites exhibited advantageous adsorption compared with traditional adsorbents such as activated carbon; in addition, they can also provide abundant reaction sites for various molecules. The chemical composition, structural acidity, and distribution of pore size can distinctly affect the efficiency of the reaction. The modification of zeolite structure, the development of novel and efficient preparation methods, as well as the improvement of reaction efficiency, have always been the focus of research for zeolites.

Keywords: zeolite; catalyst; pore structure; redox; applications

1. Introduction

Adsorption and catalytic oxidation technology have extensive uses in energy, chemical engineering, and environmental industries, such as the cracking of fossil fuels, conversion of biomass, synthesis of non-petroleum products, low-cost production of hydrogen, catalytic treatment of industrial exhaust gas, etc. [1,2]. The efficiency of catalytic reactions is strongly influenced by adsorbents and catalysts. Under normal conditions, adsorbents are porous materials with huge specific surface areas that can provide enough space for redox reactions. Activated carbons (ACs), porous alumina, metal-organic frameworks (MOFs), and molecular sieves are constantly used as adsorbents because of their pore structure. More specifically, considering the thermal stability and the modification of pore structure, molecular sieves exhibited better performance than activated carbons.

Zeolite has generally been utilized in various synthetic processes in industrial products since its development (as shown in Figure 1). For zeolites, the basic structural unit (TO_4) is connected by bridging oxygen atoms to form a secondary building structure (SBU), which can be further combined to form composite building structures (CBUs), and multiple rings (n) are simultaneously formed [3]. Zeolites can be classified as different types according to the pore size and the number of rings: (1) small pore zeolites with $n = 8$, (2) medium pore zeolites with $n = 10$, (3) large pore zeolites with $n = 12$, (4) extra-large pore zeolites with $n > 12$.

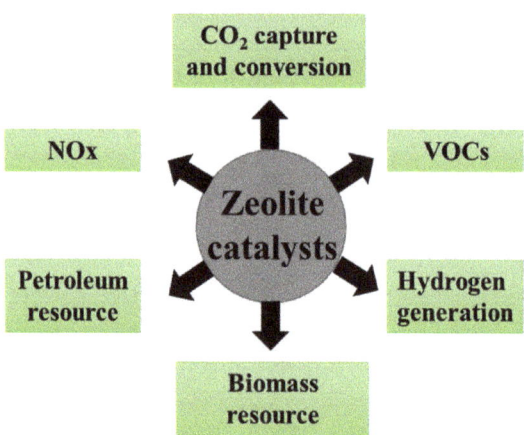

Figure 1. Various applications of zeolites in sustainable energy.

Due to the different combinations of basic units, there are significant differences in the pore structure and size of different types of zeolite molecular sieves. So far, A type, X type, Y type, β-zeolite, and ZSM-5 zeolite have been the most widely used zeolites.

A type (Linde type A, LTA) zeolites have a cubic crystal system. It has 3-dimensional channels ranging from 3 Å to 5 Å and an 8-membered topological structure. β zeolite is composed of 12-membered rings with an average channel size of 6 Å − 8 Å. X type and Y type zeolites are classified as FAU zeolites. Both of them belong to cubic crystals and are composed of hexagonal column cages and β cages formed from 6-membered rings. The Si/Al ratio of the X type varies from 2.2 to 3, and the Y type is higher than 3. Due to the special channels, X type and Y type zeolites have a large adsorption capacity for organic pollutant molecules. However, the adsorption capacity of organic molecules is reduced in moist gases. ZSM-5 is a typical MFI zeolite with a "zigzag" shaped 10-membered ring channel and vertically oriented 10-membered ring channels. The medium pore size (~5 Å) and excellent hydrothermal stability facilitated its application in the petrochemical industry.

Additionally, catalysts are pretty active components in the reaction process, including three types of catalysts: noble metals, non-noble metal oxides, and composite metal oxides. In order to increase the efficiency of the catalytic oxidation process, the selection of adsorbents and catalysts and the regulation of their modification should be prioritized.

This article illustrates the main applications of zeolite catalysts in summary and analyzes new advances in zeolite catalytic redox chemistry. The effects of porous structure, acidity, and other parameters, including zeolite type, catalyst, and reaction temperature, on the activity, selectivity, stability, and deactivation of zeolites are also summarized. Finally, the challenges and strategies of zeolite catalytic technology are discussed further.

2. The Application of Zeolite Catalysts

2.1. Catalytic Cracking of Petroleum Resources

Fluid catalytic cracking (FCC) technology is an effective alkyl removal technology that is increasingly widely utilized in the biomass oil and petrochemical industry. FCC catalysts consist of various zeolites with multi-component active ingredients that can greatly improve the efficiency of catalytic cracking [4,5].

For crude oil, it is extremely important to ameliorate the selectivity of catalysts towards products and increase the yield of specific products. In addition, Y-type zeolite has gained widespread use as a molecular sieve. Introducing and increasing mesoporous surface area can effectively reduce heavy oil production. Liu et al. prepared ordered silica aluminate on an industrial scale using a pre-crystallization unit of Y zeolite precursor. However, when a single layer of zeolite was dispersed on the catalyst surface, the composite catalyst exhibited the best activity for heavy oil cracking [6]. Pre-cracking is the first process to be carried out in the base pores, followed by further cracking and selective generation of its different products. Although Y zeolite is commonly used in heavy oil catalytic cracking and hydrocracking, the mass transfer efficiency between active sites is lower than expected. Cui et al. obtained mesoporous materials with octahedra by synthesizing ultra-stable mesoporous high-silicon molecular sieves directly. Compared to industrial FCC catalysts, the conversion rate was increased by 7.64%, and the gasoline yield was increased by 16.37%, even after these catalysts were aged at 800 °C [7].

Due to the requirements of the sixth national standard for the olefin content of automotive fuels, it is of great necessity to reduce the olefin content in gasoline. Through hydrothermal treatment, the acidity of ZSM-5 loaded metal catalysts (La-Ni-Zn) was reduced, and the B/L ratio (B = Brønsted acid, L = Lewis acid) was increased, which finally increased the aromatic production by 5% and the isoalkane production by 16%, making it an industrial catalyst for the catalytic cracking of gasoline [8]. To enhance the steam stability of Y zeolite, Yu et al. prepared rare earth (RE)-exchanged Y zeolite. The substitution of Y^{3+} ions for counter-ions Na^+ resulted in a noticeable shrinkage of the unit cell owing to the relatively small ionic radius and high charge density of Y^{3+} ions. Therefore, this kind of zeolite exhibited better performance in steam stability and was more favorable for the generation of liquefied petroleum gas and C_{5+} gasoline products in n-dodecane cracking [9]. Except for Y^{3+}, Al-exchanged Y zeolite was also proved to be able to promote liquefied petroleum gas production as a result of its strong acidity [10].

Except for routine chemical products, Li et al. have conducted extensive research on the production of high value-added products through dealkylation of light fraction light circulating oil and found that mesoporous BEA zeolite catalysts with appropriate acidity and mass transfer ability can greatly raise the yield of value-added products [11]. It can be seen that, although FCC was regarded as an efficient technology of dealkylation, the modifying of micropores and B acidic sites of zeolite need more research to elevate mass transfer ability. Thus, the optimization of catalysts in FCC, such as product selectivity, thermal stability, and cycle life, should receive focused attention so as to increase production.

2.2. Conversion of Biomass

Biomass is a promising green and renewable organic carbon source, but its calorific value is sharply lower than that of fossil fuels due to its abundant aerobic molecules, making it difficult to utilize directly [12,13]. It is necessary to transform biomass into valuable bioproducts. Two critical processes are involved in this strategy. One is the conversion of biomass into platform molecules, and the other is the upgrading of platform molecules into valuable fuels. Usually, the biomass comprised 40–50% cellulose, 25–35% hemicellulose, 15–20% lignin, and others [14]. Levulinic acid (LeA) and 5-hydroxymethylfurfural (HMF) are two common biomass platform molecules due to the highly reactive functional groups such as carboxyls, aldehydes, hydroxymethyls, and furan rings. LeA can be obtained from cellulose via the C_5 route or hemicellulose via the C_6 route, and HMF is usually produced from cellulose via the C_6 route.

(1) LeA via C_5 and C_6 route

In the C_6 route, cellulose was hydrolyzed to form C_6 sugar, which was dehydrated to form HMF. LA was obtained after the hydrolysis of HMF. Due to the high separation cost of LA and formic acid, synthesizing LA from C_6 sugar was mainly adopted in the laboratory. For the C_5 route, intermediates such as furfuryl and furfuryl alcohol (FAL) were successively converted to LeA, which made the carbon utilization more efficient than that of the C_6 route [15]. HY zeolite mixed with ionic liquid (ionic liquid: HY = 0.5) has more Lewis acid sites than Bronsted acid sites and exhibits a LeA yield of 62.2% [16]. ZSM-5 zeolite with mesopores was treated by tandem alkaline and acid washing, and ZSM-5-$OH_{0.2}$-H zeolite was therefore obtained; the modification of pore structure and Al distribution promoted the hydrolytic efficiency of FAL to LeA, the LeA yield reached 64.5% after three cycles [17].

(2) HMF via the C_6 route

HMF from biomass can be utilized to synthesize various biochemicals. The conversion of HMF from biomass can be divided into three processes: glucose hydrolyzed from cellulose, the isomerization of glucose to fructose, and the dehydration of fructose [15]. Therefore, fructose was usually used as a model molecule in HMF preparation, and many zeolites such as ZSM, HY, H-USY, and MAPO were adopted. Scholars prepared Cu-Cr/ZSM-5 zeolite using the ion-exchange method; when the zeolite was 20 wt.% of glucose, 50.4% HMF was achieved from glucose at 140 °C for 4 h [18]. To efficiently convert carbohydrates into HMF, β zeolite doping with 0.4 wt.% Cr was prepared and exhibited a superior 72% HMF yield and 83% selectivity due to the moderate L/B (Lewis acid sites/Brønsted acid sites) [19]. Low L/B hindered the isomerization of glucose to fructose; high L/B excessively accelerated the dehydration of fructose and led to the degradation of fructose to humin.

As the most abundant part of biomass, cellulose can be hydrolyzed and further converted into various chemical substances, one of which is lactic acid. Lactic acid (LA) from biomass was also among the top 30 candidates for synthetic fossil product substitutes [20]. Many important industrial chemicals can be derived from LA molecules, such as polylactic acid (PLA) [21], acrylic acid (AA), and propanoic acid (PA) (as shown in Figure 2). While cellulose was hydrolyzed to produce lactic acid (LA), various products such as levulinic acid and 5-hydroxymethylfurfural (HMF) were also produced, which resulted in a lower LA yield (around 30%). Therefore, it is necessary to improve the LA selectivity. In recent years, catalysts with acidic sites, especially solid Lewis acid catalysts, have received extensive research [22,23]. It was found that yttrium-modified siliceous material β zeolite catalysts can effectively regulate the surface acidity of zeolite and inhibit the yield of dehydration products such as HMF and other derivatives. The results indicated that when Lewis acidity was increased, the yield of LA from cellulose reached 49.2% within 30 min, more efficient than previous research studies [24].

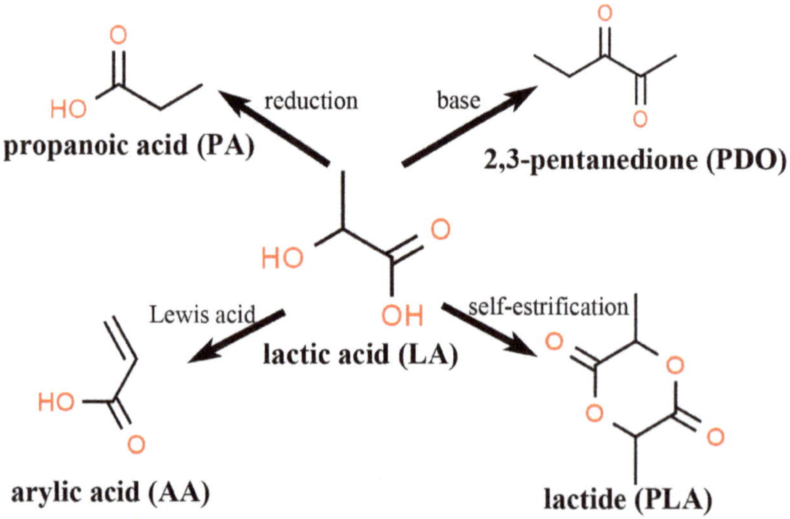

Figure 2. Reaction pathway of bio-renewable lactic acid (reproduced from reference [23]).

It was noteworthy that the composition of some organic compounds may have side effects on the catalytic performance of the catalyst. For example, phenolic substances derived from lignin not only contain O molecules but also have adverse impacts on catalysts. For this reason, the key to the catalytic cracking of bio-oil conversion is the optimization of efficient catalysts. Different catalysts have been adopted for different biomass and products.

In addition, NiMo/ASA Al$_2$O$_3$, Co Mo, HY zeolite, and ASA Al$_2$O$_3$ were simulated and found to be suitable for producing high distillate diesel [25]. The theoretical calculation results also indicated that larger mesoporous volume and surface area accelerated the diffusion rate of biomass inside the zeolite. However, this effect was continually weakened with increasing reaction time [26]. Because of the damage of by-products to catalysts and zeolites, the mass transfer efficiency was consequently decreased. To restrict the destruction of zeolites and catalysts and decrease the coke formation, an SFCC catalyst was employed in the cracking of high acid-value waste edible oil to produce biofuels. Moreover, the catalytic efficiency was attributed to doping with rare earth metals. Nguyen-Phuc et al. found that propylene and liquefied petroleum gas production were obviously improved by increasing the content of rare earth elements in ZSM-5 zeolite. Specifically, the yields of diesel, gasoline, and liquefied petroleum gas reached 29 wt.%, 42 wt.%, and 18 wt.%, respectively [27]. Additionally, to promote the conversion of biomass to valuable energy, a novel catalytic system similar to the Dylison Cycle Riser (DCR) should be designed and developed, which can produce hydrocarbon fuel intermediates from biomass-derived pyrolysis steam, such as pine and oak [28].

2.3. Preparation of Propylene

Propylene has been widely applied in diverse chemical products that are badly needed. Developing alternative methods to produce propylene using economical raw materials has attracted considerable interest, such as the hydrogenation of methanol to propylene and the dehydrogenation of propane to propylene [29–31].

2.3.1. Methanol to Propylene (MTP)

Methanol to propylene (MTP) was considered a substitutable approach to propylene production because of the extensive sources of methanol available. Due to its selectivity for propylene and the high resistance of the zeolite, ZSM-5 was preferred in the process (Table 1).

Si/Al makes a great difference to the morphology and aluminum distribution of molecular sieves. By increasing the Si/Al ratio in HZSM-5 zeolite, the selectivity of zeolite for propylene can be improved [32]. In the ZSM-5 catalyst, when the channel intersection is rich in Al pairs, the selectivity for ethylene and aromatics is higher. While the channel intersection has multiple single Al sites and lower Al pairs, it exhibits higher propylene selectivity and lower aromatic hydrocarbon selectivity [33].

Ion doping can also scale up the selectivity of zeolites towards specific products. When boron was introduced into the ZSM-5 structure through impregnation and hydrothermal synthesis, the orthogonal crystal structure of the zeolite was characterized, which retained the preferred growth orientation and high crystallinity hierarchical structure and changed the surface acidity. The selectivity of propylene was increased to 67%, and the propylene/ethylene ratio reached 8 [34]. When Mn enters the framework structure of ZSM-5, weak Brønsted acid sites increase the Al in the channel of the framework. Consequently, the carbon deposition rate was lowered, and the selectivity for propylene was improved [35].

Preventing catalyst deactivation and coke deposition from clogging pores and improving the lifespan of catalysts are still facing great challenges. Usually, modification of structural properties and catalyst acidity can efficiently solve these problems. To reduce acidity and consequently inhibit the aromatic hydrocarbon cycle, fluoride-assisted low-temperature crystallization can be applied to prepare plate-like MFI molecular sieves with similar crystal morphology and controllable acidity, which made propylene selectivity reach 52%, and catalyst lifespan 252 h [36]. In the initial process of MTP, long chain alkenes not being cracked diffused out from MFI zeolite and were found in the products (as shown in Figure 3, blue frame). Simultaneously, the long-chain alkenes propagated from olefins or alkenes were converted to polymethyl-benzenes and dienylic carbenium ions. Subsequently, lower olefins were produced in the aromatic cycle (as shown in Figure 3, red frame). Moreover, lower-density plate-like MFI can effectively inhibit the circulation of aromatic hydrocarbon groups, promote the hydrocarbon cycle, and exhibit excellent stability [37].

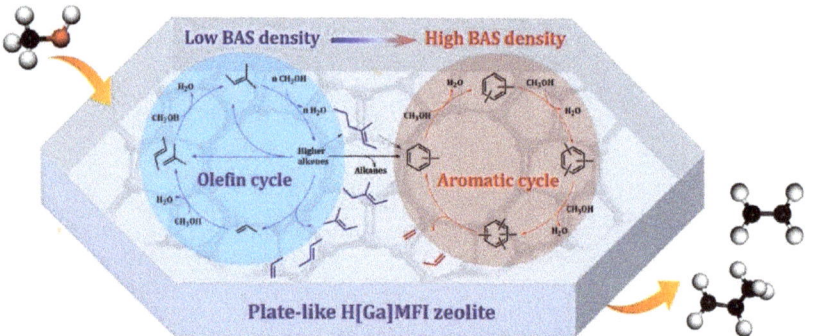

Figure 3. Dual-cycle mechanism in the MTP conversion over the plate-like H[Ga]MFI catalysts [36].

Table 1. Effects of Si/Al and specific surface area on the selectivity of propylene in MTP.

Catalysts	Method	Si/Al	S_{BET}	Selectivity of Propylene	P/E	
HZSM-5	two-stage crystallization	130	361	50.95	4.40	
		130	356	48.13	4.44	[32]
		124	352	49.68	5.89	
		139	340	50.10	7.48	
MFI		202	428	58	7.9	
B-MFI	template	205	404	66.3	7.9	[34]
Cr-MFI		201	408	59.8	2.7	
[Al/Ga]MFI	fluoride-assisted low-temperature crystallization	70–73 65–275 *	415–423 411–441	52%	-	[36]
ZSM-5	quasi-solid-phase	50	353	41.4%	-	[35]

* Si/Ga determined by ICP.

2.3.2. Propane to Propylene (PDH)

The dehydrogenation of propane (PDH) was a conventional way to produce propylene. Driven by the huge demand for propylene, Pt- or Cr- containing catalysts have drawn great attention and been applied to industrial manufacture. Although ZSM-5 is frequently used for PDH reactions due to its unique pore structure and good thermal stability, the high acidity of ZSM-5 can lead to a decrease in propylene selectivity. Through catalyst dealumination and additive regulation, the surface acidity of ZSM-5 can be reduced, and its propylene selectivity and catalytic stability in PDH can be improved. Some scholars have adopted the sequential impregnation method to introduce Ga and Mg to improve the Si/Al in HZSM-5, which can produce an effect on its stability and propylene selectivity (90.8%) [38].

Additionally, the separation of propylene and propane is another indispensable process in PDH, and it can be promoted by enhancing the propylene affinity of the adsorption site. What provided new ideas for the effective separation of propylene and propane is that scholars have prepared Ag exchanged Y zeolite (Ag-Y); this modified zeolite exhibited rapid adsorption kinetics and reversible propylene adsorption [39]. The results of molecular dynamics simulation research indicated that raising the temperature was beneficial for an increase in propane adsorption capacity, as the critical temperature of propane is higher than that of propylene. The smaller radius of non-skeletal cations in zeolite contributed to the higher efficiency of propylene adsorption [40]. After dealumination, vacant T-atom sites appeared in Si-BEA and silanol group forms. The introduced Co^{2+} can interact with silanol groups and form four coordinated structures in the zeolite (as shown in Figure 4), which consequently selectively activates the C–H bonds in propane, thereby improving the desorption efficiency after propylene generation and improving the long-term stability of the catalyst [41].

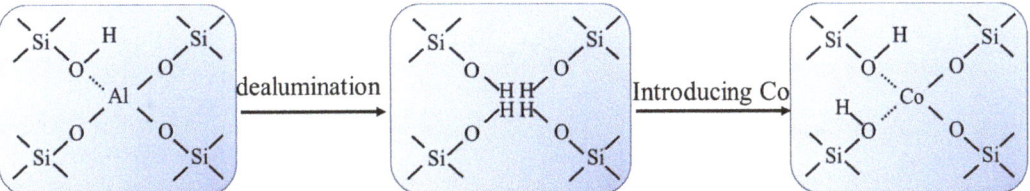

Figure 4. Local structure evolution of BEA zeolite during dealumination and Co^{2+} introduction (reproduced from reference [41]).

2.4. Selective Catalytic Reduction (SCR) of NO_x

Nitrogen oxides (NO_x) were one of the most destructive polluting gases that were mainly produced in the combustion of fossil fuels, such as car exhaust and coal combustion. In China, more than 90% of NO_x comes from vehicle exhaust emissions. Selective catalytic technology (SCR) is one efficient method to solve the problem of nitrogen oxide emissions. The catalytic reduction reactions between NO_x and NH_3 were conducted on the surface of the catalyst, and NO_x was therefore transformed into N_2. Currently, the adopted catalysts are mainly divided into metal matrix catalysts and zeolite-based catalysts. V_2O_5-WO_3/TiO_2 was once supposed to be the most effective catalyst for NO_x treatment. However, the biological toxicity, poor thermostability, and narrow working temperature range extremely restrained its application. In the 1980s, a ZSM-5 catalyst load with Cu^{2+} was discovered and exhibited excellent activity to NO_x and anti-toxicity.

Adsorption is the key process in catalytic oxidation and reduction. Zeolite catalysts exhibit good adsorption performance on NO_x owing to the large specific surface area, good thermal stability, adjustable acidity, excellent absorption, and ion exchange capabilities [42,43]. The overall efficiency of NO_x reduction can be greatly improved by metal doping and modification of zeolite catalysts. Cu-doped and Fe-doped zeolite catalysts have received more attention, such as Cu/ZSM-5, Cu/BEA, Cu/SAPO-34, and Cu/SSZ-13, which have been reported for the NH_3-SCR reaction of NO_x. The Cu-based catalyst exhibits excellent anti-propylene poisoning ability in the NH_3-SCR reaction. Researchers have found that moderately copper-doped zeolite, such as the $Cu_{7.55}$-ZK-5 catalyst, can isolate a large amount of Cu^{2+}, which is beneficial for the adsorption and activation of NO_x [44].

Based on Cu/MOR, the addition of promoting the separation of Cu^{2+} can create more Brønsted acid sites and inhibit the high-temperature ammonia oxygen reactions. This catalyst exhibited a NO_x conversion of about 88% and N_2 selectivity above 99% at 350–560 °C [45].

Hydrothermal stability is another factor influencing the application of zeolite catalysts. Yttrium-doped Cu-based zeolite catalysts can significantly reduce the fracture of Si-O-Al, improve the dispersion of Cu^{2+} active components and promote the adsorption and conversion efficiency of NO_x [46].

NO_x can be adsorbed with five reactions on Cu-ZSM-5 zeolites (as shown in Figure 5). The adsorption can be enhanced by the coordination interaction between Cu and N. Cu can be precisely dopped into H-ZSM-5 using an improved method called initial wetting impregnation microwave drying (IM). Compared with H-ZSM-5 and Na-ZSM-5, the desorption energy was reduced and the NO_x adsorption was significantly increased [47]. By designing novel pore structures, such as three dimensionally-ordered (3DOM) microporous zeolites, the contact area can be obviously extended. Some scholars have developed the steam seed-assisted colloid crystal template (SSAC) method to prepare 3DOM zeolites. After sonication, drying, and calcination, the $Pr_xMn_{1-x}O_\delta$/3DOM ZSM-5 catalyst with different Pr/Mn ratios was obtained (as shown in Figure 6) and showed an NO conversion rate of 90% [48].

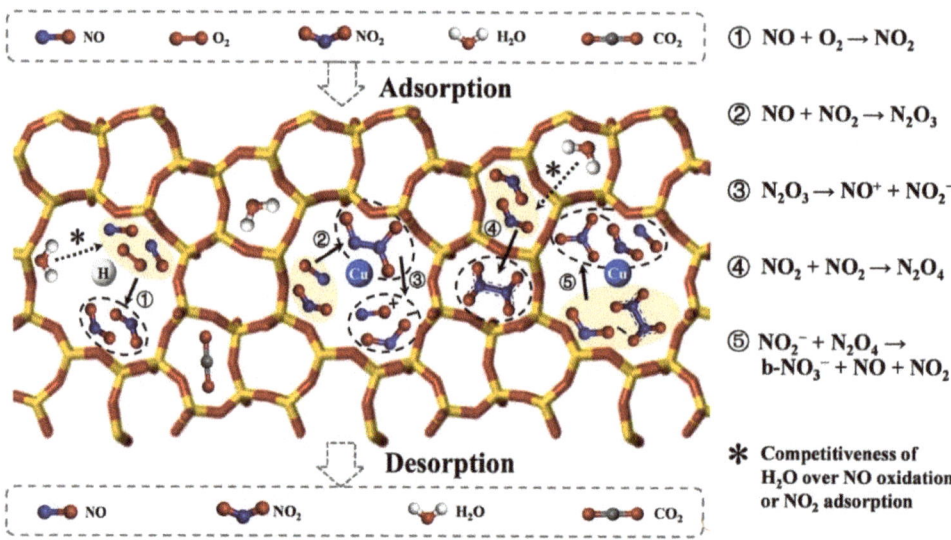

Figure 5. Proposed schematic diagram of NO$_x$ adsorption process on Cu-HZSM5_IM [47].

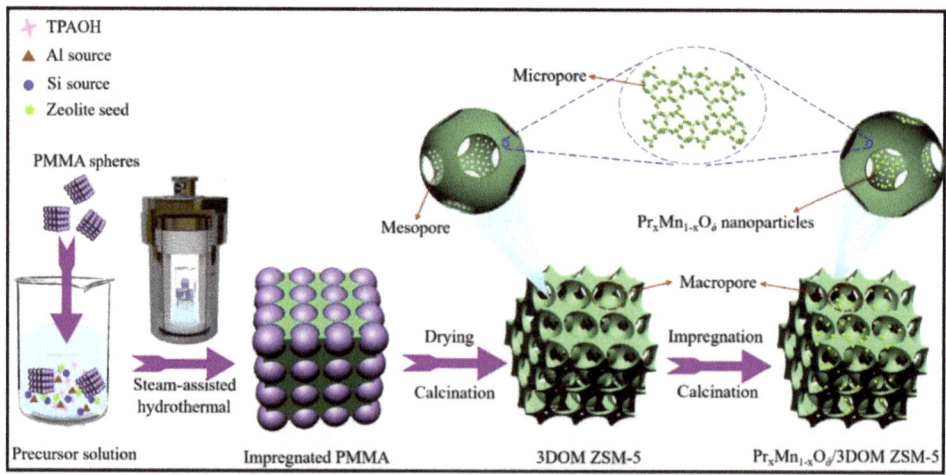

Figure 6. Schematic diagram for the preparation of Pr$_x$Mn$_{1-x}$O$_d$/3DOM ZSM-5 catalysts [48].

Except for these zeolite catalysts, scholars tried to combine metal-based catalysts (V$_2$O$_5$-WO$_3$/TiO$_2$) with zeolite catalysts (H-ZSM-5) to prepare the composite catalyst PM. The PM catalyst exhibited higher catalytic activity than the V$_2$O$_5$ WO$_3$/TiO$_2$ catalyst because the PM catalyst adsorbed more NO$_2$ and N$_2$O$_4$ [49]. Given the complexity of the exhaust purification system in SCR, some researchers adopted an H$_2$ selective catalytic reduction method using Pt/KFI molecular sieve catalysts from 150 °C to 250 °C, and the conversion rate of NO$_x$ reached 80% [50].

2.5. Hydrogen Preparation

Hydrogen energy has been viewed as a potential alternative energy in the future. However, producing hydrogen economically and environmentally is extremely challenging. Hydrogen can be prepared from natural gas (reforming of methane steam to produce hydrogen, SMR), coal gasification, electrolysis of water, and transformation of biomass. Various hydrogen production processes depend on catalysts. Due to the wide and renewable sources, the utilization of organic matter to produce hydrogen energy is gradually receiving attention. Relevant technical methods mainly include two methods. One is thermochemistry hydrogen production, which consists of high-temperature pyrolysis, liquefaction, and gasification, and the other is biomass hydrogen production which refers to the preparation of hydrogen from biomass after anaerobic fermentation, biological metabolism, and reforming. The catalysts used in hydrogen are mainly precious metals and transition metals, such as Rh, Ru, Pd, Pt, Cu, Co, Ni, etc. Molecular sieves are often used as carriers for these catalysts.

At present, ethanol steam reforming (ESR) is the most suitable industrial method for ethanol to produce hydrogen, but there are some vital problems with this technology: (1) high energy consumption; (2) expensive separation of CO_2 and H_2. Choosing appropriate catalyst carriers and catalysts can help improve catalytic activity and reduce costs. Furthermore, the Ni catalyst was relatively inexpensive and had good hydrogen generation activity. Zeolites have a large content of micropores, and the structure can be modified. However, carbon deposition during the catalytic process of Ni-loaded zeolite can easily lead to decreasing catalytic activity. A new type of quaternary ammonium cation cationic surfactant was introduced to synthesize the supported calcium-modified nickel-step classification β zeolite catalysts. The BET and pore volume of this modified zeolite were larger than that without modification. The introduction of Ca^{2+} changed the internal structure of the catalyst to some extent, which promoted the interaction between the carrier and Ni. Consequently, the yield of H_2 was increased [51,52]. When Mg was introduced into Rh/β zeolite, Rh clusters from the atomic level to the sub-nanometer scale were formed in microporous channels. It was found that the C=O formed in the reaction was the key to increasing selectivity, which can be promoted by Mg addition [53]. The DFT calculation results indicated that the stability was obviously affected by the zeolite structure. Specifically, after dealumination, the binding energy of Rh to the vacancy defect was reduced by 0.2 eV. If the Mg was added to the zeolite structure, the binding energy was additionally reduced to -1.927 eV (as shown in Figure 7a). Due to the proximity of Rh and Mg in zeolite with Mg, the energy barrier of O-H was significantly reduced, which was beneficial to the dissociation of phenol to phenoxy (as shown in Figure 7b). ITQ-6 zeolite loaded with Ni and Co can also be used in ESR. Among them, the ITQ-6 zeolite loaded with Co exhibited higher hydrogen production on account of the smaller size of Co particles. Moreover, the carbon deposition effect of Co/ITQ-6 was weaker, and the deactivation was lower [54].

In order to reduce catalyst deactivation, some scholars have tried to design catalysts with small metal nanoparticles. An encapsulated ultra-small Ni catalyst with mesoporous and hollow structure was prepared using separated metal dispersions Si-1 zeolite with high Ni dispersion (Ni@Si-1), which facilitated local mass transfer and reduced carbon deposition [55]. The research on supercritical water dynamics indicated that the gasification of catalysts such as Ni can be promoted by improving hydrogen selectivity. The pore distribution of Ni particles can be divided into micropores (1–10 nm) and mesopores (20–60 nm). When the size of the mesopores was distributed reasonably in zeolite, the active sites increased, and the catalytic activity was higher [56].

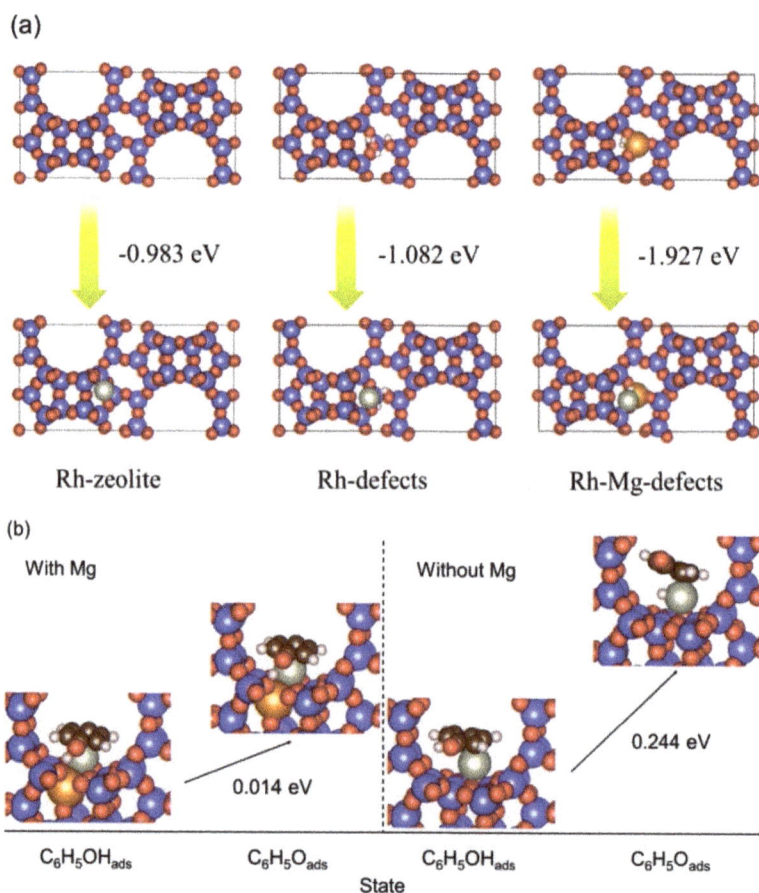

Figure 7. (a) Structural configurations of Rh located on three different supports and corresponding binding energies and (b) relative energy levels for O–H bond dissociation of phenol in zeolite Beta with and without Mg [53].

In addition to ethanol, methanol can also be used as the raw material for hydrogen production, that is, methanol steam reforming (MSR). A catalyst suitable for the catalytic reaction of MSR was prepared using an attapulgite molecular sieve as a carrier and Cu-Zr active components. The addition of Zr significantly increased the distribution of active metal particles and reaction sites, as well as the H_2 yield. At the same time, the sintering of active metals and coke were inhibited [57].

Photocatalysis is another important method for producing hydrogen gas, and zeolite loaded with TiO_2 photocatalyst was supposed to be one of the potentially efficient catalysts [58,59]. To improve the transmission efficiency of photoelectrons, some scholars have designed $Cu_2O@TiO_2@ZIF-8$, where an internal electric field was formed due to the p-n junction, which promoted the transfer of electrons to the conduction band. The photoelectrons generated from TiO_2 were transferred to the conduction band and further transferred through ZIF-8. The water can be therefore reduced by these photoelectrons (as shown in Figure 8) [52]. To provide more proton sources and active sites, $Na^+/K^+/Ca^{2+}$ can be added into zeolite, which consequently promotes the hydrogen evolution rate [60].

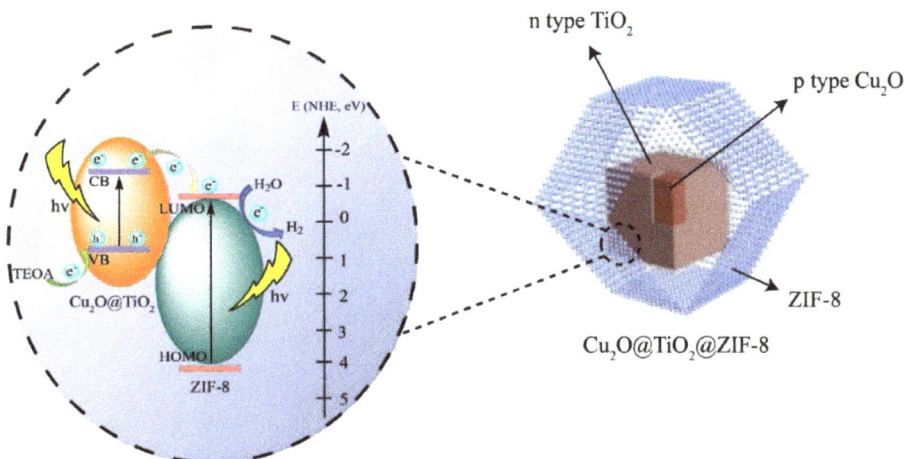

Figure 8. Proposed mechanism in the Cu$_2$O@TiO$_2$@ZIF-8 photocatalyst [58].

2.6. VOCs Abatement

Volatile organic compounds (VOCs) refer to organic compounds with a boiling temperature in the range of 50–260 °C. They were usually emitted in the petrochemical industry and the rubber industry and had serious effects on human health and the natural environment. With the rapid development of industry and increasing consumption of fossil fuel energy in vehicles, the emission of VOCs has dramatically increased in recent years. According to their boiling points, molecular structures, and molecular polarities, VOCs can be classified into different types [61]. They were hard to eliminate using direct combustion due to the low concentration of VOCs. To reduce the pollution and damage of VOCs, multiple technologies have been developed and applied, such as adsorption, a photocatalytic degradation method, plasma degradation, catalytic combustion, etc. Whether adopting an adsorption method or catalytic oxidation method, it is no wonder that developing efficient adsorbent and porous materials are principal considerations. An excellent adsorbent should have massive adsorption sites, good hydrothermal stability, and weak resistance to molecular diffusion. Worldwide, research studies show that activated carbons [62], metal-organic frameworks, and zeolites are all potential adsorption materials that have exhibited excellent adsorption performance in many experiments. However, the combustion of activated carbons and the high cost of metal-organic frameworks restrained their application in VOC treatment.

The zeolite of the 255 framework type, approved by the International Zeolite Association (IZA), was considered a promising absorbent due to its abundant micropores, adjustive chemical stability, etc. The adsorption efficiency of different zeolite structures on different VOCs molecules and the competitive adsorption mechanism of various VOCs have been a puzzle to reveal. The absorption of VOCs can be influenced by the specific surface area, pore structure, and surface functional groups. Additionally, the kinetic diameter and polarity of the VOC molecule may also affect the absorption process.

The absorption efficiency was also proved to be related to humidity. Therefore, developing novel zeolites and improving the hydrophobicity of zeolites contributed to the adsorption efficiency of zeolites under humid conditions. The hydrophobicity of zeolites can be improved through ion doping and coating. A Mo-doped MEL zeolite was developed using a self-developed recrystallization method for the adsorption and reduction of non-methane hydrocarbons (NMHC) in cooking oil fumes. The affinity for VOC molecules is significantly enhanced due to its high atomic coordination level and lack of silanol, especially under humid conditions. This Mo MEL zeolite exhibited longer adsorption saturation time, larger adsorption capacity, and better hydrophobicity, which led to the more excellent adsorption efficiency of Mo MEL than that of Si MEL and Al MEL [63]. In addition to Mo, introducing Mn (Cu or Fe) to replace Al can also improve the hydrophobicity of zeolites [64]. The addition of Mn affected the crystallization behavior of zeolites, which in turn, changed the distribution of Mn in zeolites and the properties of zeolites. Y-type zeolite, with a high specific surface area and unique microporous structure, can have its hydrophobicity and diffusion efficiency improved by being coated with mesoporous SiO_2 [65]. In general, after being coated with mesoporous hydrophobic materials, the diffusion activation energy of VOCs was decreased, and the diffusion coefficient was increased. The π-n electron pair interaction was weakened, while the dipole interaction was enhanced. The dispersion interaction plays a positive role in the adsorption of VOCs and gradually enhances with increasing relative humidity. However, the repeatability of core-shell composite materials needs to be verified. The evaporation-induced self-assembly (EISA) method was adopted to prepare NaY zeolite loaded with metal oxide nanoparticles. It strongly improved the adsorption capacity of VOCs. Li SHI et al. successfully prepared uniformly dispersed Y@M_xO_y nanoparticles, which significantly improved the adsorption effect of Y-zeolite on isopropanol and acetone in humid environments under RH = 50%. The metal oxides enhanced the adsorption capacity of VOCs, and greatly reduced the competitive adsorption between water molecules and VOCs molecules [66]. Multilevel mesoporous USY molecular sieves can be prepared using etching and surfactant template processes, during which the non-skeletal aluminum was effectively removed inside the zeolite, leading to a mesoporous size of approximately 4 nm [67]. Some scholars have also proposed a new strategy called citric acid sacrifice that can synthesize non adhesive monolithic zeolites. The mechanical strength and the water resistance were obviously improved. After the in situ dealuminization, the specific surface area and mesoporous volume were greatly improved, the adsorption capacities of toluene and acetone were also increased. When the relative humidity was 90%, the adsorption efficiency of acetone and toluene reached 90% of that attained in dry conditions [68].

Adjusting the pore structure of molecular sieves is also vital for improving the adsorption and reaction rates of VOCs. The hierarchical design and synthesis of zeolite structures are gradually receiving attention [69,70]. For example, the microporous size of ZSM-5 zeolite is generally less than 2 nm, and larger VOC molecules have a slower diffusion rate during the adsorption process, making it difficult to quickly enter the reaction active center and the generation rate of specific products is pretty low. When ZSM-5 is modified with organic functionalized silica and organic additives, the size and surface morphology of ZSM-5 nanocrystals can be greatly improved, forming a hierarchical mesoporous structure, and the adsorption capacity for toluene is greatly increased [71]. According to the hierarchical crystallization mechanism revealed in the corresponding research (as shown in Figure 9), organic silica fragments modified by organosilane PHAPTMS were combined and then gathered around MFI. After hydrolysis of -SiO_3, PHAPTMS participated in the formation of nanocrystalline frameworks in a hydrothermal process. The inter-crystalline and intra-crystalline mesopores were finally formed after calcination.

The adsorption and catalytic efficiency of zeolite can be effectively improved by regulating the active sites. Effective regulation of active sites can reduce carbon deposition in pore structures and improve the durability of catalysts. Nano Pt is usually adopted as an active ingredient, and its high dispersibility facilitates the catalytic oxidation of aromatic hydrocarbons and alkanes. Synergistically, it can interact with surface acid sites that contribute to the removal of VOCs at low temperatures [72]. Generally, adjusting the Si/Al ratio in zeolites can help regulate the acidic sites. A controllable framework modulation strategy can be adopted to prepare high-silica zeolites. In this strategy, it was very important to control the matching degree of dealuminization and Si-insertion. For Pt/β zeolite, Pt^0 has a stronger activation ability for O_2 and shows higher efficiency than $Pt^{\delta+}$. That makes it much easier to supply oxygen, which is more conducive to the ring opening reaction of the benzene ring, thus reducing the production of gaseous benzene on the high silicon catalyst. Furthermore, both high-silicon and low-silicon Pt/Beta catalysts show excellent durability and follow a similar toluene oxidation path to form intermediates, including alcohol oxide, carboxylate, and anhydride [73]. For Pt/ZSM-5, the main active oxidation site is Pt^0. For the deep catalytic oxidation of oxygen/nitrogen-containing VOCs, Pt/ZSM-5 had abundant acidic sites on its surface. Different SiO_2/Al_2O_3, Pt^0 ratio, and Pt dispersion equilibrium in the catalyst may be changed, and the synergistic effect of acidic sites and oxidation sites were promoted, which led to higher reaction activity of Pt/ZSM-5 (25), with a T90% value of only 207 °C for acetonitrile and 175 °C for ethyl acetate. It is beneficial for low-temperature catalytic degradation of specific VOCs [74].

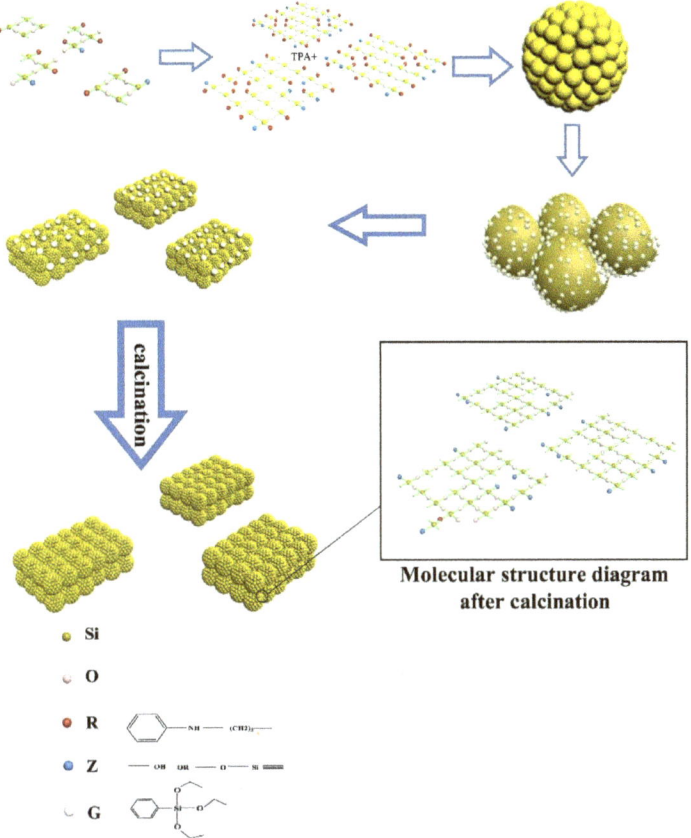

Figure 9. The crystallization mechanism diagram for the ZSM-5 zeolites [71].

In industrial activities, multiple types of VOCs are often generated simultaneously. Ideally, a single adsorbent and catalyst can simultaneously handle multiple VOCs. However, due to the various molecular structure of VOCs, the different pore structures, and the complexity of the catalyst's electronic structure, ensuring the treatment effect of VOCs is difficult. Therefore, it is particularly important to explore and understand the adsorption and oxidation mechanism of catalysts, which contributes to the development of novel catalysts and adsorbents aimed at VOC mixtures. The molecular polarity and volatility of VOCs play a crucial role in the adsorption process. For these VOC mixtures, the priority order of adsorption varies. For example, under static equilibrium conditions, the absorption order was determined as acetone > ethyl acetate > toluene. However, under dynamical conditions, molecular polarity and volatility played key roles. The absorption order was found to be ethyl acetate > toluene > acetone [75]. Complex interference effects were identified when this VOC mixture was adsorbed onto different zeolites. The adsorption effect of toluene on USY is significantly better than that on 13X. However, for acetone, the adsorption of 13X and ZSM-5 is better. ZSM-5 and β Zeolites can selectively adsorb ethyl acetate, as evidenced by DFT calculations. This competitive adsorption widely exists in the treatment of various VOCs, and the differential adsorption effect is one of the problems to be solved in the future [75]. To develop efficient catalysts, some scholars have attempted to explore the possibility of synergistic treatment of acetone and benzene by changing the amount of Mn added [76]. When Mn was introduced, the activity of the catalyst for acetone was significantly increased, while the activity for benzene was decreased. Because the introduction of Mn regulated the electronic structure of Pt, leading to the transfer of electrons to Pt, which enhanced the adsorption of acetone and gaseous oxygen and improved the performance of the acetone oxidation catalyst, and promoted the degradation of acetone. However, the high electron density of Pt inhibited the adsorption and degradation of benzene, reducing the production of formic acid products.

In addition, the treatment of VOCs in open spaces has gradually attracted the attention of researchers. For example, asphalt materials are often used in the process of road paving and continually release VOCs into the open air. Due to the limitations of open spaces, it is very difficult to collect and dispose of these VOCs. Steel slag and red mud were usually used as road materials, zeolites synthesized from these materials can reduce the volatility of asphalt and absorb VOCs in road structures [77]. Therefore, preparing efficient zeolites using in-road materials (such as steel slag) was the key approach to deal with VOC emissions in open spaces.

2.7. CO_2 Capture

CO_2 has been regarded as the main greenhouse gas resulting in global warming, which is one of the most challenging environmental issues. To prevent the damage caused by global warming, various technologies are being developed [78]. Among them, one technology called carbon capture, utilization, and storage (CCUS) has received great attention. In the CCUS system, CO_2 capture is the key stage that matures this technology. According to the manufacturing methods of CO_2, pre-combustion capture, post-combustion capture, and oxyfuel combustion were the main approaches. Regardless of different approaches, the CO_2 separation process was the critical part of CO_2 capture, such as absorption, membranes, adsorption, chemical looping combustion, and calcium looping, and adsorption has been considered one of the most promising ways to capture CO_2.

Compared with other CO_2 adsorbents, zeolite exhibited a stable cycle, huge surface area, and fast kinetics of CO_2 adsorption. X-type zeolite, Y-type zeolite, and A-type zeolite were usually studied as CO_2 capture adsorbents. Therefore, zeolite has received increasing attention in CO_2 capture over the past decades. It has been found that there are several factors influencing the capture efficiency of CO_2, such as Si/Al, distribution of pore sizes, exchangeable ions, and moisture effects.

As a kind of aluminosilicate, Al ions can replace Si ions in the TO_4 structure, which breaks the charge balance of the original structure and leads to increasing basicity. It was reported that higher basicity contributes to higher CO_2 adsorption [78]. Zeolites with a lower Si/Al ratio were more stable at high temperatures and less likely to exchange with other ions. To achieve higher CO_2 capacity, the chemical composition of zeolites can be optimized. Scholars found that the Si/Al of GIS-type zeolite strongly affected the mechanism of CO_2 adsorption. Specifically, for Na-GIS with Si/Al lower than 2.2, the CO_2 adsorption was hindered, and CO_2 uptake was negligible due to the large number of extra-framework cations near the 8-ring window [79].

The pore size of different zeolites was usually distributed in the range from 0.5 nm to 1.2 nm due to their crystalline nature. It means that only when the kinetic diameter of gas molecules is smaller than the pore size the gas can be adsorbed. To strengthen the adsorption efficiency of CO_2, it is very promising to synthesize zeolites with mesopores. Due to the hierarchical structure, the obstruction of molecular diffusion is obviously reduced, which can consequently promote CO_2 adsorption. Chen et al. prepared mesoporous LTA zeolite (Meso-LTA); it exhibited faster CO_2 adsorption kinetics (1 bar) and higher CO_2 adsorption capacities (>10 bar) at 298 K compared with microporous LTA zeolite [80].

CO_2 adsorption is essentially the interaction between CO_2 molecules and the electric field caused by ions in zeolites, which is similar to CO_2 desorption. Therefore, the CO_2 capture efficiency can be promoted by adjusting the chemical composition of zeolite, for instance, ion exchange. For instance, the simulation and experimental results of ion-exchanged 13X zeolite (FAU type) indicated that, compared with Na^+ and K^+, LiX-80 exhibited better separation of CO_2/N_2. Furthermore, the doping of Pd^{2+} and Ag^+ changed the gradient of potential and strengthened the electrostatic potential (ESP), which increased the CO_2 adsorption capacity (1.89%) and the CO_2/N_2 selectivity (85.97%) compared with LiX-80 [81].

Apart from the above factors, in flue gas, vapor is always accompanied by CO_2. The water molecules may compete with CO_2 molecules for the active sites in zeolites, which dramatically reduces the efficiency of CO_2 adsorption. Some scholars proposed a strategy that fabricating a shell around the zeolite using sol-gel coating and a poly-ethylenimine impregnation process, and the diffusion of water molecules was hindered [82].

As a kind of promising CO_2 solid adsorbent, zeolite has drawn great attention for the past decades and exhibited excellent performance. However, its application still faces a lot of challenges due to its structure characteristics and chemical composition. The modification of zeolite used for CO_2 capture will still be a research focus in the future.

3. Conclusions and Future Perspectives

Zeolite catalyst is featured in energy conversion, selective separation, pollution prevention, and treatment. Generally, it exhibited better adsorption when the pore size of zeolite was equivalent to that of the adsorbed molecule. Therefore, it was very important to select the appropriate zeolite for different catalytic reactions. Since a state of long-term hydrothermal treatment at high temperatures is a necessary condition for its synthesis, numerous problems such as long synthesis time, high energy consumption, high equipment cost, and safety issues caused by the use of high-pressure hot-pressing tanks continuously emerged. There is no doubt that shortening synthesis time and lowering synthesis temperature are desperately needed. Reducing synthesis costs (energy consumption, autoclave productivity, etc.) and environmental footprint and safety issues has always been the primary issue.

Despite the special structure and excellent properties, the promotion of the application of zeolite catalysts also faces great challenges, such as the selectivity of specific products, temperature adaptability, and the lifetime of catalysts. It was also noticed that the retention of the reaction medium leads to the generation of by-products, resulting in micropore blockage and reduced catalytic activity. However, it can be noticed that upgrading the preparation method, regulating the pore structure, and increasing acid sites are three awesome measures to improve the properties of zeolite catalysts. Zeolites are expected to show their great advantage in energy-saving catalytic processes, environmentally friendly adsorption and separation, and energy storage.

Author Contributions: Conceptualization, W.Z., D.F. and Y.Z.; methodology, W.Z. and D.F.; software, W.Z.; validation, G.H., D.F. and Y.Z.; formal analysis, D.F. and Y.Z.; investigation, W.Z. and D.F.; resources, D.L. and Y.Z.; data curation, G.H., D.L. and Y.Z.; writing—original draft preparation, W.Z.; writing—review and editing, D.F. and G.H.; visualization, D.L. and Y.Z.; supervision, D.F. and Y.Z.; project administration, D.F.; funding acquisition, D.F. All authors have read and agreed to the published version of the manuscript.

Funding: This work was financially supported by the 2023 Teaching Case Program of Graduate School for the course named contemporary analytical techniques for materials characterization (Wuhan University of Technology).

Data Availability Statement: Data will be made available on request.

Conflicts of Interest: The authors declare no conflict of interest.

References

1. Mishra, R.K.; Chistie, S.M.; Naika, S.U.; Mohanty, K. Catalytic pyrolysis of biomass over zeolites for bio-oil and chemical production: A review on their structure, porosity and acidity co-relation. *Bioresour. Technol.* **2022**, *366*, 128189. [CrossRef] [PubMed]
2. Moliner, M.; Martinez, C.; Corma, A. Multipore Zeolites: Synthesis and Catalytic Applications. *Angew. Chem. Int. Ed.* **2015**, *54*, 3560–3579. [CrossRef]
3. Li, Y.; Li, L.; Yu, J.H. Applications of Zeolites in Sustainable Chemistry. *Chemistry* **2017**, *3*, 928–949. [CrossRef]
4. Martínez, C.; Vidal-Moya, A.; Yilmaz, B.; Kelkar, C.P.; Corma, A. Minimizing rare earth content of FCC catalysts: Understanding the fundamentals on combined P-La stabilization. *Catal. Today* **2023**, *418*, 114123. [CrossRef]
5. Wang, T.; Le, T.; Ravindra, A.V.; Jue, H.; Zhang, L.; Wang, S. Enhanced regeneration of spent FCC catalyst by using oxalic acid-sulfuric acid mixture under ultrasonic irradiation. *J. Mater. Res. Technol.* **2021**, *15*, 7085–7099. [CrossRef]
6. Zhang, L.; Hu, Q.; Qin, Y.; Liu, H.; Liu, H.; Cao, G.; Gao, X.; Song, L.; Zhaolin, S. Optimizing the accessibility of zeolite Y on FCC catalyst to improve heavy oil conversion capacity. *Microporous Mesoporous Mater.* **2023**, *359*, 112627. [CrossRef]
7. Cui, W.; Zhu, D.; Tan, J.; Chen, N.; Fan, D.; Wang, J.; Han, J.; Wang, L.; Tian, P.; Liu, Z. Synthesis of mesoporous high-silica zeolite Y and their catalytic cracking performance. *Chin. J. Catal.* **2022**, *43*, 1945–1954. [CrossRef]
8. Zhang, R.; Ju, Y.; Wu, P.; Chen, J.; Lv, Z.; Zhang, Y.; Song, S.; Zhang, Z.; Ma, C.; Zhang, R.; et al. Efficiently reducing olefin content of FCC gasoline over ZSM-5 zeolite based catalyst via hydro-upgrading. *Catal. Today* **2022**, *405–406*, 57–65. [CrossRef]
9. Liu, P.; Cui, Y.; Wang, J.; Du, X.; Zhang, H.; Humphries, A.; Jia, M.; Yu, J. Structure stabilization of zeolite Y induced by yttrium and its role in promoting n-docosane conversion. *Microporous Mesoporous Mater.* **2021**, *323*, 111225. [CrossRef]
10. Yamazaki, H.; Hasegawa, H.; Tanaka, C.; Takamiya, Y.; Mitsui, T.; Mizuno, T. Al ion-exchanged USY in FCC catalyst for high LPG yield. *Catal. Commun.* **2021**, *159*, 106354. [CrossRef]
11. Miao, P.; Zhu, X.; Zhou, Z.; Feng, X.; Miao, J.; Hou, C.; Li, C. Combined dealkylation and transalkylation reaction in FCC condition for efficient conversion of light fraction light cycle oil into value-added products. *Fuel* **2021**, *304*, 121356. [CrossRef]
12. Zhong, M.; Li, X.; Chu, X.; Gui, H.; Zuo, S.; Yao, C.; Li, Z.; Chen, Y. Solar driven catalytic conversion of cellulose biomass into lactic acid over copper reconstructed natural mineral. *Appl. Catal. B Environ.* **2022**, *317*, 121718. [CrossRef]
13. Perego, C.; Bosetti, A. Biomass to fuels: The role of zeolite and mesoporous materials. *Microporous Mesoporous Mater.* **2011**, *144*, 28–39. [CrossRef]
14. Ong, H.C.; Yu, K.L.; Chen, W.-H.; Pillejera, M.K.; Bi, X.; Tran, K.-Q.; Pétrissans, A.; Pétrissans, M. Variation of lignocellulosic biomass structure from torrefaction: A critical review. *Renew. Sustain. Energy Rev.* **2021**, *152*, 111698. [CrossRef]
15. Yan, P.; Wang, H.; Liao, Y.; Wang, C. Zeolite catalysts for the valorization of biomass into platform compounds and biochemicals/biofuels: A review. *Renew. Sustain. Energy Rev.* **2023**, *178*, 113219. [CrossRef]
16. Abu Zarin, M.A.; Zainol, M.M.; Ramli, N.A.S.; Amin, N.A.S. Zeolite immobilized ionic liquid as an effective catalyst for conversion of biomass derivatives to levulinic acid. *Mol. Catal.* **2022**, *528*, 112506. [CrossRef]
17. Yan, P.; Wang, H.; Liao, Y.; Sun, P.; Wang, C. Introducing mesopore and regulating Al distribution for improving catalytic performances of ZSM-5 in furfuryl alcohol to levulinic acid. *Fuel* **2022**, *329*, 125213. [CrossRef]
18. Chung, N.H.; Oanh, V.T.; Thoa, L.K.; Hoang, P.H. Catalytic Conversion of Glucose into 5-Hydroxymethyl Furfural Over Cu-Cr/ZSM-5 Zeolite. *Catal. Lett.* **2020**, *150*, 170–177. [CrossRef]
19. Xu, S.Q.; Pan, D.H.; Hu, F.; Wu, Y.F.; Wang, H.Z.; Chen, Y.; Yuan, H.; Gao, L.J.; Xiao, G.M. Highly efficient Cr/beta zeolite catalyst for conversion of carbohydrates into 5-hydroxymethylfurfural: Characterization and performance. *Fuel Process. Technol.* **2019**, *190*, 38–46. [CrossRef]
20. Werpy, T.; Petersen, G. *Top Value Added Chemicals from Biomass: Volume I—Results of Screening for Potential Candidates from Sugars and Synthesis Gas*; Medium: Boston, MA, USA, 2004; 76p.
21. Venschott, M.; Hoelderich, W.F.; Eisenacher, M. 2nd generation PLA. Lactide formation directly from aqueous lactic acid. *Catal. Commun.* **2023**, *177*, 106636. [CrossRef]
22. Ma, H.; Tingelstad, P.; Chen, D. Lactic acid production by catalytic conversion of glucose: An experimental and techno-economic evaluation. *Catal. Today* **2023**, *408*, 2–8. [CrossRef]
23. Kim, J.; Bang, J.; Choi, J.-S.; Lim, D.-H.; Guk, D.; Jae, J. Selective conversion of lactic acid to renewable acrylic acid over SDA-free Na-ZSM-5: The critical role of basic sites of sodium oxide. *J. Catal.* **2023**, *421*, 271–284. [CrossRef]
24. Ye, J.; Chen, C.; Zheng, Y.; Zhou, D.; Liu, Y.; Chen, D.; Ni, L.; Xu, G.; Wang, F. Efficient conversion of cellulose to lactic acid over yttrium modified siliceous Beta zeolites. *Appl. Catal. A Gen.* **2021**, *619*, 118133. [CrossRef]
25. Koyunoğlu, C.; Gündüz, F.; Karaca, H.; Çınar, T.; Soyhan, G.G. Developing an adaptive catalyst for an FCC reactor using a CFD RSM, CFD DPM, and CFD DDPM–EM approach. *Fuel* **2023**, *334*, 126550. [CrossRef]
26. Silva, J.M.; Ribeiro, M.F.; Graça, I.; Fernandes, A. Bio-oils/FCC co-processing: Insights into the adsorption of guaiacol on Y zeolites with distinct acidity and textural properties. *Microporous Mesoporous Mater.* **2021**, *323*, 111170. [CrossRef]
27. Le-Phuc, N.; Tran, T.V.; Phan, T.T.; Ngo, P.T.; Ha, Q.L.M.; Luong, T.N.; Tran, T.H.; Phan, T.T. High-efficient production of biofuels using spent fluid catalytic cracking (FCC) catalysts and high acid value waste cooking oils. *Renew. Energy* **2021**, *168*, 57–63. [CrossRef]

28. Magrini, K.; Olstad, J.; Peterson, B.; Jackson, R.; Parent, Y.; Mukarakate, C.; Iisa, K.; Christensen, E.; Seiser, R. Feedstock and catalyst impact on bio-oil production and FCC Co-processing to fuels. *Biomass Bioenergy* **2022**, *163*, 106502. [CrossRef]
29. Zabihpour, A.; Ahmadpour, J.; Yaripour, F. Strategies to control reversible and irreversible deactivation of ZSM-5 zeolite during the conversion of methanol to propylene (MTP): A review. *Chem. Eng. Sci.* **2023**, *273*, 118639. [CrossRef]
30. Qiu, B.; Lu, W.-D.; Gao, X.-Q.; Sheng, J.; Ji, M.; Wang, D.; Lu, A.-H. Boosting the propylene selectivity over embryonic borosilicate zeolite catalyst for oxidative dehydrogenation of propane. *J. Catal.* **2023**, *417*, 14–21. [CrossRef]
31. Zhao, Y.-N.; Fan, S.-B.; Ma, Q.-X.; Zhang, J.-L.; Zhao, T.-S. Methanol converting to propylene on weakly acidic and hierarchical porous MFI zeolite. *J. Fuel Chem. Technol.* **2022**, *50*, 210–217. [CrossRef]
32. Feng, R.; Zhou, P.; Liu, B.; Yan, X.; Hu, X.; Zhou, M. Direct synthesis of HZSM-5 zeolites with enhanced catalytic performance in the methanol-to-propylene reaction. *Catal. Today* **2022**, *405–406*, 299–308. [CrossRef]
33. Feng, R.; Liu, B.; Zhou, P.; Yan, X.; Hu, X.; Zhou, M.; Yan, Z. Influence of framework Al distribution in HZSM-5 channels on catalytic performance in the methanol to propylene reaction. *Appl. Catal. A Gen.* **2022**, *629*, 118422. [CrossRef]
34. Kalantari, N.; Bekheet, M.F.; Delir Kheyrollahi Nezhad, P.; Back, J.O.; Farzi, A.; Penner, S.; Delibaş, N.; Schwarz, S.; Bernardi, J.; Salari, D.; et al. Effect of chromium and boron incorporation methods on structural and catalytic properties of hierarchical ZSM-5 in the methanol-to-propylene process. *J. Ind. Eng. Chem.* **2022**, *111*, 168–182. [CrossRef]
35. Tuo, J.; Lv, J.; Fan, S.; Li, H.; Yang, N.; Cheng, S.; Gao, X.; Zhao, T. One-pot synthesis of [Mn,H]ZSM-5 and the role of Mn in methanol-to-propylene reaction. *Fuel* **2022**, *308*, 121995. [CrossRef]
36. Zhang, L.; Yang, L.; Liu, R.; Shao, X.; Dai, W.; Wu, G.; Guan, N.; Guo, Z.; Zhu, W.; Li, L. Design of plate-like H[Ga]MFI zeolite catalysts for high-performance methanol-to-propylene reaction. *Microporous Mesoporous Mater.* **2022**, *333*, 111767. [CrossRef]
37. Dai, W.; Zhang, L.; Liu, R.; Huo, Z.; Dai, W.; Guan, N. Facile fabrication of a plate-like ZSM-5 zeolite as a highly efficient and stable catalyst for methanol to propylene conversion. *Mater. Today Sustain.* **2023**, *22*, 100364. [CrossRef]
38. Yang, G.; Yan, X.; Chen, Y.; Guo, X.-J.; Lang, W.-Z.; Guo, Y.-J. Improved propylene selectivity and superior catalytic performance of Ga-xMg/ZSM-5 catalysts for propane dehydrogenation (PDH) reaction. *Appl. Catal. A Gen.* **2022**, *643*, 118778. [CrossRef]
39. Xiong, Y.; Tian, T.; L'Hermitte, A.; Méndez, A.S.J.; Danaci, D.; Platero-Prats, A.E.; Petit, C. Using silver exchange to achieve high uptake and selectivity for propylene/propane separation in zeolite Y. *Chem. Eng. J.* **2022**, *446*, 137104. [CrossRef]
40. Moradi, H.; Azizpour, H.; Mohammadi, M. Study of adsorption of propane and propylene on CHA zeolite in different Si/Al ratios using molecular dynamics simulation. *Powder Technol.* **2023**, *419*, 118329. [CrossRef]
41. Wei, S.; Dai, H.; Long, J.; Lin, H.; Gu, J.; Zong, X.; Yang, D.; Tang, Y.; Yang, Y.; Dai, Y. Nonoxidative propane dehydrogenation by isolated Co^{2+} in BEA zeolite: Dealumination-determined key steps of propane CH activation and propylene desorption. *Chem. Eng. J.* **2023**, *455*, 140726. [CrossRef]
42. Liu, Y.; Tao, H.; Yang, X.; Wu, X.; Li, J.; Zhang, C.; Yang, R.T.; Li, Z. Adsorptive purification of NO_x by HZSM-5 zeolites: Effects of Si/Al ratio, temperature, humidity, and gas composition. *Microporous Mesoporous Mater.* **2023**, *348*, 112331. [CrossRef]
43. Sunil Kumar, M.; Alphin, M.S.; Manigandan, S.; Vignesh, S.; Vigneshwaran, S.; Subash, T. A review of comparison between the traditional catalyst and zeolite catalyst for ammonia-selective catalytic reduction of NO_x. *Fuel* **2023**, *344*, 128125. [CrossRef]
44. Zang, Y.; Bi, Y.; Liu, C.; Zhang, Y.; Li, Q.; Wang, Y.; Zhang, M.; Liu, Q.; Zhang, Z. The study on Cu-ZK-5 catalyst of copper content regulation and anti-propylene poisoning mechanism in NH_3-SCR reaction. *Fuel* **2023**, *340*, 127442. [CrossRef]
45. Jin, P.; Yang, L.; Sheng, Z.; Chu, X.; Chen, D. Selective catalytic reduction of NO_x with NH_3 and tolerance to H_2O & SO_2 at high temperature over zeolite supported indium-copper bimetallic catalysts for gas turbine. *J. Environ. Chem. Eng.* **2023**, *11*, 109218. [CrossRef]
46. Gao, L.; Gao, W.; Wang, H.; Xu, S.; Tian, X.; Cao, J.; Chen, J.; Zhang, Q.; Ning, P.; Hao, J. Boosting low-temperature and high-temperature hydrothermal stability of Cu/SAPO-34 for NO_x removal via yttrium decoration. *Chem. Eng. J.* **2023**, *455*, 140520. [CrossRef]
47. Liu, Y.; Wu, X.; Yang, X.; Tao, H.; Li, J.; Zhang, C.; Yang, R.T.; Li, Z. Enhancement of NO_x adsorption performance on zeolite via a facile modification strategy. *J. Hazard. Mater.* **2023**, *443*, 130225. [CrossRef]
48. Wang, L.; Ren, Y.; Yu, X.; Peng, C.; Yu, D.; Zhong, C.; Hou, J.; Yin, C.; Fan, X.; Zhao, Z.; et al. Novel preparation method, catalytic performance and reaction mechanisms of $Pr_xMn_{1-x}O_\delta$/3DOM ZSM-5 catalysts for the simultaneous removal of soot and NO_x. *J. Catal.* **2023**, *417*, 226–247. [CrossRef]
49. Kang, T.H.; Kim, H.S.; Lee, H.; Kim, D.H. Synergistic effect of V_2O_5-WO_3/TiO_2 and H-ZSM-5 catalysts prepared by physical mixing on the selective catalytic reduction of NO_x with NH3. *Appl. Surf. Sci.* **2023**, *614*, 156159. [CrossRef]
50. Zhang, L.; Shan, Y.; Yan, Z.; Liu, Z.; Yu, Y.; He, H. Efficient Pt/KFI zeolite catalysts for the selective catalytic reduction of NO_x by hydrogen. *J. Environ. Sci.* **2024**, *138*, 102–111. [CrossRef]
51. Wang, S.; He, B.; Wang, Y.; Wu, X.; Duan, H.; Di, J.; Yu, Z.; Liu, Y.; Xin, Z.; Jia, L.; et al. Hydrogen production from the steam reforming of bioethanol over novel supported Ca/Ni-hierarchical Beta zeolite catalysts. *Int. J. Hydrogen Energy. Energy* **2021**, *46*, 36245–36256. [CrossRef]
52. Husin, H.; Erdiwansyah, E.; Ahmadi, A.; Nasution, F.; Rinaldi, W.; Abnisa, F.; Mamat, R. Efficient hydrogen production by microwave-assisted catalysis for glycerol-water solutions via NiO/zeolite-CaO catalyst. *South Afr. J. Chem. Eng.* **2022**, *41*, 43–50. [CrossRef]

53. Zhang, H.; Zhong, L.; Bin Samsudin, I.; Okumura, K.; Tan, H.-R.; Li, S.; Jaenicke, S.; Chuah, G.-K. Mg-stabilized subnanometer Rh particles in zeolite Beta as highly efficient catalysts for selective hydrogenation. *J. Catal.* **2022**, *405*, 489–498. [CrossRef]
54. Da Costa-Serra, J.F.; Miralles-Martínez, A.; García-Muñoz, B.; Maestro-Cuadrado, S.; Chica, A. Ni and Co-based catalysts supported on ITQ-6 zeolite for hydrogen production by steam reforming of ethanol. *Int. J. Hydrogen Energy. Energy* **2022**, *48*, 26518–26525. [CrossRef]
55. Ismaila, A.; Chen, H.; Fan, X. Nickel encapsulated in silicalite-1 zeolite catalysts for steam reforming of glycerol (SRG) towards renewable hydrogen production. *Fuel Process. Technol.* **2022**, *233*, 107306. [CrossRef]
56. Yu, L.; Zhang, R.; Cao, C.; Liu, L.; Fang, J.; Jin, H. Hydrogen production from supercritical water gasification of lignin catalyzed by Ni supported on various zeolites. *Fuel* **2022**, *319*, 123744. [CrossRef]
57. Chen, M.; Sun, G.; Wang, Y.; Liang, D.; Li, C.; Wang, J.; Liu, Q. Steam reforming of methanol for hydrogen production over attapulgite-based zeolite-supported Cu-Zr catalyst. *Fuel* **2022**, *314*, 122733. [CrossRef]
58. Zhou, Y.; Wang, P.; Qin, L.; Kang, S.-Z.; Li, X. Double shell composite nanoarchitectonics of Cu_2O core with TiO_2/metal-organic frameworks for efficient hydrogen generation. *Int. J. Hydrogen Energy. Energy* **2023**, *48*, 629–639. [CrossRef]
59. Saka, C. Highly active hydrogen generation from sodium borohydride methanolysis and ethylene glycolysis reactions using protonated chitosan-zeolite hybrid metal-free particles. *Appl. Catal. B Environ.* **2023**, *325*, 122335. [CrossRef]
60. Sun, T.; Wei, J.; Zhou, C.; Wang, Y.; Shu, Z.; Zhou, J.; Chen, J. Facile preparation and enhanced photocatalytic hydrogen evolution of cation-exchanged zeolite LTA supported TiO_2 photocatalysts. *Int. J. Hydrogen Energy. Energy* **2023**, *48*, 13851–13863. [CrossRef]
61. Shen, Y. Biomass-derived porous carbons for sorption of Volatile organic compounds (VOCs). *Fuel* **2023**, *336*, 126801. [CrossRef]
62. Yang, F.; Li, W.; Ou, R.; Lu, Y.; Dong, X.; Tu, W.; Zhu, W.; Wang, X.; Li, L.; Yuan, A.; et al. Superb VOCs capture engineering carbon adsorbent derived from shaddock peel owning uncompromising thermal-stability and adsorption property. *Chin. J. Chem. Eng.* **2022**, *47*, 120–133. [CrossRef]
63. Zhang, Y.; Yu, Q.; Yuan, Y.; Tang, X.; Zhao, S.; Yi, H. Adsorption behavior of Mo-MEL zeolites for reducing VOCs from cooking oil fumes. *Sep. Purif. Technol.* **2023**, *322*, 124059. [CrossRef]
64. Yu, Q.; Feng, Y.; Wei, J.; Tang, X.; Yi, H. Development of Mn-Si-MEL as a bi-functional adsorption-catalytic oxidation material for VOCs elimination. *Chin. Chem. Lett.* **2022**, *33*, 3087–3090. [CrossRef]
65. Liu, H.; Wei, K.; Long, C. Enhancing adsorption capacities of low-concentration VOCs under humid conditions using NaY@meso-SiO_2 core–shell composite. *Chem. Eng. J.* **2022**, *442*, 136108. [CrossRef]
66. Yin, T.; Meng, X.; Wang, S.; Yao, X.; Liu, N.; Shi, L. Study on the adsorption of low-concentration VOCs on zeolite composites based on chemisorption of metal-oxides under dry and wet conditions. *Sep. Purif. Technol.* **2022**, *280*, 119634. [CrossRef]
67. Feng, A.; Yu, Y.; Mi, L.; Cao, Y.; Yu, Y.; Song, L. Synthesis and VOCs adsorption performance of surfactant-templated USY zeolites with controllable mesopores. *Chem. Phys. Lett.* **2022**, *798*, 139578. [CrossRef]
68. Lu, S.; Han, R.; Wang, H.; Song, C.; Ji, N.; Lu, X.; Ma, D.; Liu, Q. Three birds with one stone: Designing a novel binder-free monolithic zeolite pellet for wet VOC gas adsorption. *Chem. Eng. J.* **2022**, *448*, 137629. [CrossRef]
69. Chen, L.H.; Sun, M.H.; Wang, Z.; Yang, W.M.; Xie, Z.K.; Su, B.L. Hierarchically Structured Zeolites: From Design to Application. *Chem. Rev.* **2020**, *120*, 11194–11294. [CrossRef]
70. Ivanova, I.I.; Knyazeva, E.E. Micro-mesoporous materials obtained by zeolite recrystallization: Synthesis, characterization and catalytic applications. *Chem. Soc. Rev.* **2013**, *42*, 3671–3688. [CrossRef]
71. Zhao, C.; Hu, X.; Liu, C.; Chen, D.; Yun, J.; Jiang, X.; Wei, N.; Li, M.; Chen, Z. Hierarchical architectures of ZSM-5 with controllable mesoporous and their particular adsorption/desorption performance for VOCs. *J. Environ. Chem. Eng.* **2022**, *10*, 106868. [CrossRef]
72. Wang, J.; Shi, Y.; Kong, F.; Zhou, R. Low-temperature VOCs oxidation performance of Pt/zeolites catalysts with hierarchical pore structure. *J. Environ. Sci.* **2023**, *124*, 505–512. [CrossRef] [PubMed]
73. Li, D.; Wang, L.; Lu, Y.; Deng, H.; Zhang, Z.; Wang, Y.; Ma, Y.; Pan, T.; Zhao, Q.; Shan, Y.; et al. New insights into the catalytic mechanism of VOCs abatement over Pt/Beta with active sites regulated by zeolite acidity. *Appl. Catal. B Environ.* **2023**, *334*, 122811. [CrossRef]
74. Kong, F.; Li, G.; Wang, J.; Shi, Y.; Zhou, R. Promoting effect of acid sites in hierarchical porous Pt/ZSM-5 catalysts for low-temperature removal of VOCs. *Appl. Surf. Sci.* **2022**, *606*, 154888. [CrossRef]
75. Yu, B.; Deng, H.; Lu, Y.; Pan, T.; Shan, W.; He, H. Adsorptive interaction between typical VOCs and various topological zeolites: Mixture effect and mechanism. *J. Environ. Sci.* **2024**, *136*, 626–636. [CrossRef]
76. Zheng, Y.; Han, R.; Yang, L.; Yang, J.; Shan, C.; Liu, Q. Revealing opposite behaviors of catalyst for VOCs Oxidation: Modulating electronic structure of Pt nanoparticles by Mn doping. *Chem. Eng. J.* **2023**, *465*, 142807. [CrossRef]
77. Wu, R.; Xiao, Y.; Zhang, P.; Lin, J.; Cheng, G.; Chen, Z.; Yu, R. Asphalt VOCs reduction of zeolite synthesized from solid wastes of red mud and steel slag. *J. Clean. Prod.* **2022**, *345*, 131078. [CrossRef]
78. Yu, C.-H.; Huang, C.-H.; Tan, C.-S. A Review of CO_2 Capture by Absorption and Adsorption. *Aerosol Air Qual. Res.* **2012**, *12*, 745–769. [CrossRef]
79. Choi, H.J.; Hong, S.B. Effect of framework Si/Al ratio on the mechanism of CO_2 adsorption on the small-pore zeolite gismondine. *Chem. Eng. J.* **2022**, *433*, 133800. [CrossRef]
80. Chen, C.; Ahn, W.-S. CO_2 adsorption on LTA zeolites: Effect of mesoporosity. *Appl. Surf. Sci.* **2014**, *311*, 107–109. [CrossRef]

81. Chen, S.J.; Zhu, M.; Tang, Y.C.; Fu, Y.; Li, W.L.; Xiao, B. Molecular simulation and experimental investigation of CO_2 capture in a polymetallic cation-exchanged 13X zeolite. *J. Mater. Chem. A* **2018**, *6*, 19570–19583. [CrossRef]
82. Liu, X.; Gao, F.; Xu, J.; Zhou, L.; Liu, H.; Hu, J. Zeolite@Mesoporous silica-supported-amine hybrids for the capture of CO_2 in the presence of water. *Microporous Mesoporous Mater.* **2016**, *222*, 113–119. [CrossRef]

Disclaimer/Publisher's Note: The statements, opinions and data contained in all publications are solely those of the individual author(s) and contributor(s) and not of MDPI and/or the editor(s). MDPI and/or the editor(s) disclaim responsibility for any injury to people or property resulting from any ideas, methods, instructions or products referred to in the content.

Article

Silver-Doped Zeolitic Imidazolate Framework (Ag@ZIF-8): An Efficient Electrocatalyst for CO_2 Conversion to Syngas

Muhammad Usman * and Munzir H. Suliman

Interdisciplinary Research Center for Hydrogen and Energy Storage (IRC-HES),
King Fahd University of Petroleum & Minerals (KFUPM), Dhahran 31261, Saudi Arabia;
munzir.suliman@kfupm.edu.sa
* Correspondence: muhammadu@kfupm.edu.sa; Tel.: +966-013-860-8539

Abstract: To enable the reuse of carbon dioxide (CO_2), electrocatalytic reduction of CO_2 (CO_2RR) into syngas with a controllable H_2/CO ratio is considered a cost-effective and intriguing approach. Here, a number of silver (Ag)-doped, zeolitic imidazole framework composites were prepared by a facile method. The outcomes demonstrate that CO_2 electroreduction on Ag-doped ZIF-8 catalysts produces just CO and H_2, without having any liquid fuel, resulting in a total faradaic efficiency approaching 100%. The most optimal Ag-Zn-ZIF-8 (10% Ag, 90% Zn) demonstrates good selectivity for syngas (CO and H_2) that can be easily adjusted from 3:1 to 1:3 (H_2/CO) by changing the applied voltage during the CO_2 conversion process.

Keywords: metal–organic frameworks; bimetallic; electrocatalysts; CO_2 utilization; syngas

1. Introduction

Syngas, known as synthetic gas, comprises carbon monoxide (CO) and diatomic hydrogen (H_2). It can be prepared using a number of processes, such as steam-reforming methane, partial oxidation of hydrocarbons, and gasification of biomass. Syngas is a fuel gas mixture that can be used in many ways [1]. It has the ability to make several products, including chemicals, fuels, and electricity. Compositions of syngas that consist of varying volumetric ratios of H_2/CO in quantities of 33.33/66.67, 50/50, 66.67/33.33, 80/20, and 100/0 can be used as precursors to make synthetic fuels such as natural gas, methanol, and dimethyl ether via a Fischer–Tropsch (F–T) process [2]. In addition, syngas can be used as an alternative to fossil fuels because it can be made from different feedstocks, such as biomass and waste feedstock, and can be further used in gas turbines to produce electricity, which generates less greenhouse gas overall [3].

One such greenhouse gas is carbon dioxide (CO_2), a colorless and odorless compound that is a vital part of the carbon cycle. One of the main causes of the world's climate change is the rising level of CO_2 in the atmosphere. The current concentration of CO_2 in the atmosphere is at 414 ppm [4]. This high level of CO_2 in air is a detriment to the environment. In order to preserve human health and safety, monitoring and managing CO_2 levels in our environment is crucial. Net zero emissions is the desired outcome of minimizing the worst effects of climate change and limiting the rise in global temperatures. The term "net zero emission" describes a situation in which the amount of greenhouse gas emissions created is equal to the amount of those emissions that are removed from the atmosphere. Numerous nations, businesses, and organizations have established goals to reach net zero emissions by a target year, in most cases, either 2050 or 2060. This goal can be accomplished in a number of ways by following the four Rs—rethink, reduce, reuse, and recycle. Examples of such approaches include the utilization of renewable energy sources, improvements in energy efficiency, and the transition to low-carbon modes of transportation. The above strategies can also be combined with the removal of carbon dioxide from the atmosphere through practices such as reforestation and afforestation, as well as carbon capture, storage,

Citation: Usman, M.; Suliman, M.H. Silver-Doped Zeolitic Imidazolate Framework (Ag@ZIF-8): An Efficient Electrocatalyst for CO_2 Conversion to Syngas. *Catalysts* **2023**, *13*, 867. https://doi.org/10.3390/catal13050867

Academic Editors: De Fang and Yun Zheng

Received: 28 March 2023
Revised: 5 May 2023
Accepted: 6 May 2023
Published: 10 May 2023

Copyright: © 2023 by the authors. Licensee MDPI, Basel, Switzerland. This article is an open access article distributed under the terms and conditions of the Creative Commons Attribution (CC BY) license (https://creativecommons.org/licenses/by/4.0/).

and utilization [5]. Reusing CO_2 as a feedstock for various compounds, such as methane, formic acid, alcohol, and hydrocarbons, is a promising approach. Moreover, producing syngas by the CO_2 reduction reaction (CO_2RR) method can address the above issues while reducing the greenhouse effect [6].

Electrochemical CO_2 reduction is a promising method for lowering greenhouse gas emissions and creating useful products from CO_2 due to its high selectivity, durability, energy efficiency, and versatility. However, there are still a number of improvements required before widespread industrial use, including raising the effectiveness and longevity of catalysts, enhancing reaction conditions, and lowering costs [7–9].

Metal–organic frameworks (MOFs) are crystalline porous materials made up of metal ions or nodes joined by organic linkers [10,11]. They possess favourable properties, including high porosity, substantial surface area, variable pore size, exceptional tunability, and good stability, which have drawn much research attention. A number of applications take advantage of the distinctive characteristics of MOFs, such as gas storage and separation, catalysis, drug delivery, and sensing [10,12–16]. One such application is CO_2 conversion—a critical step in the effort to lower the emissions of greenhouse gases. In CO_2 conversion operations, MOFs can be utilized as catalysts to transform CO_2 efficiently into usable products or fuels. The zeolitic imidazolate framework (ZIF) is an important subclass of MOFs, with the majority of series including Zn or Co as the metal core and imidazole as linkers. Among ZIFs, the Zn-based zeolitic imidazole framework (ZIF-8) shows strong thermal and chemical stability that distinguishes it from other MOFs.

Ag and its composite materials are usually used as promising catalysts due to their inexpensive cost, high catalytic activity, and stability [17–20]. Ag-anchored ZIFs have shown promising synergetic applications [21–23]. Research has demonstrated that Ag is capable of converting CO_2 to CO with a high faradaic efficiency (FE), although the current density remains low [24–31]. Therefore, this work aimed to use Ag and ZIF-8 composite materials for the E(electrochemical) CO_2RR. ZIF-8 was used to anchor doped silver nanoparticles by a simple, low-temperature chemical deposition technique. Characterization and application of the resulting catalysts for electrochemical CO_2 reduction reaction (ECO_2RR) to syngas were evaluated in aqueous solutions of 0.1 M $KHCO_3$ at room temperature in an H-cell. In addition, the materials were examined in flow cells to assess their electrocatalytic properties in depth to produce pure syngas at various potentials.

2. Results and Discussion

The overall strategy in this work relied on using ZIF-8 to disperse the Ag nanoparticles in the framework. ZIF-8 contains Zn atoms as metal centers, which are considered as promising active sites for CO_2 electrocution to CO [32]. In addition, the ZIF-8 contains homogenous pores that can act as nano-reactors to confine the growth of Ag^+ nanoparticles [33]. Finally, a mild reducing agent such as ascorbic acid is used to reduce Ag^+ into metallic Ag nanoparticles within the pores of the ZIF-8. The washing step prior to reduction was crucial in removing most of the surface Ag ions to avoid the agglomeration and the growth of large Ag particles, which could lead to a significant drop in the surface area. The as-prepared Ag-ZIF-8 was characterized with several techniques.

X-ray diffraction (XRD) was carried out to investigate the phase formation and purity of the prepared material. As can be observed in Figure 1a, the ZIF-8 sample exhibited a sharp and intense diffraction pattern, which confirms the formation of high crystalline material and is in agreement with the simulated reference pattern [34]. The samples of 5% Ag-ZIF-8 and 10% Ag-ZIF-8 showed similar diffraction patterns, with the exception of a noticeable peak at 38.0° in 10% Ag-ZIF-8 corresponding to phase 111 of metallic Ag [35]. The elemental composition was confirmed by the energy dispersive X-ray (EDX) in Figure 1b,c and showed the existence of Ag and ZIF-8 elements (Zn, C, N and O) for 5% Ag-ZIF-8 and 10% Ag-ZIF-8. The calculated Ag ratio in Figure S1 is in good agreement with the theoretical one, 4.6 and 8.5% for the samples of 5% Ag-ZIF-8 and 10% Ag-ZIF-8, respectively. Conversely, due to the small Ag loading in 5% Ag-ZIF-8, the same peak was not as clear. The elemental composition was studied by the energy dispersive X-ray (EDX) in Figure 1b, which confirmed the presence of Ag, Zn, and O atoms in the 10% Ag-ZIF-8 sample.

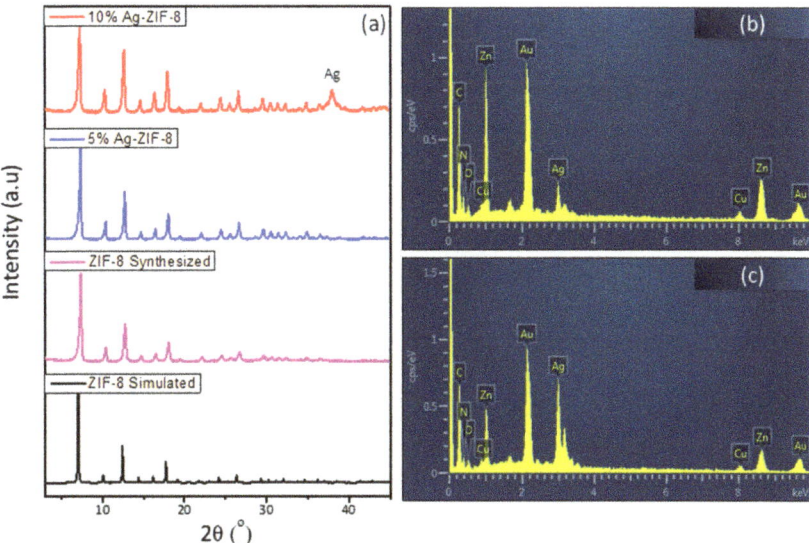

Figure 1. (a) The XRD of ZIF-8 and Ag-ZIF-8, (b) the EDX of 5% Ag-ZIF-8, (c) the EDX of 10% Ag-ZIF-8.

The morphological and structural properties of the prepared material were investigated by the SEM. Figure 2a shows the SEM of ZIF-8, which reveals uniform dodecahedron crystals. In comparison of ZIF-8 with the Ag-loaded sample, there was no significant difference in the morphology and no agglomeration of Ag nanoparticles was observed. The SEM results were supported after carrying out the TEM (Figure 2g–i). The TEM showed uniform crystals corresponding to the ZIF-8 frame work. In the high-resolution image, no Ag particles agglomerated were observed on the surface of the MOF. The elemental mapping (Figure 3) confirmed the uniform dispersion of the elements (C, N, Zn, O, and Ag). The phase of Zn in the ZIF-8 was Zn^{2+} due to its binding with the nitrogen atoms in the organic imidazole linkers, while the doped Ag phase was metallic within the framework as confirmed by XRD, and had no covalent bond with the organic linker or the metal node in the framework. These results were also supported by XPS analysis [34,36,37].

The surface area of the ZIF-8 and Ag-ZIF-8 was investigated with the aid of the BET surface analyzer (Figure 4a). The pristine ZIF-8 showed a high surface area characteristic of most MOFs, which is about 1500 $m^2\ g^{-1}$. Upon the loading of the silver nanoparticles, a decrease in the surface area was observed to ~1200 and 1000 $m^2\ g^{-1}$ for 5% Ag-ZIF-8 and 10% Ag-ZIF-8, respectively, which further confirmed the loading of the nanoparticles into the framework of the ZIF-8, as shown in Figure 4a. Moreover, FTIR was carried out for ZIF-8, 5% Ag-ZIF-8, and 10% Ag-ZIF-8, revealing identical IR spectra as shown in Figure 4b. The peak for 2-methylimidazole (MeIm) was seen at 694 cm^{-1} (C-H bend).

Figure 2. SEM of (**a,d**) ZIF-8, (**b,e**) 5% Ag-ZIF-8, and (**c,f**) 10% Ag-ZIF-8. TEM of (**g–i**) 10% Ag-ZIF-8.

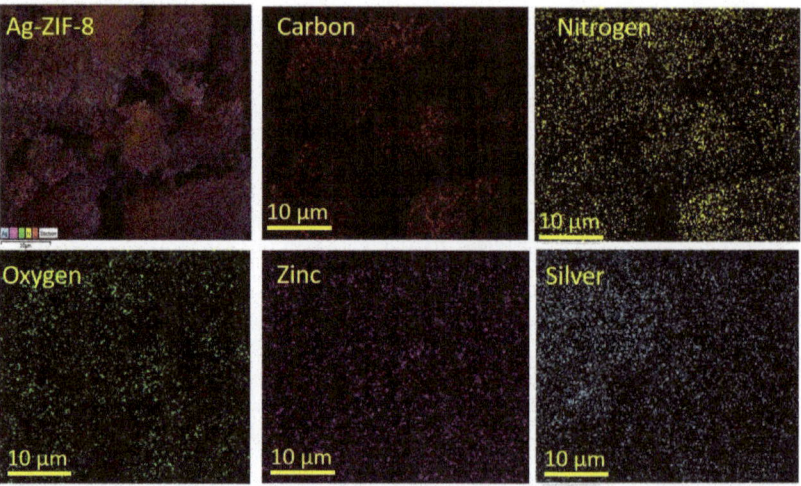

Figure 3. The elemental mapping of 10% Ag-ZIF-8.

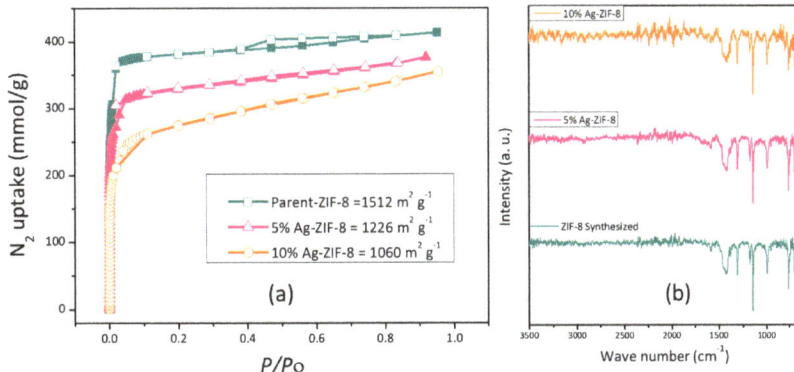

Figure 4. (a) BET isotherm of ZIF-8, 5% Ag-ZIF-8, and 10% Ag-ZIF-8; (b) FTIR spectra of ZIF-8, 5% Ag-ZIF-8 and 10% Ag-ZIF-8.

The in-plane deformation vibration of (=C-H) was confirmed by the peak at 1147 cm^{-1}. The CH$_2$ wagging transmits light at 1313 cm^{-1}, and the =C-H in-plane bending happens at 995 cm^{-1}. Both CH$_3$ and CH$_2$ responded at 1384 cm^{-1} and 1427 cm^{-1}, respectively, due to their asymmetric bends. Stretches of the carbon–carbon double bond (C=C stretch) and the carbon-nitrogen double bond (C=N stretch) had peaks at 1456 cm^{-1} and 1585 cm^{-1}, respectively. The (C-H) symmetric stretch and (=C-H) stretch were both confirmed by the two tiny narrow peaks at 2931 cm^{-1} and 3137 cm^{-1}, respectively. The IR spectra in Figure 4b confirm the preservation of the framework even after Ag loading [38], which further supports the XRD results.

In order to demonstrate the overall electrocatalytic performance, the electrocatalysts (ZIF-8, 5% Ag-ZIF-8, 10% Ag-ZIF-8) were analyzed using linear sweep voltammetry (LSV) in N$_2$- or CO$_2$-saturated 0.1 M KHCO$_3$ electrolytes (Figure 5a). The faradaic currents found in an N$_2$-saturated electrolyte are inextricably linked to the hydrogen evolution reaction (HER), whereas the faradaic currents in the presence of CO$_2$ are ascribed to contributions from both the HER and CO$_2$RR. In all cases, the CO$_2$-saturated electrolyte exhibited higher current densities than the N$_2$-saturated electrolyte. It can be observed clearly that pristine ZIF-8 showed the lowest current density. Upon the loading of the ZIF-8 with 5% Ag, the current density increased significantly (from 5 to 10 mA cm^{-2}). The current density also increased with a greater Ag loading of 10% to 14.5 mA cm^{-2}.

To gain insight about the mechanism and the kinetics of the ECO$_2$RR, the Tafel slope was estimated from the polarization curves (Figure 5b). Tafel values of 420, 221, and 154 mV dec^{-1} were calculated for the electrodes ZIF-8, 5% Ag-ZIF-8, and 10% Ag-ZIF-8, respectively. The lowest Tafel slope value suggests faster reaction kinetics and facilitated adsorption of the CO$_2$•$^-$ intermediate [39].

Another important factor is the electrochemical active surface area (ESCA), which can be estimated from the double-layer capacitance (C$_{dl}$). The C$_{dl}$ was evaluated by recording cyclic voltammograms (Figure S2) at different scan rates (50, 100, 150, 200, and 250 mV s^{-1}) and plotting the capacitive current vs. the scan rate. As shown in Figure 5c, 10% Ag-ZIF-8 exhibited the highest C$_{dl}$ value (1.25 mF), followed by 5% Ag-ZIF-8 (1.00 mF), and the pristine ZIF-8 showed the lowest C$_{dl}$ value (0.90 mF). Electrochemical impedance spectroscopy (EIS) was used to study the interaction between the electrode and electrolyte interface through charge transfer resistance (R$_{ct}$), which was obtained from Nyquist plot [40] (Figure 5d). The high frequency is ascribed to the CO$_2$RR to CO (mass transport), whereas the low frequency could be related to the electrolysis process (HER). The R$_{ct}$ values were 205, 148, and 140 Ω cm^2 for the electrodes ZIF-8, 5% Ag-ZIF-8, and 10% Ag-ZIF-8, respectively. It can be seen that doping the ZIF-8 with Ag enhances the charge transfer rate dramatically, and the 10% loading showed the lowest R$_{ct}$.

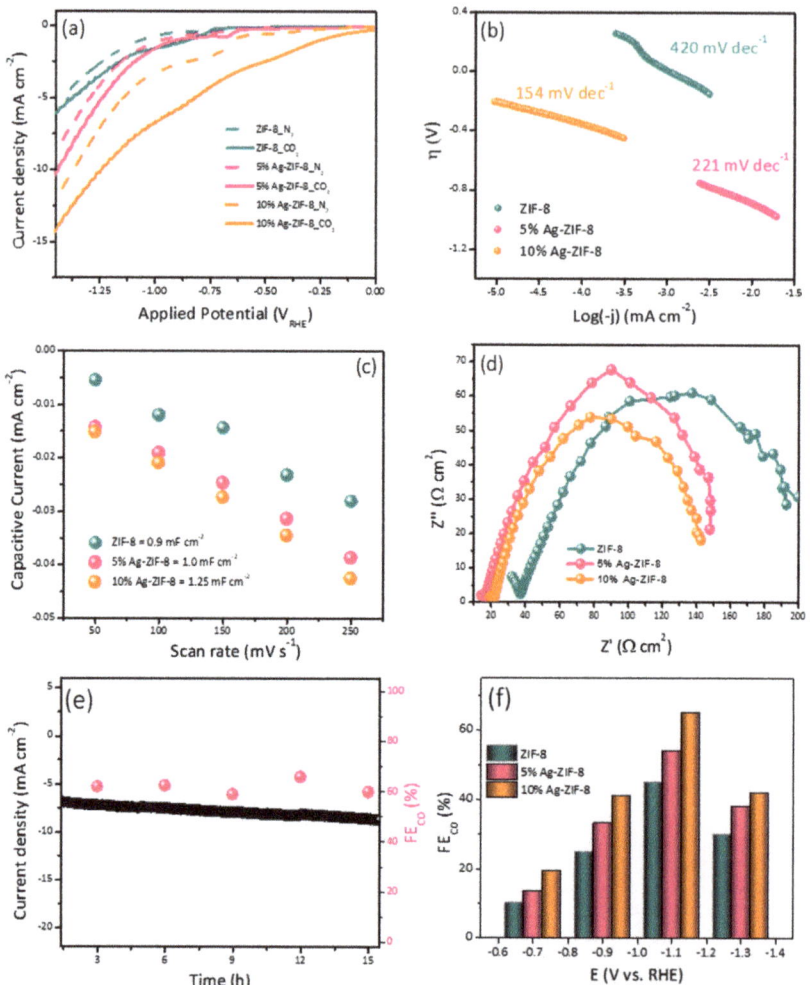

Figure 5. (**a**) The polarization curves of ZIF-8, 5% Ag-ZIF-8, and 10% Ag-ZIF-8 in N_2 and CO_2-saturated 0.1 M $KHCO_3$. (**b**) Tafel slope of ZIF-8, 5% Ag-ZIF-8, 10% Ag-ZIF-8. (**c**) C_{dl} slopes of ZIF-8, 5% Ag-ZIF-8, and 10% Ag-ZIF-8. (**d**) Nyquist plots of ZIF-8, 5% Ag-ZIF-8, and 10% Ag-ZIF-8 (**e**) chronoamperometry of 10% Ag-ZIF-8 in CO_2-saturated 0.1 M $KHCO_3$. (**f**) FE of ZIF-8, 5% Ag-ZIF-8, and 10% Ag-ZIF-8 in CO_2-saturated 0.1 M $KHCO_3$.

The electrode durability was investigated using chronoamperometry (Figure 5e) in 0.1 M $KHCO_3$. The long-term current time curve revealed a good stability for 16 h. The electrochemical performance toward CO_2 reduction was investigated for the three electrodes (ZIF-8, 5% Ag-ZIF-8, and 10% Ag-ZIF-8) using the chronoamperometry at different applied potentials for 1 h and the products were quantified using online connected GC-BID. The CO faradic efficiency (FE) of the electrodes is compared in Figure 5f. The three electrocatalysts showed a similar trend in the CO production. At low applied potential, a low CO yield was observed. Increasing the potential led to a significant increase in the CO FE%. The applied potential -1.1 V_{RHE} exhibited the highest FE%. A further increase in potential led to a decrease in the FE%. The lower CO FEs at greater negative potentials (-1.3 V versus RHE) may result from the limited CO_2 and polarization losses. The applied voltage increased the current densities of the electrocatalysts, the CO FE gradually reduced, and the CO current densities remained constant due to the competing HER on the electrode surface, which is consistent with the results reported previously. The 5% Ag-ZIF-8 showed higher FE% than the ZIF-8 at these applied potentials. The 10% Ag-ZIF-8 showed the highest FE (70% for CO and 30% for H_2) at −1.1 V.

The optimized electrode (10% Ag-ZIF-8) in the H-cell was also investigated in the flow cell system, which is a more practical setup. The LSV and FE obtained from the flow cell were compared with the results obtained from the H-cell. As can be observed in Figure 6a, the current density in the flow cell was significantly higher than that in the H-cell. This can be attributed to the GDE, which allows the diffusion of more CO_2 gas into the catalyst surface. Additionally, the flow system provides fresh electrolyte to the electrode surface, which facilitates the overall reaction. Since the electrolyte compartments in the flow system are separated, unlike the H-cell, two different electrolytes can be used. KOH in the anolyte serves as a proton source due to the oxygen evolution reaction in the anode, which is more efficient than $KHCO_3$. The FE% trend was different in the case of the flow cell. The highest FE% values (69.0 and 80%) were observed at relatively lower potentials (−0.7 and −0.9 V_{RHE}). Moreover, when the FEs were compared (Figure 6b,c), the flow cell showed higher conversion rate at lower applied potential (80% at −0.9 V_{RHE}) compared to (70% at −1.1 V_{RHE}) for the H-cell. In addition to the higher FE in the flow cell, its higher efficiency can be noted by calculating the partial current density (the current utilized in the CO_2 conversion). As shown in Figure 6d, the current was almost 5 times the one used in the H-cell.

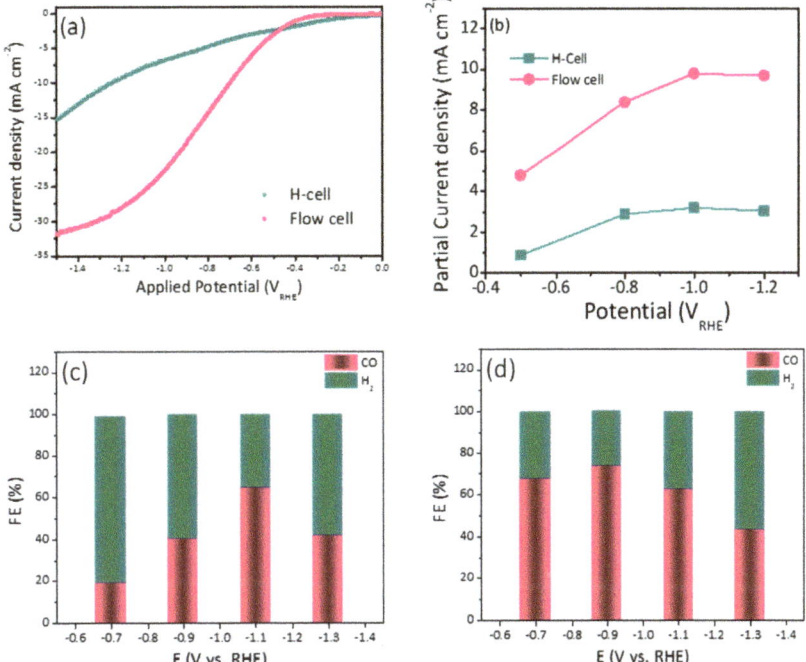

Figure 6. (a) Comparative LSV for 10% Ag-ZIF-8 using an H-cell and flow cell. (b) The partial current density for 10% Ag-ZIF-8 using the H-cell and the flow cell and the FE using (c) the H-cell (d) the flow cell.

The electrochemical performance and conversion efficiency of the current findings are shown in (Table 1), which compares Ag-based, ZIF-8, and Ag-ZIF-8 composites for the electroreduction of CO_2 into usable liquid chemicals. The 10% Ag-ZIF-8 produced syngas with a FE% ratio of 70:30 (CO:H$_2$) at a potential of −1.1 V vs. RHE and 80:20 at a potential of −0.9 V vs. RHE, which is a marked improvement over previous reports.

Table 1. Comparison of the catalytic performances of 10% Ag-ZIF-8 and the similar electrocatalysts reported in literature for the reduction of CO$_2$.

Electrocatalyst	Electrolyte	Main Product	FE% [CO:H$_2$]	Current Density [mA cm^{-2}]	Potential [V] vs. RHE	Cell Type	Ref.
ZIF-8	0.5 M NaCl	CO	65:35	3	−1.14	H-cell	[32]
ZIF-8	0.25 M K$_2$SO$_4$	CO	81:15	8.5	−1.1	H-cell	[41]
ZIF-108	0.25 M K$_2$SO$_4$	CO	52:48	24.6	−1.3	H-cell	[41]
ligand-doped ZIF-8	0.1 M KHCO$_3$	CO	90:10	10.1	−1.2	H-cell	[42]
Ag$_2$O/layered ZIF	0.25 M K$_2$SO$_4$	CO	80:20	32	−1.3	H-cell	[28]
Ag nanosheets	0.5 M KHCO$_3$	CO	91:9	6.48	−0.9	H-cell	[29]
Ag/carbon paper	0.5 M K$_2$HPO$_4$ + 0.5 M KH$_2$PO$_4$ at pH 10	CO	80:20	51	3 V (E$_{Cell}$)	Flow cell	[43]
Ag/carbon paper	0.5 M KHCO$_3$	CO	60:20	50	−1.45	Flow cell	[44]
Ag/ZIF-8	0.5 M KHCO$_3$	CO	70:30	2.6	−1.1 V	H-cell	This work
Ag/ZIF-8	1 M KOH	CO	80:20	28	−0.9 V	Flow cell	This work

3. Experimental

3.1. Materials

Zinc nitrate (ZnNO$_3$) (99.95%), silver nitrate (AgNO$_3$), ascorbic acid (99.0%) 2-methyl imidazole (99.0%), potassium bicarbonate (99.9), and potassium hydroxide(99.5%) were purchased from Sigma Aldrich, St. Louis, MO, USA. Methanol (CH$_3$OH) (99.8%) was procured from Sharlu (Sharjah, United Arab Emirates).

3.2. Preparation of ZIF-8

ZIF-8 was prepared according to the procedure reported by Lee et al. [45]. Briefly, 1.31 g of Zn(NO$_3$)$_2$ was dissolved in 45 mL of methanol in a 100 mL beaker. In another beaker, 2.87 g of 2-methyl imidazole was dissolved until a clear solution appeared. The two clear solutions were mixed in a 150 mL round bottom flask and stirred at room temperature for 1 h. The synthesized white ZIF-8 suspension was separated by centrifugation at 8000 rpm for 10 min. The white crystal was washed three times with methanol.

3.3. Preparation of Silver Nanoparticles Decorated on ZIF-8

The 5% Ag-ZIF-8 was prepared as follows (as shown in Figure 7). Briefly, 10 mg of silver nitrate (AgNO$_3$) was dissolved in 20 mL of methanol, then 100 mg the as-prepared ZIF-8 was added to the solution. The solution was sonicated for 10 min and stirred for 2 h. The white crystals were separated and washed with methanol three times to remove the surface Ag$^+$. The crystals were re-dispersed in 20 mL methanol and 20 mg of ascorbic acid was added and the solution was stirred for 30 min. Then, the crystals were separated and washed several times with DI water and methanol and dried at 50 °C under vacuum for 5 h.

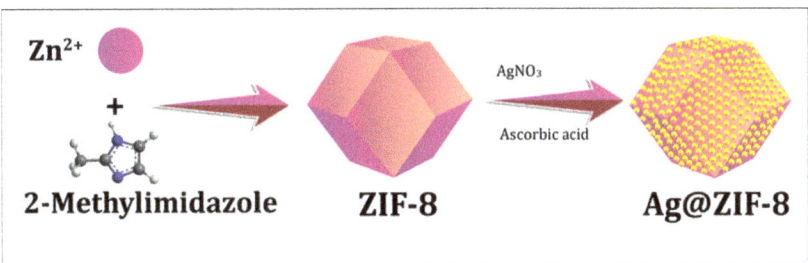

Figure 7. The schematic presentation of the synthesis of Ag-doped ZIF-8 particles.

3.4. Preparation of the Electrocatalyst

3.4.1. Electrode Fabrication of H-Cell

Ten milligrams of the ZIF-8 or Ag-ZIF-8 catalyst was dispersed in 1 mL mixture of 750 µL isopropanol, 200 µL DI water, and 50 µL Nafion (5%). The mixture was sonicated for 20 min. Then, 100 µL of the suspension was drop-cast onto 1 cm^2 conductive carbon paper and dried at room temperature.

3.4.2. Electrode Fabrication of Flow Cell

The spray painting method was applied to prepare the working electrodes for the flow cell. This process is efficient and cost-effective for creating gas diffusion electrode (GDE) with superior electrochemical performance. Twenty milligrams of the ZIF-8 or Ag-ZIF-8 catalyst was dispersed in a 2 mL mixture of 1500 µL isopropanol, 400 µL DI water, and 100 µL Nafion (5%). The mixture was homogenized using a magnetic stirrer or ultrasonication. Next, 200 µL of the prepared ink was kept under constant air flow and pressure and sprayed on GDE using the spray gun method. Using a high-pressure air spray gun, the catalyst ink was sprayed onto the carbon support material's surface. The spray nozzle is normally operated at 1.5 to 3 bar of pressure, with a 2 to 5 cm gap between it and the GDE. The GDE coated with the catalyst ink was then dried for 30 to 60 min at a temperature of 60 to 80 °C before its application in the flow cell.

3.5. Characterization

Morphological and detailed microstructural attributes of the materials were discerned by field emission scanning electron microscopy (FESEM, Tescan Lyra-3, Kohoutovice, Czech Republic). The sample was gold-coated for 30 s before SEM and EDS analysis. Another technique employed for the characterization of the samples was X-ray diffraction (XRD, Rigaku MiniFlex, Austin, TX, USA) to reveal the crystal structure of the materials. To begin, the samples were finely powdered using a mill and pestle to ensure consistency and remove any big pieces. The next step was to put the powdered samples into a sample holder. The sample was then prepared for analysis by having its surface flattened and leveled using a glass slide. This helps to ensure that reliable results are obtained. After that, the sample holder was transferred into the XRD machine, and the analysis commenced. The obtained XRD was matched with the simulated XRD of the materials. Fourier-transform infrared (FT-IR Thermo, Waltham, MA, USA) was used to identify the functional groups in the materials. BET surface analyzer (Triplex) was used to calculate the porous nature of a material. This method is based on gas adsorption as a function of pressure. Typically the sample is first degassed at 120 °C for 6 h. Under a nitrogen environment, the gas pressure is measured vs. the amount of gas adsorbed by the sample to calculate the porosity. A gas chromatographer (GC) equipped with barrier ion discharge detector (Shimadzu, Kyoto, Japan), and potentiostat (Gammray 620, Warminster, UK) were used.

3.6. The Electrochemical Studies

The ECO$_2$RR performance was investigated with the aid of H-cell and flow cell systems. The H-cell consisted of a silver silver-chloride electrode (Ag/AgCl) as a reference electrode. A platinum mesh was used as a counter electrode. As-prepared Cu-NP@NC film on conductive carbon paper was used as the working electrode. A potentiostat (Gammray 620) was connected to the electrodes in the cell.

The flow cell setup consisted of three main components. The first one was the electrolyte compartments, one compartment containing the catholyte, which was 0.5 M KHCO$_3$, and the second compartment containing the anolyte, which was 1.0 M KOH. The second component was the cell, which consisted of the cathode part (where the CO$_2$ gas passes on one side of the GDE and the catholyte passes on the other side) and the anode part connected to the anolyte. The two parts of the cell were separated by a proton permeable membrane to allow the produced H$^+$ to pass from the anode to the cathode. The third component of the flow cell was the pump, which controlled flows and circulated the catholyte and anolyte between the electrolyte compartments to the cell. Similar to the H-cell, the reference electrode was connected with working electrode (GDE) on the cathode side and counter electrode on the anode side and all were connected to a potentiostat workstation (Gammray 620).

The ECO$_2$RR performance was evaluated by carrying out linear sweep voltammetry (LSV) techniques, and calculation of the overpotential at different current densities (current normalized to the geometric surface area of the electrode). The cyclic voltammetry (CV) and LSV experiments were performed in 0.1 M potassium bicarbonate (KHCO$_3$).

$$E_{RHE} = E_{Ag/AgCl} + 0.059 \times pH + E^\circ_{Ag/AgCl} \qquad (1)$$

where $E^\circ_{Ag/AgCl} = 0.199$ V [46].

The potential was swept from (0.0 to −1.4 V vs. RHE). The electrochemical impedance spectroscopy (EIS) was performed by varying the frequency from 10^5 to 0.1 Hz under identical electrolyte and electrodes to the LSV.

4. Conclusions

In conclusion, we prepared silver-doped ZIF-8 (Ag@ZIF-8) that acted as a good catalyst for the production of syngas (CO and H$_2$) at various loading and at various potentials. The experimental findings show that Ag-doped ZIF materials had a higher current density than ZIF-8. Furthermore, the flow cell had a higher current density than the H-cell. The product analysis revealed 100% FE for the gas products. Based on qualitative and quantitative analyses, the products contained syngas at various ratios of H$_2$ and CO, and could be influenced by the applied potential. These findings reveal that the Ag-ZIF-8 platform offers promising materials for effective CO$_2$ conversion to syngas.

Supplementary Materials: The following supporting information can be downloaded at https://www.mdpi.com/article/10.3390/catal13050867/s1, Figure S1: (a) EDX of 5% Ag-ZIF-8 and (b) 10% Ag-ZIF-8; Figure S2: CVs of (a) ZIF-8, (b) 5% Ag-ZIF-8 and (c) 10% Ag-ZIF-8 at scan rates of 50, 100, 150, 200 and 250 mV s^{-1}.

Author Contributions: Conceptualization, M.U., and M.H.S.; methodology, M.U. and M.H.S.; investigation, M.U.; resources, M.U. and M.H.S.; writing—original draft preparation, M.U.; writing—review and editing, M.U., and M.H.S.; supervision, M.U.; formal analysis, M.U., and M.H.S.; funding acquisition, M.U. All authors have read and agreed to the published version of the manuscript.

Funding: This research was funded by King Fahd University of Petroleum and Minerals: ORCP2390.

Data Availability Statement: The data can be found in the main text and Supplementary Materials.

Acknowledgments: We acknowledge the Interdisciplinary Research Center for Hydrogen and Energy Storage (IRC-HES) for its continued support. The Saudi Aramco Chair Professor Project at KFUPM ORCP2390 supported this research.

Conflicts of Interest: The authors declare no conflict of interest.

References

1. Jahangiri, H.; Bennett, J.; Mahjoubi, P.; Wilson, K.; Gu, S. A review of advanced catalyst development for Fischer–Tropsch synthesis of hydrocarbons from biomass derived syn-gas. *Catal. Sci. Technol.* **2014**, *4*, 2210–2229. [CrossRef]
2. Sattarzadeh, M.; Ebrahimi, M.; Jazayeri, S.A. A detail study of a RCCI engine performance fueled with diesel fuel and natural gas blended with syngas with different compositions. *Int. J. Hydrogen Energy* **2022**, *47*, 16283–16296. [CrossRef]
3. Mohanty, U.S.; Ali, M.; Azhar, M.R.; Al-Yaseri, A.; Keshavarz, A.; Iglauer, S. Current advances in syngas ($CO + H_2$) production through bi-reforming of methane using various catalysts: A review. *Int. J. Hydrogen Energy* **2021**, *46*, 32809–32845. [CrossRef]
4. Garba, M.D.; Usman, M.; Khan, S.; Shehzad, F.; Galadima, A.; Ehsan, M.F.; Ghanem, A.S.; Humayun, M. CO_2 towards fuels: A review of catalytic conversion of carbon dioxide to hydrocarbons. *J. Environ. Chem. Eng.* **2021**, *9*, 104756. [CrossRef]
5. Li, M.; Irtem, E.; Iglesias van Montfort, H.-P.; Abdinejad, M.; Burdyny, T. Energy comparison of sequential and integrated CO_2 capture and electrochemical conversion. *Nat. Commun.* **2022**, *13*, 5398. [CrossRef] [PubMed]
6. Joshi, G.; Pandey, J.K.; Rana, S.; Rawat, D.S. Challenges and opportunities for the application of biofuel. *Renew. Sustain. Energy Rev.* **2017**, *79*, 850–866. [CrossRef]
7. Usman, M.; Humayun, M.; Garba, M.D.; Ullah, L.; Zeb, Z.; Helal, A.; Suliman, M.H.; Alfaifi, B.Y.; Iqbal, N.; Abdinejad, M.; et al. Electrochemical Reduction of CO_2: A Review of Cobalt Based Catalysts for Carbon Dioxide Conversion to Fuels. *Nanomaterials* **2021**, *11*, 2029. [CrossRef]
8. Usman, M.; Li, D.; Li, C.; Zhang, S. Highly selective and stable hydrogenation of heavy aromatic-naphthalene over transition metal phosphides. *Sci. China Chem.* **2015**, *58*, 738–746. [CrossRef]
9. Masel, R.I.; Liu, Z.; Yang, H.; Kaczur, J.J.; Carrillo, D.; Ren, S.; Salvatore, D.; Berlinguette, C.P. An industrial perspective on catalysts for low-temperature CO_2 electrolysis. *Nat. Nanotechnol.* **2021**, *16*, 118–128. [CrossRef]
10. Usman, M.; Iqbal, N.; Noor, T.; Zaman, N.; Asghar, A.; Abdelnaby, M.M.; Galadima, A.; Helal, A. Advanced strategies in Metal-Organic Frameworks for CO_2 Capture and Separation. *Chem. Rec.* **2022**, *22*, e202100230. [CrossRef]
11. Trickett, C.A.; Helal, A.; Al-Maythalony, B.A.; Yamani, Z.H.; Cordova, K.E.; Yaghi, O.M. The chemistry of metal–organic frameworks for CO_2 capture, regeneration and conversion. *Nat. Rev. Mater.* **2017**, *2*, 17045. [CrossRef]
12. Usman, M.; Zeb, Z.; Ullah, H.; Suliman, M.H.; Humayun, M.; Ullah, L.; Shah, S.N.A.; Ahmed, U.; Saeed, M. A review of metal-organic frameworks/graphitic carbon nitride composites for solar-driven green H_2 production, CO_2 reduction, and water purification. *J. Environ. Chem. Eng.* **2022**, *10*, 107548. [CrossRef]
13. Usman, M.; Helal, A.; Abdelnaby, M.M.; Alloush, A.M.; Zeama, M.; Yamani, Z.H. Trends and Prospects in UiO-66 Metal-Organic Framework for CO_2 Capture, Separation, and Conversion. *Chem. Rec.* **2021**, *21*, 1771–1791. [CrossRef] [PubMed]
14. Ghanem, A.S.; Ba-Shammakh, M.; Usman, M.; Khan, M.F; Dafallah, H.; Habib, M.A.; Al-Maythalony, B.A. High gas permselectivity in ZIF-302/polyimide self-consistent mixed-matrix membrane. *J. Appl. Polym. Sci.* **2020**, *137*, 48513. [CrossRef]
15. Helal, A.; Sanhoob, M.A.; Hoque, B.; Usman, M.; Zahir, M.H. Bimetallic Metal-Organic Framework Derived Nanocatalyst for CO_2 Fixation through Benzimidazole Formation and Methanation of CO_2. *Catalysts* **2023**, *13*, 357. [CrossRef]
16. Helal, A.; Cordova, K.E.; Arafat, M.E.; Usman, M.; Yamani, Z.H. Defect-engineering a metal–organic framework for CO_2 fixation in the synthesis of bioactive oxazolidinones. *Inorg. Chem. Front.* **2020**, *7*, 3571–3577. [CrossRef]
17. Sekine, K.; Yamada, T. Silver-catalyzed carboxylation. *Chem. Soc. Rev.* **2016**, *45*, 4524–4532. [CrossRef]
18. Abdinejad, M.; Ferrag, C.; Hossain, M.N.; Noroozifar, M.; Kerman, K.; Kraatz, H.B. Capture and electroreduction of CO_2 using highly efficient bimetallic Pd–Ag aerogels paired with carbon nanotubes. *J. Mater. Chem. A* **2021**, *9*, 12870–12877. [CrossRef]
19. Abdinejad, M.; Irtem, E.; Farzi, A.; Sassenburg, M.; Subramanian, S.; Iglesias van Montfort, H.-P.; Ripepi, D.; Li, M.; Middelkoop, J.; Seifitokaldani, A.; et al. CO_2 Electrolysis via Surface-Engineering Electrografted Pyridines on Silver Catalysts. *ACS Catal.* **2022**, *12*, 7862–7876. [CrossRef]
20. Abdinejad, M.; Santos da Silva, I.; Kraatz, H.B. Electrografting amines onto silver nanoparticle-modified electrodes for electroreduction of CO_2 at low overpotential. *J. Mater. Chem. A* **2021**, *9*, 9791–9797. [CrossRef]
21. Shi, J.; Zhang, L.; Sun, N.; Hu, D.; Shen, Q.; Mao, F.; Gao, Q.; Wei, W. Facile and Rapid Preparation of Ag@ZIF-8 for Carboxylation of Terminal Alkynes with CO_2 in Mild Conditions. *ACS Appl. Mater. Interfaces* **2019**, *11*, 28858–28867. [CrossRef] [PubMed]
22. Wu, J.-K.; Tan, P.; Lu, J.; Jiang, Y.; Liu, X.-Q.; Sun, L.-B. Fabrication of Photothermal Silver Nanocube/ZIF-8 Composites for Visible-Light-Regulated Release of Propylene. *ACS Appl. Mater. Interfaces* **2019**, *11*, 29298–29304. [CrossRef]
23. Chen, J.; Gu, A.; Miensah, E.D.; Liu, Y.; Wang, P.; Mao, P.; Gong, C.; Jiao, Y.; Chen, K.; Zhang, Z.; et al. Silver-decorated ZIF-8 derived ZnO concave nanocubes for efficient photooxidation-adsorption of iodide anions: An in-depth experimental and theoretical investigation. *J. Solid State Chem.* **2021**, *297*, 122039. [CrossRef]
24. Hori, Y.; Ito, H.; Okano, K.; Nagasu, K.; Sato, S. Silver-coated ion exchange membrane electrode applied to electrochemical reduction of carbon dioxide. *Electrochim. Acta* **2003**, *48*, 2651–2657. [CrossRef]

25. Hoshi, N.; Kato, M.; Hori, Y. Electrochemical reduction of CO_2 on single crystal electrodes of silver Ag(111), Ag(100) and Ag(110). *J. Electroanal. Chem.* **1997**, *440*, 283–286. [CrossRef]
26. Hatsukade, T.; Kuhl, K.P.; Cave, E.R.; Abram, D.N.; Jaramillo, T.F. Insights into the electrocatalytic reduction of CO_2 on metallic silver surfaces. *Phys. Chem. Chem. Phys.* **2014**, *16*, 13814–13819. [CrossRef]
27. Daiyan, R.; Lu, X.; Ng, Y.H.; Amal, R. Highly Selective Conversion of CO_2 to CO Achieved by a Three-Dimensional Porous Silver Electrocatalyst. *ChemistrySelect* **2017**, *2*, 879–884. [CrossRef]
28. Jiang, X.; Wu, H.; Chang, S.; Si, R.; Miao, S.; Huang, W.; Li, Y.; Wang, G.; Bao, X. Boosting CO_2 electroreduction over layered zeolitic imidazolate frameworks decorated with Ag_2O nanoparticles. *J. Mater. Chem. A* **2017**, *5*, 19371–19377. [CrossRef]
29. Yan, S.; Chen, C.; Zhang, F.; Mahyoub, S.A.; Cheng, Z. High-density Ag nanosheets for selective electrochemical CO_2 reduction to CO. *Nanotechnology* **2021**, *32*, 165705. [CrossRef]
30. Lu, Q.; Rosen, J.; Zhou, Y.; Hutchings, G.S.; Kimmel, Y.C.; Chen, J.G.; Jiao, F. A selective and efficient electrocatalyst for carbon dioxide reduction. *Nat. Commun.* **2014**, *5*, 3242. [CrossRef]
31. Liu, M.; Liu, M.; Wang, X.; Kozlov, S.M.; Cao, Z.; De Luna, P.; Li, H.; Qiu, X.; Liu, K.; Hu, J.; et al. Quantum-Dot-Derived Catalysts for CO_2 Reduction Reaction. *Joule* **2019**, *3*, 1703–1718. [CrossRef]
32. Wang, Y.; Hou, P.; Wang, Z.; Kang, P. Zinc Imidazolate Metal–Organic Frameworks (ZIF-8) for Electrochemical Reduction of CO_2 to CO. *Chemphyschem* **2017**, *18*, 3142–3147. [CrossRef] [PubMed]
33. Suliman, M.H.; Baroud, T.N.; Siddiqui, M.N.; Qamar, M.; Giannelis, E.P. Confined growth and dispersion of FeP nanoparticles in highly mesoporous carbons as efficient electrocatalysts for the hydrogen evolution reaction. *Int. J. Hydrogen Energy* **2021**, *46*, 8507–8518. [CrossRef]
34. Pan, Y.; Liu, Y.; Zeng, G.; Zhao, L.; Lai, Z. Rapid synthesis of zeolitic imidazolate framework-8 (ZIF-8) nanocrystals in an aqueous system. *Chem. Commun.* **2011**, *47*, 2071–2073. [CrossRef]
35. Khani, M.; Sammynaiken, R.; Wilson, L.D. Electrocatalytic Oxidation of Nitrophenols via Ag Nanoparticles Supported on Citric-Acid-Modified Polyaniline. *Catalysts* **2023**, *13*, 465. [CrossRef]
36. Veeramani, V.; Van Chi, N.; Yang, Y.-L.; Hong Huong, N.T.; Van Tran, T.; Ahamad, T.; Alshehri, S.M.; Wu, K.C.W. Decoration of silver nanoparticles on nitrogen-doped nanoporous carbon derived from zeolitic imidazole framework-8 (ZIF-8) via in situ auto-reduction. *RSC Adv.* **2021**, *11*, 6614–6619. [CrossRef] [PubMed]
37. Tian, F.; Cerro, A.M.; Mosier, A.M.; Wayment-Steele, H.K.; Shine, R.S.; Park, A.; Webster, E.R.; Johnson, L.E.; Johal, M.S.; Benz, L. Surface and Stability Characterization of a Nanoporous ZIF-8 Thin Film. *J. Phys. Chem. C* **2014**, *118*, 14449–14456. [CrossRef]
38. Ahmad, A.; Iqbal, N.; Noor, T.; Hassan, A.; Khan, U.A.; Wahab, A.; Raza, M.A.; Ashraf, S. Cu-doped zeolite imidazole framework (ZIF-8) for effective electrocatalytic CO_2 reduction. *J. CO2 Util.* **2021**, *48*, 101523. [CrossRef]
39. Hu, C.; Bai, S.; Gao, L.; Liang, S.; Yang, J.; Cheng, S.-D.; Mi, S.-B.; Qiu, J. Porosity-Induced High Selectivity for CO_2 Electroreduction to CO on Fe-Doped ZIF-Derived Carbon Catalysts. *ACS Catal.* **2019**, *9*, 11579–11588. [CrossRef]
40. Suliman, M.H.; Yamani, Z.H.; Usman, M. Electrochemical Reduction of CO_2 to C1 and C2 Liquid Products on Copper-Decorated Nitrogen-Doped Carbon Nanosheets. *Nanomaterials* **2023**, *13*, 47. [CrossRef]
41. Jiang, X.; Li, H.; Xiao, J.; Gao, D.; Si, R.; Yang, F.; Li, Y.; Wang, G.; Bao, X. Carbon dioxide electroreduction over imidazolate ligands coordinated with Zn(II) center in ZIFs. *Nano Energy* **2018**, *52*, 345–350. [CrossRef]
42. Dou, S.; Song, J.; Xi, S.; Du, Y.; Wang, J.; Huang, Z.-F.; Xu, Z.J.; Wang, X. Boosting Electrochemical CO_2 Reduction on Metal–Organic Frameworks via Ligand Doping. *Angew. Chem. Int. Ed.* **2019**, *58*, 4041–4045. [CrossRef]
43. Kim, B.; Ma, S.; Molly Jhong, H.-R.; Kenis, P.J.A. Influence of dilute feed and pH on electrochemical reduction of CO_2 to CO on Ag in a continuous flow electrolyzer. *Electrochim. Acta* **2015**, *166*, 271–276. [CrossRef]
44. Li, Y.C.; Zhou, D.; Yan, Z.; Gonçalves, R.H.; Salvatore, D.A.; Berlinguette, C.P.; Mallouk, T.E. Electrolysis of CO_2 to Syngas in Bipolar Membrane-Based Electrochemical Cells. *ACS Energy Lett.* **2016**, *1*, 1149–1153. [CrossRef]
45. Lee, Y.-R.; Jang, M.-S.; Cho, H.-Y.; Kwon, H.-J.; Kim, S.; Ahn, W.-S. ZIF-8: A comparison of synthesis methods. *Chem. Eng. J.* **2015**, *271*, 276–280. [CrossRef]
46. Wang, Y.; Niu, C.; Zhu, Y. Copper–Silver Bimetallic Nanowire Arrays for Electrochemical Reduction of Carbon Dioxide. *Nanomaterials* **2019**, *9*, 173. [CrossRef]

Disclaimer/Publisher's Note: The statements, opinions and data contained in all publications are solely those of the individual author(s) and contributor(s) and not of MDPI and/or the editor(s). MDPI and/or the editor(s) disclaim responsibility for any injury to people or property resulting from any ideas, methods, instructions or products referred to in the content.

Review

Research Progress in Gas Separation and Purification Based on Zeolitic Materials

Kai Qi [1,2,3,*], Lili Gao [1,2,3], Xuelian Li [1,2,3] and Feng He [4,*]

1. College of Environmental Science and Engineering, Taiyuan University of Technology, Taiyuan 030024, China
2. Research Center of CO_2 Capture, Utilization and Storage, Taiyuan University of Technology, Taiyuan 030024, China
3. Shanxi Key Laboratory of Compound Air Pollutions Identification and Control, Taiyuan University of Technology, Taiyuan 030024, China
4. State Key Laboratory of Silicate Materials for Architectures, Wuhan University of Technology, Wuhan 430070, China
* Correspondence: qikai@tyut.edu.cn (K.Q.); he-feng2002@163.com (F.H.)

Abstract: The characteristics and preparation methods of zeolite-based adsorbents and membranes were reviewed and their applications in gas separation and purification were introduced according to classification. The effects of framework structure, equilibrium cations and pore size of zeolites as well as temperature and pressure of the system on gas adsorption and separation were discussed, and the separation mechanisms were also summarized. The main defects and improved methods of zeolite-based adsorbents and membranes were briefly described, and their future trend for gas separation and purification was finally prospected.

Keywords: zeolites; zeolitic membranes; gas separation and purification

1. Introduction

Nitrogen (N_2), oxygen (O_2), carbon dioxide (CO_2), hydrogen (H_2), light hydrocarbons (C_xH_y, $x \leq 4$) and other common industrial gases are mainly derived from petrochemical and natural gas processing industries, and these are the basic energy resources and raw materials for the production of important industrial chemicals [1]. However, the separation and purification of these gases, which usually exist in the form of mixtures, is a very critical and challenging industrial process. With people's attention turned towards green production, the living environment and their own health, the importance of adsorption and separation technology has become increasingly apparent.

At present, the main separation technologies for common light gases (H_2, CO_2, O_2, N_2, lower olefins and alkynes) include cryogenic distillation, pressure swing adsorption (PSA) and membrane separation, among which the most effective and traditional method is cryogenic distillation process with phase change and high energy consumption [2]. In order to save energy and reduce consumption, PSA and membrane technologies without phase change gradually expanded their markets to become the technology leader, and these are also characterized by simple operation and less investment for facilities. The key of these two technologies for gas separation, furthermore, lies in the research and development of high-performance adsorbents and membrane materials.

In recent years, zeolites as advanced functional materials have attracted extensive attention in more and more research fields, especially in the petroleum and chemical industries. Different from disordered porous materials such as activated carbon (including carbon molecular sieve), silica gel, porous alumina and diatomite, zeolites are ordered crystalline and porous materials constructed by TO_4 tetrahedrons (T = Al, Si, P, Ti, etc.) as building blocks connected to each other by oxygen atoms, with regular pore structure, high micropore volume and specific surface area, as well as outstanding thermal, mechanical and chemical stability [3]. The negative charged TO_4 tetrahedrons which are balanced by counter-ions (such as Na^+, K^+, or Ca^{2+}) make the surface of zeolites highly polarized to easily adsorb molecules with strong polarity. Moreover, the channel dimensions of zeolites generally fall in the sub-nanometric scale, comparable with the kinetic diameters of small molecules, which can be adjusted by ion-exchange properties and by modifying the Si/Al ratio during synthesis [4]. Moreover, artificially-synthesized zeolites made with cheap and easily available raw materials also ensure high purity and consistency of products [5]. Zeolitic membrane is a kind of inorganic porous membrane, which refers to the continuous defect-free membrane formed by the growth of zeolites. It has the unique properties of zeolite itself, such as pore size uniformity, ion exchange, acid-base resistance and high-temperature resistance, which has been widely adopted by virtue of its advantages over organic polymer membrane in high-temperature resistance and easy control of microporous structure in nanoscale. Thereupon, zeolite-based adsorbents and membranes have extensive application value for gas separation. As emphasized in Figure 1, the number of publications concerning on zeolite-based adsorbents or membranes in gas separation and purification has experienced a considerable growth in recent decades. Although its overall research is still in its infancy and in its rapid development stage, zeolite-based adsorbents or membranes exhibit great potential in the field of gas separation and purification from the perspective of a very limited number of research results showing excellent separation effects.

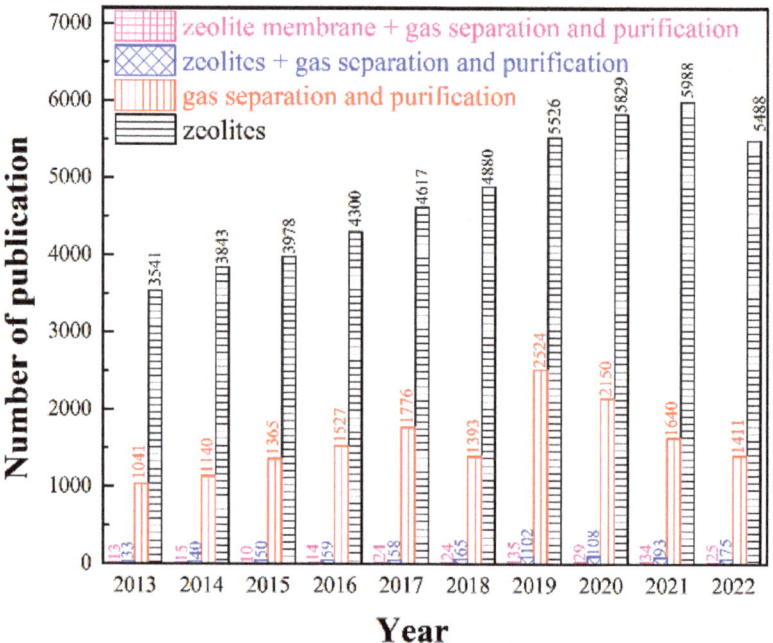

Figure 1. Number of publications per year for the period of 2012–2022 based on Web of Science database using the keywords: (Black columns) zeolites; (Red columns) gas separation and purification; (Blue columns) zeolites, gas separation and purification; (Pink columns) zeolite membrane, gas separation and purification.

2. Synthesis Strategies toward Zeolite-Based Adsorbents and Membranes

The early discovery and use of zeolites began with natural zeolites, of which clinoptilolite, chabazite, erionite, mordenite, phillipsite and ferrierite are suitable types for gas adsorption and separation [6]. However, due to the limitations of the pore structure of natural zeolites, they are far from meeting the industrial demands with the increase of their application volume and scope. Therefore, people began to research and synthesize artificial zeolites. In recent years, with the advancement of techniques, synthesis methods have become increasingly diverse so that zeolites with complex structures, stable performance and high catalytic efficiency have been extensively synthesized, greatly expanding their application fields.

2.1. Synthesis Methods of Powder Zeolites

The synthetic chemistry of zeolites is extremely complex. After long-term exploration, researchers at home and abroad have continuously improved and innovated on traditional methods. At present, the main synthesis methods include: hydrothermal synthesis, solvothermal synthesis, ionothermal synthesis, vapor-phase transport (VPT), dry gel conversion, green synthesis, etc. [7–9]. Various synthesis methods have their own advantages and disadvantages. A broad prospect will surely be opened for the green and efficient synthesis of zeolites by integrating several different synthesis routes.

Hydrothermal synthesis is designed to fully mix alkali (NaOH, KOH, etc.), Al_2O_3, SiO_2 and H_2O in a certain proportion under heating in a closed vessel, followed by nucleation and growth to form zeolites. As the most traditional synthesis method, it has the advantages of uniform dissolution of reactants, mild reaction conditions, simple operation, low cost and pollution. Different from hydrothermal synthesis, non-aqueous solvent instead of water is applied in solvothermal synthesis to avoid the interference of H_2O and break the boundary of water solvent. However, the large use of organic solvent causes the problem of high cost in actual production and greatly increases the production risk. Among them, ion-thermal synthesis, which uses ionic liquids with low steam pressure, low volatility and strong designability as both solvent and template, has gradually developed. However, the high cost of ionic liquids is not suitable for the large-scale production of zeolites.

In vapor-phase transport (VPT), the gel reactants are heated at a certain temperature in a specific reactor, without contact with organic amine and water in liquid phase. Derived from VPT, dry gel conversion mixes the non-volatile directing agent (e.g., quaternary ammonium base/salt) in gel reactants rather than in aqueous solvent, which has the advantages of less pollution, saving raw materials and reducing costs, but limited practical application due to the fussy operation process and impure products.

As the concept of "green chemistry" has taken root, the shortcomings of traditional synthesis methods have also been exposed. Researchers have studied greening the synthesis process of zeolites from various perspectives, including the application of natural minerals (e.g., kaolinite and diatomite) and green organic templates as well as novel green synthesis methods. Among them, zeolites synthesized by seeding growth which employs crystal seeds instead of organic templates occupies a significant part, which plays an important role in effectively shortening the nucleation and growth period, inhibiting the generation of hybrid crystals and regulating the grain size. In solventless synthesis, raw materials in solid phase are ground and mixed uniformly for high-temperature crystallization during the reaction process without adding other solvents, which significantly increases available space of reactor and reduces environmental pollution and resource waste with simple operating procedures and lower reaction pressure. However, this method is still in the exploration and research stage, and has not yet achieved large-scale industrial production. Microwave-assisted synthesis can quickly transfer energy to the reaction system and accelerate nucleation of microcrystalline and dissolution of reaction gel, however, harmful radiation and high pressure generated under closed conditions should be highly emphasized due to high microwave energy. Hydroxyl radical-assisted synthesis can also effectively accelerate the crystallization of zeolites, reduce the amount of alkali and organic templates, and minimize energy consumption by promoting the depolymerization and polymerization of Si-O-Si bonds in initial aluminosilicate gel.

2.2. Fabrications of Zeolite Membranes

For zeolitic membranes, the main synthesis methods include hydrothermal synthesis (including in-situ and secondary growth), pore plugging, microwave heating, pulsed laser deposition, and electrophoretic deposition [10]. Different from in-situ hydrothermal synthesis in which zeolitic membranes nucleate and grow directly on the surface of support, zeolitic membranes synthesized by secondary growth are formed based on crystal seeds precoated on the surface of support, showing the characteristics of easy control of crystal growth and membrane microstructure, as well as excellent permeability. Currently, most research has focused on the synthesis of zeolitic membranes on porous carriers (e.g., Al_2O_3, stainless steel, TiO_2, SiO_2, mullite, glass, etc.) for conferring mechanical strength using in-situ hydrothermal synthesis or secondary growth, because the self-standing layer is very brittle. The secondary growth, as mentioned above, is more reproducible than the one-step method.

During the synthesis of zeolitic membranes, different methods of seed laying, hydrothermal and calcination conditions will have impacts on structure, density, gas permeability, and separation selectivity of the prepared membranes to varying degrees [11]. Currently, the main methods of seed laying include pull-up, wetting-rubbing seeding, vacuum seeding, electrophoretic deposition, spray coating, electrostatic adsorption, cross flow filtration, pulsed laser deposition, graded seeds dip-coating, steam-assisted conversion seeding, Langmuir-Blogett (L-B) assembly, etc. With the innovation of the crystallization process, various novel methods such as dynamic heating, variable solution synthesis, variable temperature synthesis, electric field-assisted crystallization, oil-bath heating, gel-free steam-assisted conversion, two-step hydrothermal synthesis plus dry gel conversion, and ultrasonic-assisted crystallization are gradually emerging in terms of accelerating nucleation and crystallization, shortening synthesis time, reducing film thickness, improving film flux, and reducing synthesis costs. Furthermore, recent developments designed to avoid the formation of thermally induced defects include multi-stage calcination, ozone oxidation, plasma-assisted calcination, rapid thermal processing (RTP), optimized rapid thermal processing (O-RTP), etc. In the RTP or O-RTP, a composite membrane is instantly heated by IR illumination to 600–900 °C without programmed heating for several minutes and then quickly cooled to room temperature to yield a zeolite membrane with far less inter-crystalline defects, thus improving separation performance as compared with membranes calcined using conventional ramp rates. In addition, post treatment of the synthesized zeolitic films can not only repair the film defects, but also improve the surface properties. For example, chemical vapor deposition (CVD) of silane compounds (usually tetraethoxysilane) enables plugging of nanometer-scale defects, but it is inefficient for big defects because a large quantity of silane compounds is required. Sol-gel or a polymeric solution (e.g., silicon rubbers, mainly polydimethylsiloxane (PDMS)) can also be used for plugging the defects of the membranes.

Over the years, some research groups have prepared membranes with oriented crystals for improving the molecular sieving effect. Taking a zeolitic membrane with an MFI-type topological structure (sinusoidal pore structure (a-axis orientation, 0.51 nm × 0.55 nm) and straight channels (b-axis orientation, 0.53 nm × 0.56 nm)) as an example, the pore structure within membrane varies greatly when crystals grow in different directions, which will have a significant impact on its mass transfer characteristics [12]. Compared to MFI zeolitic membranes with arbitrary orientation, the inter-crystalline defects within the membrane will be significantly reduced alongside the membrane thickness and mass transfer resistance when crystals grow perpendicular to the support surface in b-axis direction, which would play a key role in improving permeability and separation selectivity [13]. Another route explored for reducing defects is using very thin zeolite nanosheets during the seeding step. Two-dimensional zeolitic nanosheets have unique structural advantages in constructing high-performance oriented zeolitic membranes due to their nanoscale thickness and high aspect ratio. Dakhchoune et al. reported the fabrication of high-performance H_2/CO_2 separation by zeolitic membranes prepared by a reactive assembly of sodalite nanosheets that host hydrogen-sieving six-membered rings (6-MRs) of SiO_4 tetrahedra, which effectively blocked CO_2 transport and led to a H_2/CO_2 ideal selectivity of above 100 at 250–300 °C [14].

2.3. General Modification Strategies of Zeolite-Based Adsorbents and Membranes

Several common modification strategies for improving the adsorption and separation performance of zeolite-based adsorbents and membranes are as follows:

(1) Adjustment of Si/Al ratios. To promote gas adsorption and separation of zeolites/zeolitic membranes by adjusting Si/Al ratios, there are two primary aspects, namely, enhancing surface adsorption and internal diffusion. In terms of enhancing surface adsorption, the lower the Si/Al ratios are, the more charge-compensated the cations will be, which is more conducive to the adsorption of gas molecules with large dipole or quadrupole moments and can promote the separation of gas mixtures with large polarity differences (e.g., CO_2/CH_4); In terms of enhancing internal diffusion, the higher the Si/Al ratios, the less the cations, which can reduce the steric hindrance of internal diffusion and promote the separation of gas mixtures with large dynamic diameter differences (e.g., N_2/CH_4). The study by Sun et al. demonstrated that the combination of X zeolites with a lower Si/Al ratio and a lower number of binders is significantly effective in improving the adsorption and separation properties of X zeolites in purification for LNG production with low CO_2 concentration in the feed gas and at high pressure [15]. Guo et al. successfully synthesized ZSM-11 zeolites with high Si/Al ratios ($n_{(Si/Al)}$ = 100, 500), the adsorption capacity of the two samples for CH_4 was higher than that of N_2, and the selectivity of CH_4/N_2 reached more than 4.0 at 25 °C and 500 kPa, which is much higher than that of commercial low-silica zeolites [16].

(2) Cation exchange. The exchange of various cations can not only change the adsorption behaviors of gas molecules on/in the surface/pore of zeolites/zeolitic membranes, but also affect the permeation behavior of different gases. The higher the basicity of the cations (low electronegativity), the stronger the binding force with weakly-acidic molecules; the higher the charge density of the cations, the stronger the electrostatic interaction with gases. In addition, the pore characteristics of zeolites will also change after cation exchange, thus affecting the internal diffusion of gas molecules. The Li^+, K^+, Mg^{2+} and Ca^{2+}-exchanged X zeolites were prepared by Sun et al. as efficient adsorbents for purification of liquefied natural gas (LNG), and the interaction between CO_2 and zeolites were reported to not only depend on the types of cations, but also on the amounts of cations, which significantly affected the adsorption at low pressure [17].

(3) Heteroatom substitution. Taking the substitution of Al by Ti as an example, the substitution of Al atom in the framework of zeolites by Ti atom can make the zeolites/zeolitic membranes have higher thermal stability, acidic stability and hydrophobicity, thus promoting the gas adsorption and separation under wet conditions. CHA-type titanosilicate (Ti-CHA) zeolite prepared by Araki et al. exhibited a relatively high CO_2 permeance compared with that of previously reported CHA-type zeolite membranes, and the CO_2 permeance and selectivity were only marginally reduced as a result of the highly hydrophobic pore structure in the presence of 1 vol.% H_2O [18].

(4) Surface/pore modification. Based on the covalent bonding of Si-O-Si groups, specific functional organic groups (including amino, hydroxy, hydrophobic groups, etc.) are grafted on/in the surface or pore of zeolites/zeolitic membranes by covalent grafting method to enhance the surface/pore polarity or acid-base properties, further promoting adsorption and separation. The study by Ilyas et al. focused on achieving highly selective membranes for CO_2 by chemically grafting 4A zeolites with methoxy groups containing cation and acetate anion based ionic liquid in order to enhance the CO_2 solubility owing to the molecular interaction of CO_2 with methoxy moieties and acetate as anion due to its hydrogen bond basicity [19].

(5) Advance techniques utilization. Other than commonly using chemical and/or physical modification methods, various advance techniques (i.e plasma technology, microwave,

irradiation, ultrasonic, etc.) have also been proven to be used to make zeolitic modifications potentially for gas separation/adsorption. Wahono et al. prepared amine adsorbent through a simple plasma polymerization and deposition on physicochemically modified natural mordenite-clinoptilolite zeolite, resulting in a significant increase in surface area-weighted CO_2 adsorption capacity [20]. Tang et al. reported that ultrasound-assisted method represents a rapid and controllable means of synthesizing nano/micro-scale zeolites with enhanced mass transfer and adsorption capacity of CO_2, CH_4, N_2, and O_2 [21]. The conjunction of e fastvolumetric microwave heating with a unique counter diffusion of metal and linker solutions by Hillman et al. enables the unprecedented rapid synthesis of wellintergrown ZIF-7-8 membranes with tunable molecular sieving properties, showing prospects for the commercial gas separation applications of ZIF membranes [22].

3. The Applications in Gas Separation and Purification

As a matter of fact, zeolites/zeolitic membranes have been employed in various separation processes for over two centuries and they are still the most important candidates for material-based separations. Generally, micropores play a decisive role in gas separation and adsorption, while mesopores and macropores mainly serve as transport channels for gas molecules. The comparison of pore size of typical zeolites with kinetic diameters of common industrial gases is shown in Figure 2.

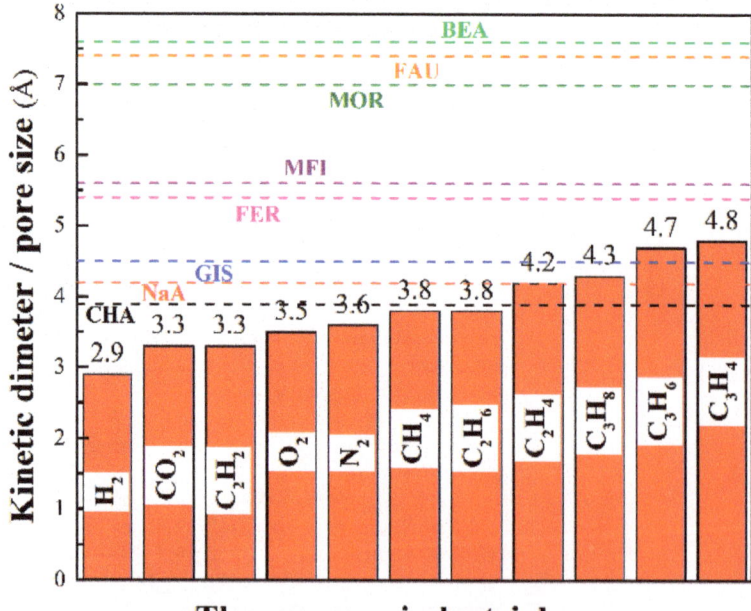

Figure 2. The comparison of pore size of typical zeolites with kinetic diameters of common industrial gases.

3.1. Air Separation

Oxygen (O_2) is not only known as the source of life, but also very important in industrial production and scientific research. Oxygen-enriched combustion is considered an effective method to improve energy efficiency, so O_2/N_2 separation is of great significance. N_2 with a lone pair of electrons has a larger quadrupole moment compared to O_2, inducing stronger interaction with cations in zeolites/zeolitic membranes and enriched O_2 in outflow gases. Researchers have carried out a large amount of research on O_2/N_2 separation based on the interaction between N_2 and zeolitic cations. In addition, the Si/Al ratio is also an important factor affecting O_2/N_2 adsorption and separation. Due to low Si/Al ratio of A-type and X-type zeolites with a large number of ion-exchange sites, research on O_2/N_2 separation mainly focuses on A- and X-type zeolites/zeolitic membranes (Table 1).

Table 1. O_2/N_2 separation performances for various zeolites/zeolitic membranes.

Zeolites/ Zeolitic Membranes (Support)	Rings	Operating Conditions	O_2 Permeability $\times 10^7$ /[mol·(m²·s·Pa)⁻¹] or Adsorption Capacity/ <mmol/g>	Separation Factor for O_2/N_2	Ref.
4A (PDMS)	8	T = 25 °C, P = 1–2.50 bar	[1052 Barrer]	2.6	[23]
CeY	12	T = 25 °C, P = 7 bar	[0.67]	0.9	[24]
Li-RHO	8	T = 25 °C, P = 6 bar	[0.05]	2447.72	[25]
CeLiX	12	T = 25 °C, P = 1 bar	<0.35>	1.8	[26]
3A + ZSM-5 (Polyurethane-)	8, 5	T = 30 °C, P = 4 bar	[21.5 Barrer]	2.7	[27]
NaA (Si_3N_4)	12	T = 25 °C, P = 1 bar	[117.9]	4.54	[28]

3.2. Hydrogen Purifification

Hydrogen (H_2), which is mainly generated from natural gas reforming, coal gasification and industrial by-products (e.g., coke oven gas (COG, 54–59% H_2)), is widely used as an important industrial raw material and zero-emission energy resource, while attracting more and more attention for acquiring sufficient pure hydrogen in continuous long-term industrial reprocessing (hydrogen production, purification and separation). In Table 2, H_2 separation performances for various zeolites/zeolitic membranes are given. Taken together, it is highly desired to prepare small pore zeolites/zeolitic membranes (e.g., SAPO-34) based on the view of size-exclusion.

Table 2. H_2 separation performances for various zeolites/zeolitic membranes.

Zeolites/ Zeolitic Membranes (Support)	Rings	Operating Conditions	H_2 Permeability $\times 10^7$/ [mol·(m²·s·Pa)⁻¹]	Separation Factor	Ref.
SSZ-13 (Mullite tube)	8	T = 25 °C, P = 2 bar, equimolar H_2/CH_4	-	43	[29]
Si-CHA (α-Al_2O_3 tube)	8	T = 40 °C, P = 1 bar	11	34 (H_2/CH_4)	[30]
Cs-LTA (α-Al_2O_3 tube)	8	T = 50 °C, P = 2 bar	-	8 (H_2/CO_2) 23.4 (H_2/CH_4) 176.8 (H_2/C_3H_8)	[31]
STT (α-Al_2O_3 hollow fiber)	7, 9	T = 25 °C, P = 2 bar, equimolar H_2/CH_4	0.28	49.6	[32]
SAPO-34 (α-Al_2O_3 hollow fiber)	8	T = 600 °C, equimolar H_2/C_3H_8	3.1	41	[33]
SAPO-34 (α-Al_2O_3 tube)	8	T = 25 °C, P = 2 bar, equimolar H_2/CH_4	14.5	42.2	[34]
$AlPO_4$-14 (Mullite tube)	8	T = 25 °C, P = 1 bar, equimolar H_2/CH_4	6.3	28	[35]
Natural zeolite (Hydroxyethyl cellulose mixed-matrix membrane)	-	T = 25 °C, P = 4 bar	699.01 Barrer	8.85 (H_2/CO_2)	[36]

3.3. Separation of Light Hydrocarbons

Light hydrocarbons are basic organic chemical raw materials, and their production capacity is an important standard to measure a country's chemical level. Generally, light hydrocarbons are mostly obtained through naphtha cracking and methanol to olefins, among which there are many by-products (e.g., acetylene in trace amounts) seriously affecting their further processing and utilization. Therefore, the separation and purification of light hydrocarbons plays an important role in industry. As shown in Table 3, various zeolites/zeolitic membranes are constantly reported for light hydrocarbons separation in recent years, among which unsaturated hydrocarbon-selective adsorbents basically realized with thermodynamic, kinetic, equilibrium-kinetic synergetic, gate-opening, and molecular sieving separations are the most common. However, the development of alkane-selective adsorbents is highly desired for industrial purification of unsaturated hydrocarbons in an effort to simplify the separation process and save energy consumption.

Table 3. Separation performances of light hydrocarbons for various zeolites/zeolitic membranes.

Zeolites/Zeolitic Membranes (Support)	Rings	Operating Conditions	Permeability × 10^7/[mol·(m^2·s·Pa)$^{-1}$] or Adsorption Capacity/<mmol/g>	[Separation Factor] or <Purity>	Ref.
SSZ-13 (Mullite tube)	8	T = 30 °C, P = 2 bar equimolar C_2H_4/C_2H_6	[0.029, C_2H_4]	[11]	[37]
Nano-sized K-Chabazite	8	T = 25 °C, P = 1 bar equimolar CH_4/N_2	<1.79, CH_4>	[4.7] <84%, CH_4>	[38]
Ni@FAU	12	T = 25 °C, P = 1 bar equimolar C_2H_2/C_2H_4	<1.72, C_2H_2>	[~100] <C_2H_2 < 1 ppm>	[39]
		T = 25 °C, P = 1 bar equimolar C_3H_4/C_3H_6	<1.59, C_3H_4>	[92] <C_2H_2 < 1 ppm>	
		T = 25 °C, P = 1 bar equimolar $C_4H_6/1,3$-C_4H_6	<1.58, C_4H_6>	[83] <C_2H_2 < 1 ppm>	
		T = 25 °C, P = 1 bar equimolar C_2H_2/C_3H_4	<1.80, C_2H_2>	[109] <C_2H_2 < 1 ppm>	
Ag-ZK-5	8	T = 25 °C, P = 1 bar 20% CH_4/N_2	< 1.6, CH_4	[11.8]	[40]

3.4. Capture and Separation of Carbon Dioxide

Carbon dioxide (CO_2) is a common gas that is also a major greenhouse gas. The large amount of CO_2 released into the atmosphere brings about more than 60% of global warming changes. About 80% this carbon dioxide comes from fuel combustion, industrial processes and biogas/natural/shale gas processing. Capturing and separating CO_2 effectively from flue gas is of great importance to mitigate climate change resulting from greenhouse gas emissions, which has attracted wide attention all over the world. Table 4 lists the CO_2 separation performances for various zeolites/zeolitic membranes. In the adsorption process, high adsorption capacity and selectivity are two key parameters for effective removal of carbon dioxide from gas stream. While in membrane separation, size screening plays a major role, which allows the passing of CO_2 and exclude other components from the mixtures to be separated, and kinetic separation based on differences in the diffusion rates also accounts for a portion. Generally, introduction of amine containing groups including ionic liquids to enhance CO_2 affinity is usually adopted.

3.5. Denitration for Flue Gas Purification

Nitrogen oxides (NO_x) exhausted from stationary engines (thermal power, cement, and steel industries, etc.) are one of the major sources of air contamination. At present, selective catalytic reduction of NO_x (NH_3-SCR) technology using NH_3 as a reductant is considered to be the most effective NO_x purification method, and catalyst is the key of NH_3-SCR process. Among various catalysts, zeolites with large specific area, strong gas adsorption on abundant acid sites, wide temperature window, high thermal stability and good SO_2 durability play an indispensable role as efficient and stable catalysts in the petrochemical, energy and environmental fields. As shown in Table 5, the NH_3-SCR performances of various zeolites/zeolitic membranes are listed. Nowadays, the most studied denitration catalysts are Fe, Cu zeolites, Mn, Ce zeolites and ZSM-5, BEA, SAPO-n, SSZ-13 topological zeolites are also the current research hotspots due to their excellent catalytic activity, selectivity and hydrothermal stability.

Table 4. CO_2 separation performances for various zeolites/zeolitic membranes.

Zeolites/ Zeolitic Membranes (Support)	Rings	Operating Conditions	CO_2 Permeability/[$\times 10^{-7}\cdot$mol$\cdot$(m$^2\cdot$s$\cdot$Pa)$^{-1}$] or Adsorption Capacity/<mmol/g>	[Separation Factor] or <CO_2 Purity>	Ref.
SSZ-13 (α-Al$_2$O$_3$ tube)	8	T = 20 °C, P = 1.4 bar, equimolar CO_2/CH_4	[39]	[162]	[41]
SAPO-34 (α-Al$_2$O$_3$ tube)	8	T = 25 °C, P = 1.4 bar, equimolar CO_2/CH_4	[44.6]	[155]	[42]
SSZ-13 (Mullite tube)	8	T = 25 °C, P = 2 bar, equimolar CO_2/N_2	-	[32]	[29]
HS-CHA (α-Al$_2$O$_3$ tube)	8	T = 40 °C, P = 4 bar, equimolar CO_2/CH_4	[15]	[115]	[43]
5A	8	T = 130 °C, P = 3.5, 0.3 bar, 10% CO_2/N_2	<5.2>	<95.2%>	[44]
Na-X	12	T = 20 °C, P = 1 bar, 400 ppm CO_2/air	<0.45>	<99.5%>	[45]
K13X	12	T = 30 °C, P = 40 bar, 3% CO_2/CH_4	<4.00>	[46]	[17]
Amine-impregnated natural zeolite (clinoptilolite) (Polysulfone mixed matrix membranes)	8, 10	T = 35 °C, P = 3.45 bar, 5% CO_2/CH_4	[22.79 Barrer]	[45.78]	[46]
Mordenite-clinoptilolite natural zeolites	8, 10, 12	T = 0 °C, P = 30 bar	<5.22>	-	[47]

Table 5. NH_3-SCR performances for NO_x removal by various zeolites/zeolitic membranes.

Zeolites/ Zeolitic Membranes (Synthesis Methods)	T [a] (X [b] > 80%)	X_{max} (T (X_{max}))	S [c] [Gas Permeation]	Reaction Conditions	SO_2 Tolerance	Ref.
Cu-TNU-9, Cu-ZSM-5 (Ion-exchange)	150–450 °C	100% (200–350 °C)	~100% in the whole temperature range	500 ppm NO, 575 ppm NH_3, 4 Vol.% O_2, 5 Vol.% H_2O, GHSV = 30,000 h^{-1}	-	[48]
Monolithic MnO$_x$/Cu-SSZ-13 (Ion-exchange & 3D printing)	160–540 °C	100% (200–350 °C)	~100% in the whole temperature range	[NO] = [NH_3] = 500 ppm, 5 Vol.% O_2, 2 Vol.% H_2O, GHSV = 30,000 h^{-1}	-	[49]
Fe/Cu-SSZ-13 (Hydrothermal method)	160–600 °C	100% (180–570 °C)	-	490 ppm NO, 10 ppm NO_2, 500 ppm NH_3, 5 Vol.% O_2, GHSV = 50,000 h^{-1}	Remain at 100% for the first 5 h with 50 ppm SO_2, then decline severely	[50]
Cu$_{0.01}$Fe$_{0.1}$SAPO-34	200–375 °C	100% (250–375 °C)	~100% within 175–375 °C	[NO] = [NH_3] = 600 ppm, 2.5 Vol.% O_2, GHSV = 37,500 h^{-1}	Maintain > 90% in 4 h with 100 ppm SO_2 and 5 Vol.% H_2O	[51]
Monolithic CuY-M-1 (Ion exchange)	100–360 °C	100% (200–280 °C)	>80% within 100–340 °C	715 ppm NO, 800 ppm NH_3, 3 Vol.% O_2, GHSV = 20,000 h^{-1}	<180 °C & >240 °C: Without SO_2 (100 ppm) poisoning	[52]
Pt/SAPO-34@SiC membrane (Hydrothermal method)	230–340 °C	95% (280 °C)	-	300 ppm NO, 300 ppm NH_3, 3 Vol.% O_2	-	[53]
MnO$_x$@PTFE membrane (Hydrothermal method)	120–240 °C	100% (160–210 °C)	[483.5 m$^3\cdot$m$^{-2}\cdot$h$^{-1}\cdot$kPa^{-1}]	[NO] = [NH_3] = 500 ppm, 5 Vol.% O_2, GHSV = 60,000 h^{-1}	-	[54]

[a] the reaction temperature window; [b] the NO_x conversion; [c] the N_2 selectivity.

3.6. Water Vapor Adsorption

The water adsorption/moisture removal is essential in many aspectss (e.g., adsorption heat pumps (AHPs), electric dehumidifiers, dehydration of organic solvents and gaseous industrial streams including hydrogen and natural gas), even in the storage of processed foods and moisture-sensitive materials. As shown in Table 6, water vapor adsorption of various zeolites/zeolitic membranes are displayed. On the whole, water vapor adsorption on zeolitic materials dominantly depends on specific surface area and the hydrophilicity/hydrophobicity which directly dependent on the Si/Al ratio and surface hydroxyl groups. Lower Si/Al ratio and higher surface hydroxyl groups can render the surface of zeolites to be hydrophilic and vice versa.

Table 6. Water vapor adsorption by various zeolites/zeolitic membranes.

Zeolites/Zeolitic Membranes	Rings	Operating Conditions	Si/Al Molar Ratio	Specific Surface Area/[m²/g]	H₂O Adsorption Capacity/[mmol/g]	Ref.
HP-NaY-1 (Na₂O/Al₂O₃ = 7.5)				698	17.32	
HP-NaY-2 (Na₂O/Al₂O₃ = 8)	12	T = 30 °C P = 0.7 bar	7	706	17.45	[55]
HP-NaY-3 (Na₂O/Al₂O₃ = 9)				749	17.98	
HP-ZnY-1 (Na₂O/Al₂O₃ = 7.5)				668	17.91	
HP-MgY-1 (Na₂O/Al₂O₃ = 7.5)				679	18.55	
HP-LiY-1 (Na₂O/Al₂O₃ = 7.5)				685	19.08	
Natural zeolite	-	Stored in a desiccator containing water at room temperature for two months	11	179.44	4.33	[56]
Natural zeolite with plasma activation				153.15	9.17	
Ag/5A	8	T = 26 ± 2 °C P = 0.84 bar	1	128	8.82	[57]
Ni/5A				113	5.96	

4. Separation Mechanisms

The gas separation with zeolites/zeolitic membranes is generally divided into two processes: surface adsorption and pore diffusion. Firstly, gas molecules are adsorbed on the surface of zeolites/zeolitic membranes; then, under the action of chemical potential, adsorbed molecules transition from one adsorption site to the next adsorption site or vacancy, thereby entering the zeolite channel and diffusing to the permeation side. The study of separation rules helps to determine process conditions in practical applications, while the separation mechanisms study helps to explain the differences generated by different gas separation on various zeolites/zeolitic membranes, thereby selectively selecting separation objects and types of zeolites/zeolitic membranes. The following is a brief overview of several common separation mechanisms for zeolites/zeolitic membranes:

Molecular sieving effect, also known as the steric hindrance effect, is the most common and easily-understood mechanism in adsorption and separation. It involves using the difference of kinetic diameters to achieve screening through regular pore structures at the level of kinetic diameters [58]. Two major categories in this method can be divided: one is that gas molecules with kinetic diameters smaller than the pore size of zeolites/zeolitic membranes can diffuse into the channels, while larger gas molecules are excluded; the other is that the separation of gas molecules with kinetic diameters smaller than the pore size of zeolites/zeolitic membranes may be related not only to the kinetic diameter but also to the shape and size of molecules.

For two gases with a very small size difference, the energy potential barriers of the two gases when passing through the pore of zeolites/zeolitic membranes may differ greatly due to the different diffusion rates, which is called "kinetic effect" [59]. Usually, diffusion selectivity takes place when molecules of one component are smaller and their diffusivity in zeolite micropores is much faster than that of the larger component (e.g., H_2/CH_4 in many 8MR zeolite membranes). Moreover the contribution of diffusion selectivity increases with temperature.

Guest-host interactions with different adsorption strengths, including the well-known physisorption (van der Waals interaction and electrostatic interaction) and chemisorption (electron transfer, exchange or sharing between the adsorbates and adsorbents, irreversible) have also been adopted to achieve separation. Generally, the highly-polarized pore environments originating from the local electric field make zeolites/zeolitic membranes preferentially adsorb molecules with greater quadruple and dipole moments, thereby achieving adsorptive separations. However, powerful and irreversible chemisorption does not meet the requirements of adsorption and separation, while weak chemical bonds stronger than van der Waals interactions but still in the range of reversible adsorption are also popular in adsorption and separation. Specifically, Π-complexation stronger than van der Waals can be formed among molecules with Π-electrons and adsorbents or membranes containing transition metals, resulting in higher adsorption selectivity, and is more suitable for gas purification and separation with low target gas concentrations [60].

Interestingly, the extraframework cations in zeolites/zeolitic membranes can be induced temporarily and reversibly deviating from the intrinsic position of channels by some specific guest molecules which are subsequently allowed to enter the pores of zeolites/zeolitic membranes, and the according phenomenon is called "molecular trapdoor effect" [61].

Different from molecular trapdoor effect, "breathing effect" is generally caused by the flexibility of zeolite framework, which has been frequently reported in MOFs [62]. The adsorbent hardly adsorbs the adsorbate or the adsorption amount is very low at low pressures, but specific guest molecules can enter the interior as the pressure increases to a certain extent, resulting in a rapid increase in adsorption capacity. Distinctively, the adsorption isotherm of "molecular trapdoor effect" shows a typical type-I isotherm, while for "breathing effect", a rapid upward trend can be viewed after breaking a certain pressure.

When the difference between the size of gas molecules and the micropore diameter of zeolites/zeolitic membranes is comparable with the de Broglie wavelength of gas molecules, an uncertain quantum effect occurs at low temperatures, leading to different diffusion rates of molecules [63]. At present, hydrogen isotope separation has already been achieved based on "quantum effect".

In actual adsorption and separation, the above mechanisms are often not separate existence. Compatibility of multiple mechanisms in one class of adsorption or membrane materials can make the adsorbents or membranes have better adsorption capacity and selectivity, but it also adds difficulty to the study of adsorption and separation mechanisms.

5. Challenges and Future Perspectives

A good adsorbent or membrane for separation should not only have high selectivity, but also have excellent adsorption capacity or permeability and mass transfer rate. The selectivity determines whether the adsorbent or membrane can effectively separate gases, while the adsorption capacity or permeability and mass transfer rate determine the capacity to handle gases. In a general sense, homogeneous and thin films are more favorable to harvest both high flux and high separation selectivity. The ordered microporous structure in molecular scale qualifies zeolites with valuable characteristics in dealing with small molecules and great potential for gas separations thereof. So far, zeolites/zeolitic membranes have been successfully adopted as promising candidates in various separation and purification processes like air separation, hydrogen purification, separation of light hydrocarbons, CO_2 capture and separation as well as gas drying. Zeolite separations, including membrane separation and adsorptive separation, have become a research hotspot with a bright industrial future.

At present, the research on zeolite-based adsorbents and membranes for gas separation and purification is generally developing in the direction of high selectivity, high adsorption capacity or high permeability. Although certain progress has been made and great application prospects have been shown, it still faces many difficulties and challenges. The research interests and developing trends of zeolites/zeolitic membranes in the application of gas separation and purification are prospected as follows:

(1) Small-pore zeolites with pore size close to the kinetic diameter of gas molecules will still be research hotspots;
(2) Research on hierarchical porous zeolites whereby the introduction of mesopores can significantly improve the mass transfer rate and adsorption capacity;
(3) Preparation of new zeolites and zeolitic membranes;

(4) Simplifying the preparation methods, enhancing the repeatability, reducing the preparation cost and realizing large-scale production are the common problems faced by zeolite-based adsorbents and membranes.

Author Contributions: Conceptualization, K.Q. and F.H.; methodology, K.Q.; software, K.Q.; validation, L.G., X.L. and F.H.; formal analysis, K.Q.; investigation, K.Q.; resources, K.Q.; data curation, K.Q.; writing—original draft preparation, K.Q.; writing—review and editing, L.G.; visualization, X.L.; supervision, F.H.; project administration, K.Q.; funding acquisition, K.Q. All authors have read and agreed to the published version of the manuscript.

Funding: This work was supported by the "Fundamental Research Program of Shanxi Province (20210302124003)".

Data Availability Statement: Data will be made available on request.

Conflicts of Interest: The authors declare no conflict of interest.

References

1. Wang, X.Q.; Chang, Z.Y.; Li, L.B.; Yang, J.F.; Li, J.P. Progress in metal-organic frameworks for efficient separation of gaseous light hydrocarbon. *Chem. Ind. Eng. Prog.* **2020**, *39*, 2218–2234.
2. Zhao, X.; Wang, Y.; Li, D.S.; Bu, X.; Feng, P. Metal-organic frameworks for separation. *Adv. Mater.* **2018**, *30*, 1705189. [CrossRef] [PubMed]
3. Burton, A. Recent trends in the synthesis of high-silica zeolites. *Catal. Rev.-Sci. Eng.* **2018**, *60*, 132–175. [CrossRef]
4. Algieri, C.; Barbieri, G.; Drioli, E. Zeolite membranes for gas separation. In *Membrane Engineering for the Treatment of Gases: Gas-Separation Problems Combined with Membrane Reactors*; Drioli, E., Barbieri, G., Eds.; Royal Society of Chemistry: London, UK, 2011; Volume 2, pp. 223–252.
5. Feng, C.; Khulbe, K.C.; Matsuura, T.; Farnood, R.; Ismail, A. Recent progress in zeolite/zeotype membranes. *J. Membr. Sci. Res.* **2015**, *1*, 49–72.
6. Wang, S.B. Utilisation of natural zeolites for air separation and pollution control. In *Handbook of Natural Zeolites*; Bentham Science Publishers: Soest, The Netherlands, 2012; Volume 481, pp. 569–587.
7. Li, K.; Cheng, H.F. Progress in synthesis and application of zeolite molecular sieves. *China Non-Met. Miner. Ind.* **2019**, *3*, 38–41.
8. Chen, Y.H.; Ren, T.H.; Liu, G.Q.; Liu, X.X.; Ni, X.; Yin, X.L.; Li, G.Z. Development of green routes for synthesis of zeolite. *Ind. Catal.* **2020**, *28*, 17–21.
9. Wu, C.Y.; Wang, Z.W.; Zhao, Y.; Zhang, H.D.; Cheng, P. Research progress of green and efficient synthesis of zeolite. *Chem. Reag.* **2022**, *44*, 1543–1550.
10. Algieri, C.; Drioli, E. Zeolite membranes: Synthesis and applications. *Sep. Purif. Technol.* **2022**, *278*, 119295. [CrossRef]
11. Liu, Y.; Lyu, C.M.; Zhu, L.; Wang, W.L.; Li, Z.Y.; Bai, Y. Recent synthesis methods of zeolite membranes. *Membr. Sci. Technol.* **2020**, *40*, 145–150.
12. Lai, Z.; Tsapatsis, M.; Nicolich, J.P. Siliceous ZSM-5 membranes by secondary growth of b-oriented seed layers. *Adv. Funct. Mater.* **2004**, *14*, 716–729. [CrossRef]
13. Peng, Y.; Wang, Z.B. Zeolite MFI membranes for separation of ethanol/water mixture. *Prog. Chem.* **2013**, *25*, 2178–2188.
14. Dakhchoune, M.; Villalobos, L.F.; Semino, R.; Liu, L.M.; Rezaei, M.; Schouwink, P.; Avalos, C.E.; Baade, P.; Wood, V.; Han, Y. Gas-sieving zeolitic membranes fabricated by condensation of precursor nanosheets. *Nat. Mater.* **2021**, *20*, 362–369. [CrossRef] [PubMed]
15. Sun, Y.B.; Tang, J.F.; Li, G.Y.; Hua, Y.H.; Li, H.; Hu, S.Y. Experimental investigation of adsorption and CO_2/CH_4 separation properties of 13X and JLOX-500 zeolites during the purification of liquefied natural gas. *ACS Omega* **2022**, *7*, 18542–18551. [CrossRef] [PubMed]
16. Guo, W.J.; Li, Y.; Li, S.S.; Wu, X.L.; Tan, J.C.; Li, W.; Yang, J.F.; Li, J.P. Study on adsorption and separation performance of CH_4/N_2 by high-silica ZSM-11 zeolite. *Nat. Gas Chem. Ind.* **2022**, *47*, 67–72.
17. Sun, Y.B.; Tang, J.F.; Li, G.Y.; Hua, Y.H.; Sun, Y.B.; Hu, S.Y.; Wen, X.J. Adsorption, separation and regeneration of cation-exchanged X zeolites for LNG purification: Li^+, K^+, Mg^{2+} and Ca^{2+}. *Microp. Mesopor. Mat.* **2022**, *340*, 112032. [CrossRef]
18. Araki, S.; Ishii, H.; Imasaka, Y.; Yamamoto, H. Synthesis and gas permeation properties of chabazite-type titanosilicate membranes synthesized using nano-sized seed crystals. *Microp. Mesopor. Mat.* **2020**, *292*, 109798. [CrossRef]
19. Ilyas, A.; Muhammad, N.; Gilani, M.A.; Vankelecom, I.F.J.; Khan, A.L. Effect of zeolite surface modification with ionic liquid [APTMS][Ac] on gas separation performance of mixed matrix membranes. *Sep. Purif. Technol.* **2018**, *205*, 176–183. [CrossRef]
20. Wahono, S.K.; Dwiatmoko, A.A.; Cavallaro, A.; Indirathankam, S.C.; Addai-Mensah, J.; Skinner, W.; Vinu, A.; Vasilev, K. Amine-functionalized natural zeolites prepared through plasma polymerization for enhanced carbon dioxide adsorption. *Plasma Process. Polym.* **2021**, *18*, 2100028. [CrossRef]
21. Tang, X.; Liu, J.; Shang, H.; Wu, L.; Yang, J. Gas diffusion and adsorption capacity enhancement via ultrasonic pretreatment for hydrothermal synthesis of K-KFI zeolite with nano/micro-scale crystals. *Micropor. Mesopor. Mat.* **2020**, *297*, 110036. [CrossRef]
22. Hillman, F.; Brito, J.; Jeong, H.K. Rapid one-pot microwave synthesis of mixed-linker hybrid zeolitic-imidazolate framework membranes for tunable gas separations. *ACS Appl. Mater. Interfaces* **2018**, *10*, 5586–5593. [CrossRef]

23. Prajapati, P.K.; Kansara, A.M.; Aswal, V.K.; Singh, P.S. High oxygen permeable Zeolite-4A poly(dimethylsiloxane) membrane for air separation. *J. Appl. Polym. Sci.* **2019**, *136*, 48047. [CrossRef]
24. Liu, H.B.; Yuan, D.H.; Yang, L.P.; Xing, J.C.; Zeng, S.; Xu, S.T.; Xu, Y.P.; Liu, Z.M. Directly decorated CeY zeolite for O_2-selective adsorption in O_2/N_2 separation at ambient temperature. *Mater. Horiz.* **2022**, *9*, 688–693. [CrossRef]
25. Xia, H.Y.; Hu, Y.F.; Bao, Q.; Zhang, J.; Sun, P.L.; Liang, D.; Wang, B.X.; Qiao, X.; Wang, X.Y. Adsorption separation of O_2/N_2 by Li-RHO zeolite with high oxygen selectivity. *Micropor. Mesopor. Mater.* **2023**, *350*, 112442. [CrossRef]
26. Jiang, M.M.; Zhu, M.F.; Deng, C.; Zhao, L.; Ma, J.; Shi, M.S.; Gao, W.Y. Modification of LiX zeolite molecular sieve via cerium ions. *Appl. Chem. Ind.* **2017**, *46*, 332–334, 346.
27. Marani, H.T.; Sadeghi, M.; Moheb, A.; Esfahani, E.N. Optimization of the gas separation performance of polyurethane-zeolite 3A and ZSM-5 mixed matrix membranes using response surface methodology. *Chin. J. Chem. Eng.* **2019**, *927*, 110–129.
28. Hu, Y.D. Fabrication and Properties of NaA Zeolite Membranes on Porous Si_3N_4 Supports. Master's Thesis, University of Chinese Academy of Sciences, Beijing, China, 2021.
29. Wang, B.; Zheng, Y.H.; Zhang, J.F.; Zhang, W.J.; Zhang, F.; Xing, W.H.; Zhou, R.F. Separation of light gas mixtures using zeolite SSZ-13 membranes. *Micropor. Mesopor. Mater.* **2019**, *275*, 191–199. [CrossRef]
30. Kida, K.; Maeta, Y.; Yogo, K. Preparation and gas permeation properties on pure silica CHA-type zeolite membranes. *J. Membr. Sci.* **2017**, *522*, 363–370. [CrossRef]
31. Xu, C.; Wei, W.C.; He, Y. Enhanced hydrogen separation performance of Linde Type-A zeolite molecular sieving membrane by cesium ion exchange. *Mater. Lett.* **2022**, *324*, 132680. [CrossRef]
32. Zhou, T.; Shi, M.Y.; Chen, L.J.; Gong, C.; Zhang, P.; Xie, J.X.; Wang, X.R.; Gu, X.H. Fluorine-free synthesis of all-silica STT zeolite membranes for H_2/CH_4 separation. *Chem. Eng. J.* **2022**, *433*, 133567. [CrossRef]
33. Wang, Z.G.; Xu, J.; Pati, S.; Chen, T.J.; Deng, Y.Z.; Dewangan, N.; Meng, L.; Lin, J.Y.S.; Kawi, S. High H_2 permeable SAPO-34 hollow fiber membrane for high temperature propane dehydrogenation application. *AIChE J.* **2020**, *66*, e16278. [CrossRef]
34. Mei, W.L.; Du, Y.; Wu, T.Y.; Gao, F.; Wang, B.; Duan, J.G.; Zhou, J.J.; Zhou, R.F. High-flux CHA zeolite membranes for H_2 separations. *J. Membr. Sci.* **2018**, *565*, 358–369. [CrossRef]
35. Shao, G.Q.; Zhu, H.; Ma, W.; Yan, P.; Ma, J.L. Synthesis of high-performance $AlPO_4$-14 zeolite membranes for gas separation. *Chem. J. Chin. Univ.* **2019**, *40*, 2265–2273.
36. Ünügül, T.; Nigiz, F.U. Hydrogen purification using natural zeolite-loaded hydroxyethyl cellulose membrane. *Int. J. Energ. Res.* **2022**, *46*, 1826–1836. [CrossRef]
37. Zheng, Y.H.; Hu, N.; Wang, H.M.; Bu, N.; Zhang, F.; Zhou, R.F. Preparation of steam-stable high-silica CHA (SSZ-13) membranes for CO_2/CH_4 and C_2H_4/C_2H_6 separation. *J. Membr. Sci.* **2015**, *475*, 303–310. [CrossRef]
38. Yang, J.F.; Liu, J.Q.; Liu, P.X.; Li, L.B.; Tang, X.; Shang, H.; Li, J.P.; Chen, B.L. K-Chabazite zeolite nanocrystal aggregates for highly efficient methane separation. *Angew. Chem. Int. Ed.* **2022**, *61*, e202116850.
39. Chai, Y.C.; Han, X.; Li, W.Y.; Liu, S.S.; Yao, S.K.; Wang, C.; Shi, W.; Da-silva, I.; Manuel, P.; Cheng, Y.Q.; et al. Control of zeolite pore interior for chemoselective alkyne/olefin separations. *Science* **2020**, *368*, 1002–1006. [CrossRef]
40. Kencana, K.S.; Min, J.G.; Kemp, K.C.; Hong, S.B. Nanocrystalline Ag-ZK-5 zeolite for selective CH_4/N_2 separation. *Sep. Purif. Technol.* **2022**, *282*, 120027. [CrossRef]
41. Qiu, H.E.; Zhang, Y.; Kong, L.; Kong, X.; Tang, X.X.; Meng, D.N.; Xu, N.; Wang, M.Q.; Zhang, Y.F. High performance SSZ-13 membranes prepared at low temperature. *J. Membr. Sci.* **2020**, *603*, 118023. [CrossRef]
42. Zhang, Y.; Wang, M.Q.; Qiu, H.E.; Kong, L.; Xu, N.; Tang, X.X.; Meng, D.N.; Kong, X.; Zhang, Y.F. Synthesis of thin SAPO-34 zeolite membranes in concentrated gel. *J. Membr. Sci.* **2020**, *603*, 118451. [CrossRef]
43. Imasaka, S.; Itakura, M.; Yano, K.; Fujita, S.; Okada, M.; Hasegawa, Y.; Abe, C.; Araki, S.; Yamamoto, H. Rapid preparation of high-silica CHA-type zeolite membranes and their separation properties. *Sep. Purif. Technol.* **2018**, *199*, 298–303. [CrossRef]
44. Mendes, P.A.P.; Ribeiro, A.M.; Gleichmann, K.; Ferreira, A.F.P.; Rodrigues, A.E. Separation of CO_2/N_2 on binderless 5A zeolite. *J. CO_2 Util.* **2017**, *20*, 224–233. [CrossRef]
45. Wilson, S.M.W. High purity CO_2 from direct air capture using a single TVSA cycle with Na-X zeolites. *Sep. Purif. Technol.* **2022**, *294*, 121186. [CrossRef]
46. Castruita-de, L.G.; Yeverino-Miranda, C.Y.; Montes-Luna, A.D.J.; Meléndez-Ortiz, H.I.; Alvarado-Tenorio, G.; García-Cerda, L.A. Amine-impregnated natural zeolite as filler in mixed matrix membranes for CO_2/CH_4 separation. *J. Appl. Polym. Sci.* **2020**, *137*, 48286.
47. Wahono, S.K.; Stalin, J.; Addai-Mensah, J.; Skinner, W.; Vinu, A.; Vasilev, K. Physicochemical modification of natural mordenite-clinoptilolite zeolites and their enhanced CO_2 adsorption capacity. *Micropor. Mesopor. Mater.* **2020**, *294*, 109871. [CrossRef]
48. Tarach, K.A.; Jabłońska, M.; Pyra, K.; Liebau, M.; Reiprich, B.; Gläser, R.; Góra-Marek, K. Effect of zeolite topology on NH_3-SCR activity and stability of Cu-exchanged zeolites. *Appl. Catal. B-Environ.* **2021**, *284*, 119752. [CrossRef]
49. Wei, Y.Z.; Chen, M.Y.; Ren, X.Y.; Wang, Q.F.; Han, J.F.; Wu, W.Z.; Yang, X.G.; Wang, S.; Yu, J.H. One-pot three-dimensional printing robust self-supporting MnO_x/Cu-SSZ-13 zeolite monolithic catalysts for NH_3-SCR. *CCS Chem.* **2022**, *4*, 1708–1719. [CrossRef]
50. Wan, J.; Chen, J.W.; Zhao, R.; Zhou, R.X. One-pot synthesis of Fe/Cu-SSZ-13 catalyst and its highly efficient performance for the selective catalytic reduction of nitrogen oxide with ammonia. *J. Environ. Sci.* **2021**, *100*, 306–316. [CrossRef]

51. Xiao, Y.D.; Jia, X.H.; He, D.; Huang, B.C.; Yang, Y.X.; Zhang, Y.; Lu, M.J.; Tu, X.; Yu, C.L. Controllable synthesis, characterization and low-temperature NH_3-SCR activity of FeSAPO-34 and CuFeSAPO-34 molecular sieves. *Acta Sci. Circumstantiae* **2022**, *42*, 357–368.
52. Qi, K.; Yi, Q.; Fang, D.; Gong, P.J.; Shi, L.J.; Gao, L.L.; Li, X.L.; He, F.; Xie, J.L. Temperature dependence of reaction mechanisms and SO_2 tolerance over a promising monolithic CuY catalyst for NO removal. *Appl. Surf. Sci.* **2023**, *615*, 156473. [CrossRef]
53. Chen, J.H.; Pan, B.; Wang, B.; Ling, Y.J.; Fu, K.; Zhou, R.F.; Zhong, Z.X.; Xing, W.H. Hydrothermal synthesis of a Pt/SAPO-34@SiC catalytic membrane for the simultaneous removal of NO and particulate matter. *Ind. Eng. Chem. Res.* **2020**, *59*, 4302–4312. [CrossRef]
54. Feng, S.S.; Zhou, M.D.; Han, F.; Zhong, Z.X.; Xing, W.H. A bifunctional MnO_x@PTFE catalytic membrane for efficient low temperature NO_x-SCR and dust removal. *Chin. J. Chem. Eng.* **2020**, *28*, 1260–1267. [CrossRef]
55. Dabbawala, A.A.; Kumar-Reddy, K.S.; Mittal, H.; Al-Wahedi, Y.; Vaithilingam, B.V.; Karanikolos, G.N.; Singaravel, G.; Morin, S.; Berthod, M.; Alhassan, S.M. Water vapor adsorption on metal-exchanged hierarchical porous zeolite-Y. *Micropor. Mesopor. Mater.* **2021**, *326*, 111380. [CrossRef]
56. Wahono, S.K.; Suwanto, A.; Prasetyo, D.J.; Jatmiko, T.H.; Vasilev, K. Plasma activation on natural mordenite-clinoptilolite zeolite for water vapor adsorption enhancement. *Appl. Surf. Sci.* **2019**, *483*, 940–946. [CrossRef]
57. Henao-Sierra, W.; Romero-Sáez, M.; Gracia, F.; Cacua, K.; Buitrago-Sierra, R. Water vapor adsorption performance of Ag and Ni modified 5A zeolite. *Micropor. Mesopor. Mater.* **2018**, *265*, 250–257. [CrossRef]
58. Lin, J.Y.S. Molecular sieves for gas separation. *Science* **2016**, *353*, 8–10. [CrossRef]
59. Li, K.; Olson, D.H.; Seidel, J.; Emge, T.J.; Gong, H.; Zeng, H.; Li, J. Zeolitic imidazolate frameworks for kinetic separation of propane and propene. *J. Am. Chem. Soc.* **2009**, *131*, 10368–10369. [CrossRef]
60. Gao, F.; Wang, Y.Q.; Wang, S.H. Selective adsorption of CO on CuCl/Y adsorbent prepared using $CuCl_2$ as precursor: Equilibrium and thermodynamics. *Chem. Eng. J.* **2016**, *290*, 418–427. [CrossRef]
61. Shang, J.; Li, G.; Gu, Q.F.; Singh, R.; Xiao, P.; Liu, J.Z.; Webley, P.A. Temperature controlled invertible selectivity for adsorption of N_2 and CH_4 by molecular trapdoor chabazites. *Chem. Commun.* **2014**, *50*, 4544–4546. [CrossRef]
62. Georgieva, V.M.; Bruce, E.L.; Verbraeken, M.C.; Scott, A.R.; Casteel, W.J.; Brandani, S.; Wright, P.A. Triggered gate opening and breathing effects during selective CO_2 adsorption by merlinoite zeolite. *J. Am. Chem. Soc.* **2019**, *141*, 12744–12759. [CrossRef]
63. Salazar, J.M.; Lectez, S.; Gauvin, C.; Macaud, M.; Bellat, J.P.; Weber, G.; Bezverkhyy, I.; Simon, J.M. Adsorption of hydrogen isotopes in the zeolite NaX: Experiments and simulations. *Int. J. Hydrog. Energ.* **2017**, *42*, 13099–13110. [CrossRef]

Disclaimer/Publisher's Note: The statements, opinions and data contained in all publications are solely those of the individual author(s) and contributor(s) and not of MDPI and/or the editor(s). MDPI and/or the editor(s) disclaim responsibility for any injury to people or property resulting from any ideas, methods, instructions or products referred to in the content.

Article

Positive Effect of Ce Modification on Low-Temperature NH₃-SCR Performance and Hydrothermal Stability over Cu-SSZ-16 Catalysts

Yuqian Liang [1], Rui Li [2], Ruicong Liang [1], Zhanhong Li [1], Xiangqiong Jiang [1] and Jiuxing Jiang [1,3,*]

[1] MOE Key Laboratory of Bioinorganic and Synthetic Chemistry, School of Chemistry, Sun Yat-sen University, Guangzhou 510275, China
[2] State Key Laboratory of Green Chemical Engineering and Industrial Catalysis, Sinopec Shanghai Research Institute of Petrochemical Technology, Shanghai 201208, China
[3] Jiangxi Provincial Key Laboratory of Low-Carbon Solid Waste Recycling Technology, School of Geography and Environmental Engineering, Gannan Normal University, Ganzhou 341000, China
* Correspondence: jiangjiux@mail.sysu.edu.cn

Abstract: Cu-exchanged SSZ-16 zeolite catalysts exhibit outstanding NH₃-SCR activity, but their catalytic performance after hydrothermal treatments is not ideal. In order to improve the hydrothermal stability of Cu-SSZ-16, CuCe$_x$-SSZ-16 series catalysts were prepared via an ion exchange process, and the effect of Ce modification on the hydrothermal stability was investigated. In addition, increasing Ce contents significantly improved the hydrothermal stability, and CuCe$_{0.87}$-SSZ-16 showed the best hydrothermal stability. The effects of adding Ce to active species and the AFX framework were studied by various characterization measurements. The ^{27}Al MAS NMR results reveal that Ce modification can strengthen the structural stability of the CuCe$_x$-SSZ-16 catalysts. Furthermore, the combined results of XPS, H₂-TPR, and in situ DRIFTS confirm that the introduction of Ce markedly increases the active Cu^{2+}-2Z species, contributing to the remarkable hydrothermal stability.

Keywords: Cu-SSZ-16; Ce content; NH₃-SCR; low temperature; hydrothermal stability

Citation: Liang, Y.; Li, R.; Liang, R.; Li, Z.; Jiang, X.; Jiang, J. Positive Effect of Ce Modification on Low-Temperature NH₃-SCR Performance and Hydrothermal Stability over Cu-SSZ-16 Catalysts. *Catalysts* **2023**, *13*, 742. https://doi.org/10.3390/catal13040742

Academic Editors: De Fang and Yun Zheng

Received: 28 February 2023
Revised: 6 April 2023
Accepted: 7 April 2023
Published: 10 April 2023

Copyright: © 2023 by the authors. Licensee MDPI, Basel, Switzerland. This article is an open access article distributed under the terms and conditions of the Creative Commons Attribution (CC BY) license (https:// creativecommons.org/licenses/by/ 4.0/).

1. Introduction

Nitrogen oxides (NO$_x$) have been identified as a significant air pollutant that causes a large number of environmental issues and harms human health [1,2]. The primary sources of NO$_x$ in cities are emissions from power plants and automobile engines, of which diesel engines account for a large proportion [3]. Consequently, the control of NO$_x$ emitted from diesel engines is essential. Ammonia-selective catalytic reduction (NH₃-SCR) is regarded as a highly efficient denitration technique because of its excellent deNO$_x$ performance [4]. Nowadays, many Cu-exchanged zeolites have been widely considered due to their outstanding deNO$_x$ activity and hydrothermal stability. Among them, Cu-exchanged CHA, AEI, SFW, and AFX catalysts have been extensively investigated in previous studies [5–10]. However, severe high-temperature hydrothermal treatments destroy the skeleton of zeolite and reduce the active species, causing the loss of NH₃-SCR activity.

The effect of Cu species in Cu-exchanged zeolites has been extensively studied [11–13]. It is universally acknowledged that Cu^{2+} species, including Cu^{2+}-2Z and [Cu(OH)]$^+$-Z (where Z stands for a framework negative charge), provide active sites for the NH₃-SCR reaction [14]. Cu^{2+}-2Z species refer to Cu^{2+} located in the 6-ring, while the Cu^{2+} sites residing in the 8-ring are recorded as [Cu(OH)]$^+$-Z [15,16]. The two kinds of Cu^{2+} species behave differently under hydrothermal treatments. Cu^{2+}-2Z species are considered relatively stable active sites with higher hydrothermal stability, contributing to NO$_x$ removal [4,17–20]. However, [Cu(OH)]$^+$-Z is more beneficial to low-temperature (<300 °C) deNO$_x$ reactions, though it may transform to Cu^{2+}-2Z or CuO$_x$ clusters with increasing temperature [21,22]. The CuO$_x$ clusters might block the pores of zeolites, leading to a reduction in NH₃-SCR activity [23].

The Cu-SSZ-16 with an AFX structure exhibits remarkable NH_3-SCR performance, but its low-temperature activity is reduced to varying degrees after hydrothermal treatments at different temperatures [10]. Since hydrothermal stability is vital to the application of catalysts, the improvement of zeolite catalysts should also focus on hydrothermal stability. According to the literature, hydrothermal deactivation is mainly caused by a decrease in the active Cu^{2+} species and the structural instability resulting from skeleton dealumination, which can be alleviated by some means, for example, by introducing some elements [24–26]. Previous studies have demonstrated that introducing Ce to Cu-exchanged zeolites could ameliorate their hydrothermal stability. Wang et al. proposed that the addition of Ce could greatly promote the catalytic activity and hydrothermal stability of Cu-SSZ-39 catalysts [27]. Mao et al. perceived that the higher hydrothermal stability of Cu-Ce/SAPO-34 might be obtained by increasing the additional content of Ce, for Ce doping could prevent hydrothermal treatments from causing damage to the partial pore structure and a reduction in the catalyst's crystallinity [28]. Deng et al. found that Ce doping could improve the hydrothermal stability of Cu/SSZ-13 catalysts, owing to the increased framework aluminum and the more stable Cu sites [29]. Jiang and co-workers reported that the introduction of Ce might stabilize the zeolite skeleton and increase the active Cu^{2+} species, leading to the excellent hydrothermal stability of CeCu-SSZ-52 [30]. However, developing new catalysts with outstanding catalytic activity and hydrothermal stability is still crucial. Cu-SSZ-16 catalysts show superior $deNO_x$ activity, but their hydrothermal stability needs to be increased to allow commercial application.

In this study, $CuCe_x$-SSZ-16 series catalysts (x = 0.77 wt.% and 0.87 wt.%) were synthesized to study their low-temperature NH_3-SCR catalytic activity as well as their hydrothermal stability. Various characterization measurements such as XRD, ^{27}Al MAS NMR, XPS, H_2-TPR, EPR, UV-vis, and in situ DRIFTS were used to probe the influence of adding Ce to the catalysts, including the changes in the active species and zeolite framework.

2. Results and Discussion

2.1. NH_3-SCR Activity and SO_2 Resistance Test

The NO_x conversion curves of the NH_3-SCR reaction over Cu-SSZ-16-Fresh, $CuCe_{0.77}$-SSZ-16-Fresh, and $CuCe_{0.87}$-SSZ-16-Fresh (where "Fresh" represents the samples tested before the hydrothermal treatments) are displayed in Figure 1a. The NO_x reduction efficiency of Cu-SSZ-16-Fresh reaches 90% at about 215 °C and remains above 90% at 215–400 °C. Compared with Cu-SSZ-16-Fresh, $CuCe_{0.77}$-SSZ-16-Fresh and $CuCe_{0.87}$-SSZ-16-Fresh exhibit better low-temperature catalytic activity with higher NO_x conversion from 150 to 250 °C and the conversion is maintained at 95% from 250 to 400 °C. Additionally, the N_2 selectivity is slightly improved, and the selectivity of the NO_2 and N_2O byproducts declines at low temperatures (<250 °C) with the incorporation of Ce (Figure S1a–c).

In order to inquire about the changes in the hydrothermal stability with the addition of Ce, Cu-SSZ-16, $CuCe_{0.77}$-SSZ-16, and $CuCe_{0.87}$-SSZ-16 catalysts were treated at the hydrothermal temperature of 750 °C (referred to as Cu-SSZ-16-750HT, $CuCe_{0.77}$-SSZ-16-750HT, and $CuCe_{0.87}$-SSZ-16-750HT). The catalytic data in Figure 1b illustrate that the NO_x reduction efficiency of Cu-SSZ-16-750HT reaches 90% at around 265 °C, while $CuCe_x$-SSZ-16-750HT series catalysts achieve 90% NO_x conversion at about 240 °C. The results reveal that the NH_3-SCR activity is enhanced at low temperatures (<250 °C) after Ce is added. The N_2, NO_2, and N_2O selectivity of the Cu-SSZ-16-750HT and $CuCe_x$-SSZ-16-750HT series catalysts is presented in Figure S1d–f. The N_2 selectivity of Cu-SSZ-16-750HT decreases by 4%, while the $CuCe_x$-SSZ-16-750HT series catalysts have little change compared with $CuCe_x$-SSZ-16-Fresh. The selectivity of NO_2 byproducts is below 2% for $CuCe_x$-SSZ-16-750HT series catalysts in the whole temperature range. As for N_2O, the selectivity for Cu-SSZ-16-750HT is 6% at 150 °C, compared to 4% for the $CuCe_x$-SSZ-16-750HT series catalysts.

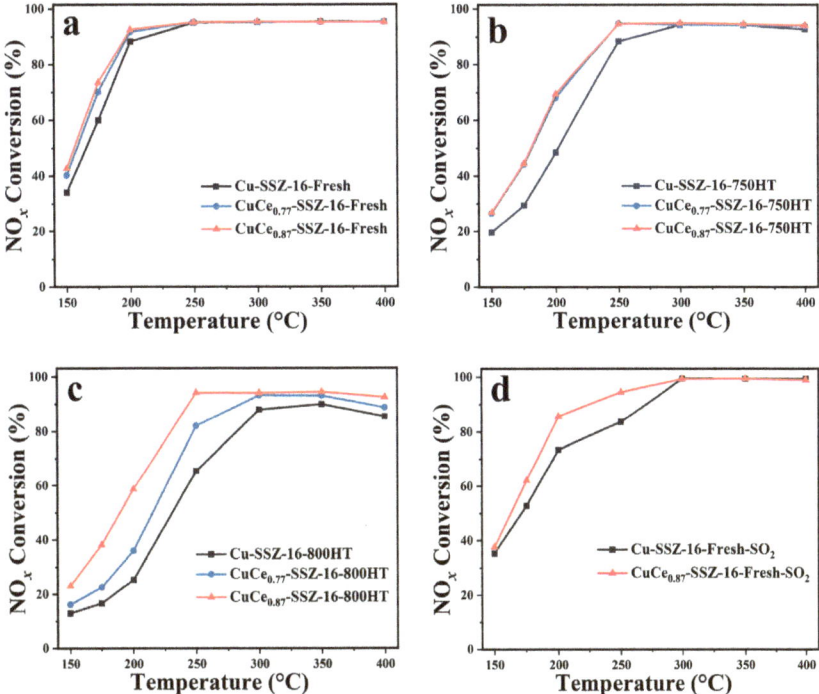

Figure 1. NO$_x$ conversion before (**a**), after hydrothermal treatments at 750 °C (**b**) and 800 °C (**c**), and in the presence of SO$_2$ (**d**) over Cu-SSZ-16, CuCe$_{0.77}$-SSZ-16, and CuCe$_{0.87}$-SSZ-16. Reaction conditions: 500 ppm NO, 500 ppm NH$_3$, 50 ppm SO$_2$ (when used), 5 vol% O$_2$, 5 vol% H$_2$O, balance N$_2$, and GHSV = 200,000 h^{-1}.

Additionally, to further investigate the effect of incorporating Ce on the hydrothermal stability of CuCe$_x$-SSZ-16, the catalysts were hydrothermally aged under more severe conditions of 800 °C. As shown in Figure 1c, the NO$_x$ conversion of Cu-SSZ-16-800HT is below 90% at 150–400 °C. After introducing Ce species, the NO$_x$ reduction efficiency of CuCe$_{0.77}$-SSZ-16-800HT and CuCe$_{0.87}$-SSZ-16-800HT is above 90% at 285–400 °C and 245–400 °C, respectively. The corresponding N$_2$ selectivity of the CuCe$_x$-SSZ-16-800HT series catalysts shows a noticeable improvement below 250 °C (Figure S1g). The N$_2$ selectivity of Cu-SSZ-16-800HT is only 85% at 150 °C, while that of the CuCe$_{0.77}$-SSZ-16-800HT and CuCe$_{0.87}$-SSZ-16-800HT catalysts is 89% and 93% at the same temperature, respectively. The three samples aged at 800 °C maintain similar low NO$_2$ selectivity (Figure S1h). Compared with Cu-SSZ-16-800HT, the N$_2$O selectivity of CuCe$_{0.77}$-SSZ-16-800HT and CuCe$_{0.87}$-SSZ-16-800HT significantly declines at low temperatures, decreasing from 10% to 7% and then to 4% at 150 °C (Figure S1i). It is suggested that the addition of Ce improves the hydrothermal stability of the catalysts and enables the aged catalysts to maintain better NH$_3$-SCR performance.

The CuCe$_{0.87}$-SSZ-16-Fresh catalyst was selected to explore the influence of SO$_2$ in the reaction mixture, and the experimental results are depicted in Figure 1d. In the presence of SO$_2$, the NH$_3$-SCR performance of CuCe$_{0.87}$-SSZ-16-Fresh-SO$_2$ (where "SO$_2$" represents 50 ppm SO$_2$ in the feed gas) is well maintained at the low-temperature range (<300 °C). The catalytic activity of Cu-SSZ-16 significantly decreases due to the toxicity of SO$_2$, compared with Cu-SSZ-16-Fresh, the NO$_x$ conversion decreases by 11% at 250 °C for Cu-SSZ-16-Fresh-SO$_2$. Meanwhile, CuCe$_{0.87}$-SSZ-16-Fresh-SO$_2$ shows 94% NO$_x$ conversion at 250 °C, which is virtually identical to that of CuCe$_{0.87}$-SSZ-16-Fresh, suggesting that the introduction of Ce improves the SO$_2$ resistance of the catalysts.

According to the above results, it can be inferred that the incorporation of Ce positively impacts the low-temperature NH$_3$-SCR performance and the SO$_2$ resistance. More importantly, it improves the hydrothermal stability of the CuCe$_x$-SSZ-16 series catalysts. Moreover, with the increase in the Ce contents, the positive effect is enhanced, and CuCe$_{0.87}$-SSZ-16 exhibits the best hydrothermal stability.

2.2. Structural Characterization

The chemical compositions of the fresh and aged catalysts are exhibited in Table 1. The fresh catalysts contain similar Si/Al ratios and Cu contents. In addition, the changes in Cu and Ce contents after hydrothermal treatment at 800 °C may be due to the destruction of the framework [29]. Figure 2a,b, shows the PXRD patterns of the fresh catalysts and the samples aged at 800 °C. As presented in Figure 2a, the PXRD patterns of the fresh catalysts exhibit the typical characteristic peaks of SSZ-16 (2θ = 7.4°, 8.6°, 11.6°, and 12.8°), implying that the AFX structure is well maintained after Cu and Ce ion exchange [31]. After hydrothermally aging at 800 °C, an amorphous structure forms in the three aged catalysts, possibly due to structural damage caused by hydrothermal treatments. Among them, the characteristic peaks of the AFX structure can be identified in the PXRD pattern of CuCe$_{0.87}$-SSZ-16-800HT. It is suggested that CuCe$_{0.87}$-SSZ-16-800HT maintains a partial AFX structure, which is important for CuCe$_{0.87}$-SSZ-16-800HT to exhibit high deNO$_x$ activity still. However, the characteristic peaks of SSZ-16 could barely be recognized for Cu-SSZ-16-800HT, indicating that the structure has collapsed after hydrothermal treatment at 800 °C. Notably, although the crystallinity decreases significantly, no characteristic peaks corresponding to CuO$_x$ and CeO$_2$ are found on all the fresh and aged catalyst samples [32,33], indicating that CuO$_x$ or CeO$_2$ particles have not formed and the Cu and Ce are distributed well in all catalysts.

Table 1. The chemical compositions and textural parameters of Cu-SSZ-16-Fresh, CuCe$_{0.77}$-SSZ-16-Fresh, CuCe$_{0.87}$-SSZ-16-Fresh, Cu-SSZ-16-800HT, CuCe$_{0.77}$-SSZ-16-800HT, and CuCe$_{0.87}$-SSZ-16-800HT.

Catalysts	Component Content [a]			S_{BET} [b] $(m^2 \cdot g^{-1})$	Pore Volume [b] $(cm^3 \cdot g^{-1})$
	Si/Al	Cu (wt.%)	Ce (wt.%)		
Cu-SSZ-16-Fresh	3.4	2.2	-	523	0.247
CuCe$_{0.77}$-SSZ-16-Fresh	3.4	2.2	0.77	577	0.251
CuCe$_{0.87}$-SSZ-16-Fresh	3.3	2.1	0.87	605	0.261
Cu-SSZ-16-800HT	3.1	2.8	-	13	0.030
CuCe$_{0.77}$-SSZ-16-800HT	3.1	2.8	1.0	30	0.035
CuCe$_{0.87}$-SSZ-16-800HT	3.1	2.7	1.1	37	0.060

[a] Measured by ICP-OES. [b] Derived from N$_2$ adsorption–desorption isotherms.

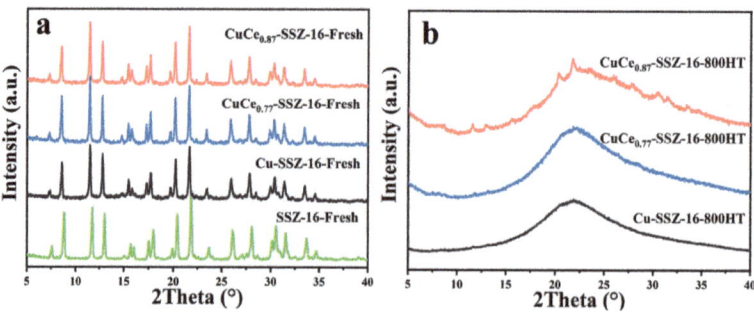

Figure 2. PXRD patterns before (a) and after hydrothermal treatment at 800 °C (b) of SSZ-16, Cu-SSZ-16, CuCe$_{0.77}$-SSZ-16, and CuCe$_{0.87}$-SSZ-16.

The N_2 adsorption–desorption analyses for Cu-SSZ-16-Fresh, $CuCe_{0.77}$-SSZ-16-Fresh, and $CuCe_{0.87}$-SSZ-16-Fresh are demonstrated in Figure 3a,b. All three fresh catalysts show type I isotherms related to typical microporous structures. Table 1 summarizes the BET surface areas (S_{BET}) and pore volumes of the fresh and aged catalysts. The table shows that the S_{BET} values are 523, 577, and 605 $m^2 \cdot g^{-1}$ for Cu-SSZ-16-Fresh, $CuCe_{0.77}$-SSZ-16-Fresh, and $CuCe_{0.87}$-SSZ-16-Fresh, respectively. Correspondingly, the pore volumes gradually increase from 0.247 $cm^3 \cdot g^{-1}$ to 0.251 $cm^3 \cdot g^{-1}$, then to 0.261 $cm^3 \cdot g^{-1}$. In general, both S_{BET} and pore volumes increase with an increase in the Ce mass fraction, and $CuCe_{0.87}$-SSZ-16-Fresh has the largest S_{BET} and pore volumes. However, the S_{BET} and pore volumes decline sharply after hydrothermal treatment at 800 °C, which may be due to the collapse of the zeolite skeleton. The S_{BET} and pore volumes of Cu-SSZ-16-800HT are only 13 $m^2 \cdot g^{-1}$ and 0.030 $cm^3 \cdot g^{-1}$, respectively, while they are 30 $m^2 \cdot g^{-1}$ and 0.035 $cm^3 \cdot g^{-1}$ for $CuCe_{0.77}$-SSZ-16-800HT and 37 $m^2 \cdot g^{-1}$ and 0.060 $cm^3 \cdot g^{-1}$ for $CuCe_{0.87}$-SSZ-16-800HT. The skeleton of Cu-SSZ-16-800HT collapses more severely, which is consistent with the PXRD results. Pore structures are retained in the aged catalysts, which may help the catalysts maintain catalytic activity.

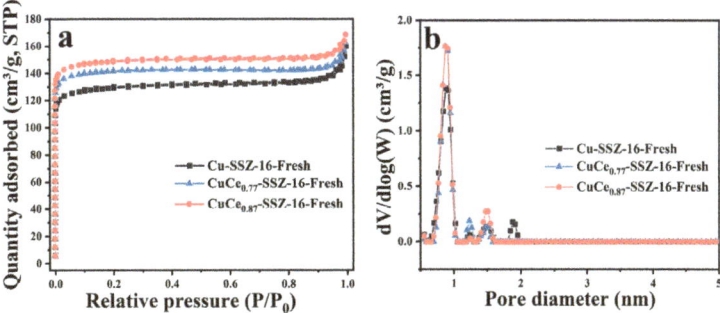

Figure 3. (**a**) N_2 adsorption–desorption isotherms and (**b**) Pore-size distribution spectra of Cu-SSZ-16-Fresh, $CuCe_{0.77}$-SSZ-16-Fresh, and $CuCe_{0.87}$-SSZ-16-Fresh.

Figure 4 displays the SEM results of the fresh catalysts at different magnifications. All the samples show similar morphologies of a double-cone prism with a similar average length of 1–2 µm. It can be concluded that the incorporation of Ce does not affect the structure or morphology of the catalysts. However, the morphology changed after hydrothermal treatment at 800 °C due to the damage to the zeolite framework (Figure S2). As shown in Figure 5, the TEM in bright and dark fields and corresponding element mapping images illustrate that both the Cu and Ce atoms are well dispersed in the catalyst samples, in agreement with the PXRD results mentioned above. Moreover, CuO_x or CeO_x clusters are not detected, leading to improved NH_3-SCR performance.

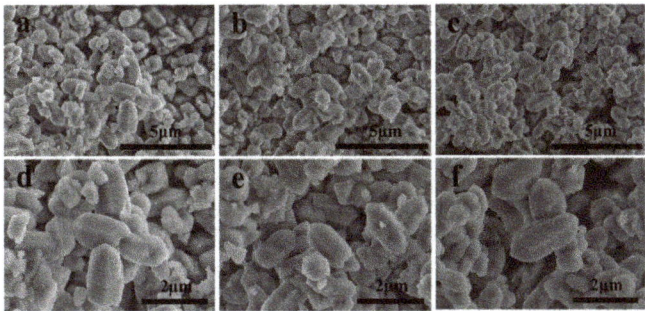

Figure 4. SEM images of Cu-SSZ-16-Fresh (**a,d**), $CuCe_{0.77}$-SSZ-16-Fresh (**b,e**), and $CuCe_{0.87}$-SSZ-16-Fresh (**c,f**).

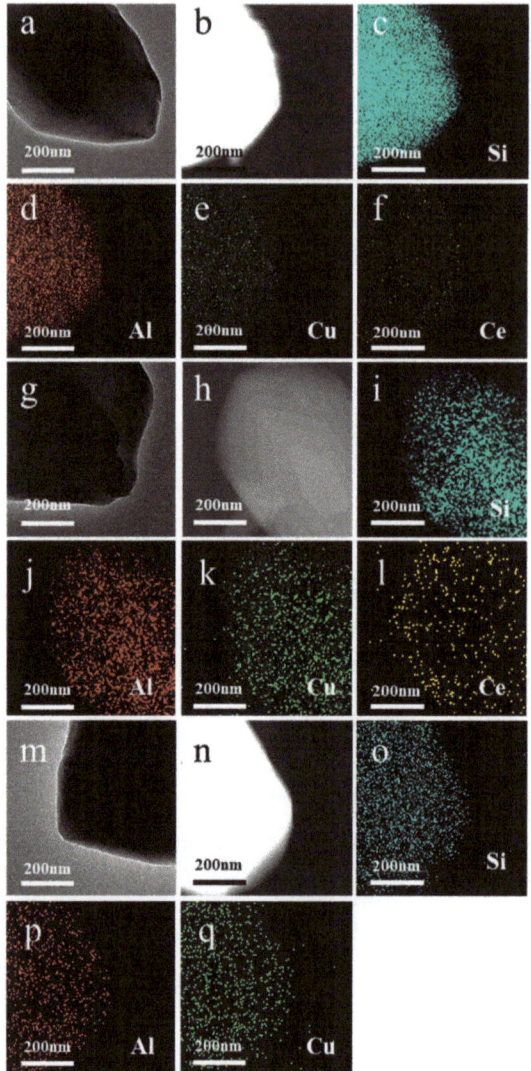

Figure 5. TEM in bright and dark fields and EDS mapping images of CuCe$_{0.87}$-SSZ-16-Fresh (**a–f**), CuCe$_{0.77}$-SSZ-16-Fresh (**g–l**), and Cu-SSZ-16-Fresh (**m–q**).

The ^{27}Al MAS NMR spectra of SSZ-16-Fresh, Cu-SSZ-16-Fresh, CuCe$_{0.77}$-SSZ-16-Fresh, and CuCe$_{0.87}$-SSZ-16-Fresh are depicted in Figure 6a. Four peaks are determined at around 57, 51, 30, and −1 ppm in the spectra of Cu-SSZ-16-Fresh, CuCe$_{0.77}$-SSZ-16-Fresh, and CuCe$_{0.87}$-SSZ-16-Fresh, respectively. For SSZ-16-Fresh, there are only three peaks, leaving out the peak at 30 ppm. The 57 ppm and 51 ppm signals are associated with two kinds of framework aluminum in zeolite; the former is attributed to tetrahedrally coordinated aluminum, and the latter corresponds to distorted aluminum [22,34]. The peak signals centered at 30 ppm and −1 ppm are characteristic of penta-coordinated and octahedral aluminum, respectively [35]. The peaks are integrally calculated and represented in Figure S3 and Table 2. The percentage of octahedral aluminum in SSZ-16-Fresh is 0.9%, which increases to 11.8% for Cu-SSZ-16-Fresh. It may be due to the distortion of the zeolite skeleton caused by Cu ion exchange, leading to the dealumination of the catalyst [34]. After Ce ion exchange, the relative content of octahedral aluminum decreases from 11.8% to 9.7% and 6.8%. The amount of tetrahedrally coordinated aluminum is markedly enhanced from 19.9% to 25.1% and 27.2%, indicating that the incorporation of Ce increases the framework Al over CuCe$_x$-SSZ-16-Fresh. The proportions of framework Al increase with an increase in Ce. After hydrothermal treatment at 800 °C, the peaks at 52 ppm and 0 ppm occupy a dominant position for Cu-SSZ-16-800HT (Figure 6b). Compared with the Cu-SSZ-16-Fresh, the non-framework Al accounts for a larger proportion of aluminum in the Cu-SSZ-16-800HT. Furthermore, the framework Al in Cu-SSZ-16-800HT is mainly composed of distorted aluminum. However, the peak at 57 ppm remains in the ^{27}Al MAS NMR spectra of CuCe$_x$-SSZ-16-800HT, demonstrating that more tetrahedrally coordinated aluminum exists in the CuCe$_x$-SSZ-16-800HT series catalysts [5]. Compared with CuCe$_x$-SSZ-16-Fresh, although the non-framework Al increases in CuCe$_x$-SSZ-16-800HT, the framework Al still accounts for the majority of aluminum in CuCe$_x$-SSZ-16-800HT. It may be one of the reasons why the CuCe$_x$-SSZ-16-800HT series catalysts can still maintain high NH$_3$-SCR catalytic activity. It is concluded that introducing Ce into the Cu-SSZ-16 catalysts reduces the dealumination reaction and improves the crystallinity, leading to outstanding hydrothermal stability.

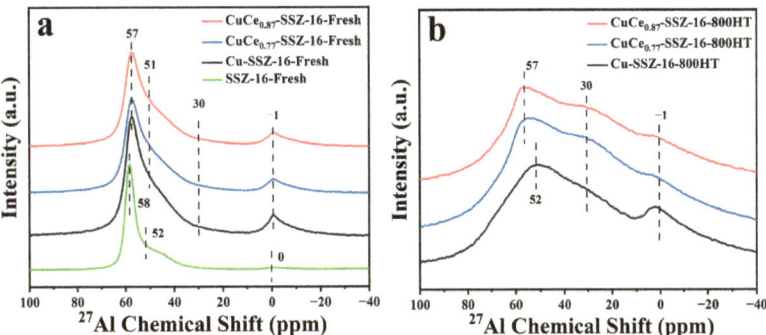

Figure 6. ^{27}Al MAS NMR spectra of (a) SSZ-16-Fresh, Cu-SSZ-16-Fresh, CuCe$_{0.77}$-SSZ-16-Fresh and CuCe$_{0.87}$-SSZ-16-Fresh, and (b) Cu-SSZ-16-800HT, CuCe$_{0.77}$-SSZ-16-800HT, and CuCe$_{0.87}$-SSZ-16-800HT.

Table 2. Quantitative analysis of the ^{27}Al NMR results of SSZ-16-Fresh, Cu-SSZ-16-Fresh, CuCe$_{0.77}$-SSZ-16-Fresh, and CuCe$_{0.87}$-SSZ-16-Fresh.

Catalysts	^{27}Al NMR Peak/ppm Relative Concentration (%)			
	57	51	30	−1
SSZ-16-Fresh	45.8	53.3	-	0.9
Cu-SSZ-16-Fresh	19.9	61.5	6.8	11.8
CuCe$_{0.77}$-SSZ-16-Fresh	25.1	58.1	7.1	9.7
CuCe$_{0.87}$-SSZ-16-Fresh	27.2	56.9	9.1	6.8

Furthermore, to investigate the change in the surface acidity, NH$_3$-TPD measurements were carried out on Cu-SSZ-16-Fresh, CuCe$_{0.77}$-SSZ-16-Fresh, and CuCe$_{0.87}$-SSZ-16-Fresh (Figure 7a). All the catalysts exhibit three desorption peaks at around 197, 318, and 505 °C (referred to as S1, S2, and S3, respectively). The signal at 197 °C is related to the weak acid sites, including physically adsorbed NH$_3$, NH$_3$ adsorbed on weak Brønsted acid sites, and NH$_3$ adsorbed by the surface hydroxyl groups [22,27,36]. The 318 °C peak is associated with moderate Lewis acid sites produced by ion exchange [30]. The peak at 505 °C is attributed to the NH$_3$ adsorbed on strong Brønsted acid sites [27]. It can be found that all the samples display similar locations and amounts of acid sites. The deconvolution areas of the fresh catalysts are shown in Figure 7b. With the incorporation of Ce, the amount of the weak acid sites (S1) decreases slightly, which may be due to the introduced Ce occupying some

Brønsted acid sites [37]. The number of moderate and strong acid sites (S2 and S3) increases with the addition of Ce, which is beneficial for NH_3 storage and NH_3-SCR performance.

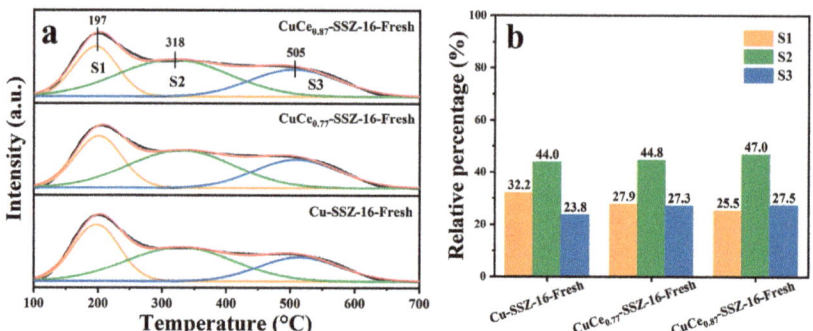

Figure 7. (**a**) Deconvolution of the NH_3-TPD curves and (**b**) the corresponding areas over Cu-SSZ-16-Fresh, $CuCe_{0.77}$-SSZ-16-Fresh, and $CuCe_{0.87}$-SSZ-16-Fresh.

2.3. Cu and Ce Species

XPS was performed to inquire about the chemical state of the two species introduced, and the XPS results of the fresh samples are exhibited in Figure 8. The Cu 2p spectrum (Figure 8a) is divided into two peaks at around 933.1 and 935.9 eV, which may correspond to the presence of Cu^+ and Cu^{2+} species [28,33,38,39]. The integral area fraction results of Cu-SSZ-16-Fresh, $CuCe_{0.77}$-SSZ-16-Fresh, and $CuCe_{0.87}$-SSZ-16-Fresh are depicted in Table 3. The relative amount of Cu^{2+} significantly increases from 32.8% to 40.4% and then to 62.6%, while the proportion of Cu^+ decreases as Ce increases, revealing that the addition of Ce facilitates the formation of surface Cu^{2+}. Moreover, the Ce 3d spectra of $CuCe_{0.77}$-SSZ-16-Fresh and $CuCe_{0.87}$-SSZ-16-Fresh are obtained (Figure 8b). According to the literature, the spectra can be recognized as having eight peaks, and the ones marked u′ and v′ are related to Ce^{3+} species; the others, which are labeled u, v, u″, v″, u‴ and v‴, are ascribed to Ce^{4+} species [40–43]. As generalized in Table 3, the $Ce^{3+}/(Ce^{3+} + Ce^{4+})$ ratios increase from 44.4% to 54.4% over the surface with an increase in the Ce contents, which is due to the formation of Cu^{2+} in the redox cycles $Cu^+ + Ce^{4+} \rightarrow Cu^{2+} + Ce^{3+}$ [37,44]. Consequently, the addition of Ce leads to the electron transfer of Cu^+, forming more Cu^{2+} and contributing to the high NH_3-SCR activity. The XPS results of the catalysts aged at 800 °C are displayed in Figure S4, and the deconvolution areas are listed in Table 3. After hydrothermal treatment at 800 °C, the percentage of Cu^{2+} decreases from 32.8% to 20.7% in Cu-SSZ-16 because hydrothermal aging transfers some of the Cu^{2+} to CuO_x [45]. The proportion of Cu^{2+} in $CuCe_{0.77}$-SSZ-16-800HT and $CuCe_{0.87}$-SSZ-16-800HT is 33.5% and 35.8%, respectively. Although the proportion decreases compared with $CuCe_{0.77}$-SSZ-16-Fresh and $CuCe_{0.87}$-SSZ-16-Fresh, it is still higher than that of Cu-SSZ-16-Fresh (32.8%). Many active Cu^{2+} species are retained in the $CuCe_x$-SSZ-16-800HT samples, which is an important reason for the low deactivation of $CuCe_x$-SSZ-16-800HT series catalysts. The ratio of $Ce^{3+}/(Ce^{3+} + Ce^{4+})$ declines, meaning that the redox ability of the aged catalysts is reduced by the hydrothermal treatment [46]. Even though the relative amounts of Cu^{2+} and Ce^{3+} decrease for $CuCe_x$-SSZ-16-800HT, many remain in the aged catalysts, inhibiting a sharp decline in catalytic activity. It is suggested that the introduction of Ce could increase the active Cu^{2+} contents and improve the resistance to hydrothermal treatments [28].

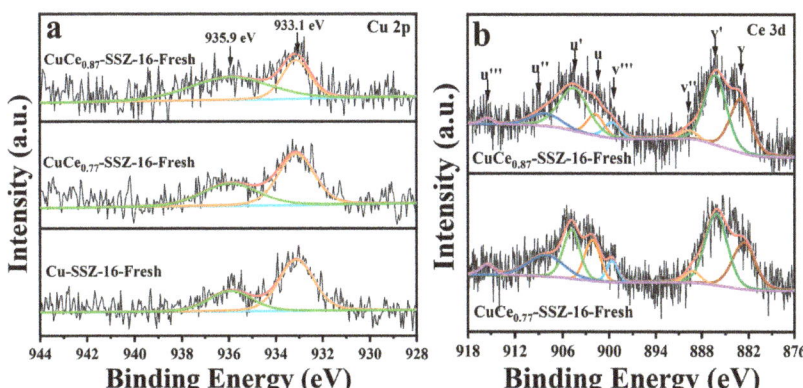

Figure 8. XPS spectra of Cu 2p (**a**) and Ce 3d (**b**) for Cu-SSZ-16-Fresh, CuCe$_{0.77}$-SSZ-16-Fresh, and CuCe$_{0.87}$-SSZ-16-Fresh.

Table 3. The distribution of Cu and Ce species revealed by XPS.

Catalysts	Cu^{2+} (%)	Cu^{+} (%)	Ce^{3+}/(Ce^{3+} + Ce^{4+}) (%)
Cu-SSZ-16-Fresh	32.8	67.2	-
CuCe$_{0.77}$-SSZ-16-Fresh	40.4	59.6	44.4
CuCe$_{0.87}$-SSZ-16-Fresh	62.6	37.4	54.4
Cu-SSZ-16-800HT	20.7	79.3	-
CuCe$_{0.77}$-SSZ-16-800HT	33.5	66.5	37.8
CuCe$_{0.87}$-SSZ-16-800HT	35.8	64.2	43.0

Figure S5 shows the UV-vis spectra of the fresh catalysts. All three catalysts display two peaks at 202 and 733 nm, related to the charge transfer from the framework oxygen to Cu^{2+} and the d-d transitions of Cu^{2+} in CuO$_x$ [12,47]. However, due to the good distribution and relatively low content of CuO$_x$, the peaks related to CuO$_x$ cannot be detected by PXRD. Additionally, a new peak appears at 297 nm for CuCe$_{0.77}$-SSZ-16-Fresh and CuCe$_{0.87}$-SSZ-16-Fresh, assigned to the charge transfer process of Ce^{3+} [48]. The intensity of the 202 nm peak for CuCe$_{0.77}$-SSZ-16-Fresh and CuCe$_{0.87}$-SSZ-16-Fresh is significantly higher than that of Cu-SSZ-16-Fresh, implying that more Cu^{2+} species exist in CuCe$_{0.77}$-SSZ-16-Fresh and CuCe$_{0.87}$-SSZ-16-Fresh. Hence, it can be inferred that adding Ce contributes to the rise in Cu^{2+} species in the catalysts, and the more pronounced effect is enhanced with an increase in the Ce contents. The peak located at 297 nm suggests the existence of Ce^{3+} in the CuCe$_x$-SSZ-16-Fresh series catalysts. Therefore, the incorporation of Ce might contribute to the formation of Cu^{2+}, which is also proven by the XPS results.

EPR was measured to evaluate the quantity and coordination environment of the Cu^{2+} species in zeolites because Cu^{+} and CuO$_x$ species could not be detected by EPR [49]. Figure S6 depicts the EPR results of Cu-SSZ-16-Fresh, CuCe$_{0.77}$-SSZ-16-Fresh, and CuCe$_{0.87}$-SSZ-16-Fresh. The three samples show similar peak features. The sharp peaks ($g_\perp$ = 2.08) can be observed in all the samples, which correspond to isolated Cu^{2+} coordinated with oxygen. After the introduction of Ce, the intensities of the $g_\perp$ = 2.08 peaks are significantly enhanced. The enhancement is greater with an increase in the Ce content, suggesting that the quantity of Cu^{2+} gradually increases with the addition of Ce. Furthermore, the hyperfine features of EPR are $g_\parallel$ = 2.37 for Cu-SSZ-16-Fresh and CuCe$_{0.77}$-SSZ-16-Fresh, demonstrating that the Cu^{2+} species are in the identical coordination environment in the two samples. For CuCe$_{0.87}$-SSZ-16-Fresh, the hyperfine feature has $g_\parallel$ = 2.33 due to the different coordination environments of Cu^{2+} after the incorporation of Ce.

Additionally, to explore the distribution and amount of Cu species in the catalysts, H$_2$-TPR was measured over the fresh catalysts (Figure 9a) and the samples aged at 800 °C (Figure 9b). The H$_2$-TPR spectrum of Cu-SSZ-16-Fresh is deconvolved into five reduction peaks at approximately 236 °C, 323 °C, 390 °C, 477 °C, and 519 °C, with each peak representing one kind of Cu species. Among these, the 236 °C peak corresponds to $[Cu(OH)]^+$-Z, while that at 390 °C is related to Cu^{2+}-2Z. Notably, Cu^{2+}-2Z species require a higher temperature to be reduced since they are situated in the 6-ring and are more stable. The peak at 323 °C is associated with CuO_x, which exerts negative effects on NH_3-SCR performance by blocking the pores of the zeolite catalysts. The signal at 477 °C is assigned to $Cu(AlO_2)_2$ in the catalysts, which is indirectly caused by the dealumination of the zeolite framework. Cu^+ species in the catalysts are reduced to Cu^0 at 519 °C [50–52]. Furthermore, the H_2-TPR curves of $CuCe_{0.77}$-SSZ-16-Fresh and $CuCe_{0.87}$-SSZ-16-Fresh show five deconvolution regions similar to those of Cu-SSZ-16-Fresh. Table 4 lists the integral calculation of the H_2-TPR profiles in the range of 100–400 °C. As presented here, the CuO_x species account for 26.2% of the Cu-SSZ-16-Fresh catalyst but 19.4% and 11.5% of the $CuCe_{0.77}$-SSZ-16-Fresh and $CuCe_{0.87}$-SSZ-16-Fresh catalysts, indicating that the modification of Ce combats the generation of CuO_x. The percentage of $[Cu(OH)]^+$-Z is 11.8%, while that of Cu^{2+}-2Z is 62.0% in Cu-SSZ-16-Fresh. After adding Ce, the proportions of $[Cu(OH)]^+$-Z and Cu^{2+}-2Z increase to 13.0% and 67.6% in the $CuCe_{0.77}$-SSZ-16-Fresh catalyst, respectively. Moreover, $[Cu(OH)]^+$-Z accounts for 14.1%, and Cu^{2+}-2Z accounts for 74.4% in the $CuCe_{0.87}$-SSZ-16-Fresh catalyst when the Ce contents increase further. The two Cu^{2+} species increase with an increase in Ce. Different from the fresh catalysts, the curves are only determined to have four peaks after hydrothermal treatment at 800 °C, namely, at 339 °C, 421 °C, 560 °C, and 698 °C, related to CuO_x, Cu^{2+}-2Z, $Cu(AlO_2)_2$, and Cu^+, respectively [52]. The peaks shift toward high temperatures, which suggests that these Cu species have become more stable during the hydrothermal treatment at 800 °C [52,53]. It is observed that the peak at about 230 °C disappears after hydrothermal aging because the $[Cu(OH)]^+$-Z is unstable and transforms to Cu^{2+}-2Z or CuO_x species at high temperatures [21]. This can also explain why the low-temperature catalytic activity of the aged samples significantly decreases. As presented in Table 4, compared with the fresh samples, the proportion of CuO_x increases and the proportion of Cu^{2+}-2Z declines in the samples aged at 800 °C. However, most Cu^{2+}-2Z species are retained in the aged catalysts, preventing a significant decrease in NH_3-SCR performance. The percentages of Cu^{2+}-2Z species in Cu-SSZ-16-800HT, $CuCe_{0.77}$-SSZ-16-800HT, and $CuCe_{0.87}$-SSZ-16-800HT are 51.4%, 54.0%, and 56.0%, respectively. Compared with Cu-SSZ-16-800HT, more active Cu^{2+}-2Z species are maintained in $CuCe_x$-SSZ-16-800HT, which helps the $CuCe_x$-SSZ-16-800HT catalysts maintain high $deNO_x$ catalytic activity. The proportion of CuO_x is 48.6%, compared to 46.0% and 44.0% for $CuCe_{0.77}$-SSZ-16-800HT and $CuCe_{0.87}$-SSZ-16-800HT, respectively, confirming that the introduction of Ce can effectively prevent the formation of CuO_x. Therefore, it can be deduced that the formation of Cu^{2+} is promoted while the generation of CuO_x is inhibited by introducing Ce, leading to higher hydrothermal stability.

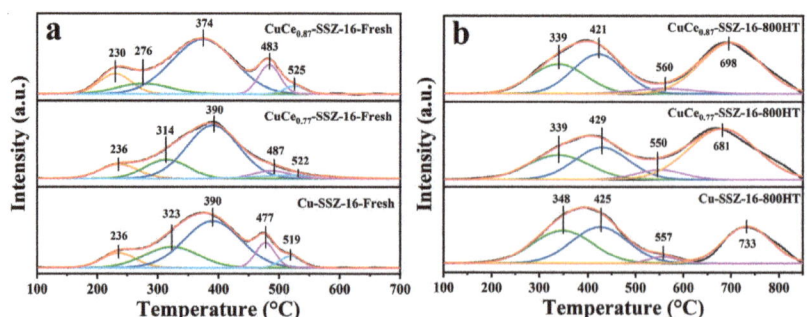

Figure 9. H_2-TPR before (**a**) and after hydrothermal treatment at 800 °C (**b**) of Cu-SSZ-16, $CuCe_{0.77}$-SSZ-16, and $CuCe_{0.87}$-SSZ-16.

Table 4. The distribution of Cu species measured by H_2-TPR.

Catalysts	$[Cu(OH)]^+$-Z (%)	Cu^{2+}-2Z (%)	CuO_x (%)
Cu-SSZ-16-Fresh	11.8	62.0	26.2
$CuCe_{0.77}$-SSZ-16-Fresh	13.0	67.6	19.4
$CuCe_{0.87}$-SSZ-16-Fresh	14.1	74.4	11.5
Cu-SSZ-16-800HT	-	51.4	48.6
$CuCe_{0.77}$-SSZ-16-800HT	-	54.0	46.0
$CuCe_{0.87}$-SSZ-16-800HT	-	56.0	44.0

In situ DRIFTS measurements under NH_3 adsorption are ideal for probing the relative contributions of the two Cu^{2+} species in the catalysts. The NH_3-DRIFTS results of Cu-SSZ-16-Fresh, $CuCe_{0.77}$-SSZ-16-Fresh, and $CuCe_{0.87}$-SSZ-16-Fresh are displayed in Figure 10. Two negative peaks appear in 860–1000 cm^{-1} wave numbers, one corresponding to $[Cu(OH)]^+$-Z at 949 cm^{-1} and the other related to Cu^{2+}-2Z at 895 cm^{-1} [13,54,55]. The two peak intensities increase effectively, demonstrating that the amounts of the two Cu^{2+} species increase through the introduction of Ce. The relative integral areas are shown in Figure S7, revealing that the relative content of the two Cu^{2+} species also changes with the addition of Ce. The percentage of Cu^{2+}-2Z in Cu-SSZ-16-Fresh is 48.7%, which increases to 51.5% and 54.3% for $CuCe_{0.77}$-SSZ-16-Fresh and $CuCe_{0.87}$-SSZ-16-Fresh, respectively. With an increase in the Ce contents, the relative proportion of Cu^{2+}-2Z increases gradually, and the percentage of $[Cu(OH)]^+$-Z decreases. It is indicated that Ce addition is conducive to forming Cu^{2+} species, especially Cu^{2+}-2Z species, which is beneficial to the hydrothermal stability of the catalysts.

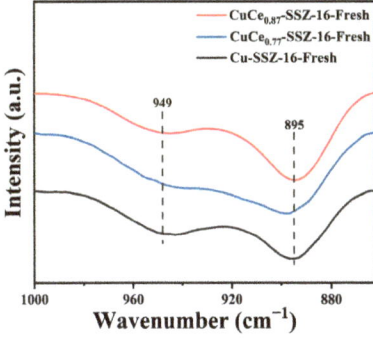

Figure 10. In situ DRIFTS spectra of Cu-SSZ-16-Fresh, $CuCe_{0.77}$-SSZ-16-Fresh, and $CuCe_{0.87}$-SSZ-16-Fresh.

From the analysis results above, we can conclude that the incorporation of Ce can stabilize the skeleton of Cu-SSZ-16 and also promote the formation of $[Cu(OH)]^+$-Z and Cu^{2+}-2Z, especially Cu^{2+}-2Z. The PXRD results show that the AFX structure is retained in the hydrothermally treated $CuCe_x$-SSZ-16, suggesting that adding Ce can improve the structural stability of the catalysts. The ^{27}Al MAS NMR results also confirm this deduction, as the dealumination is reduced and the framework Al increases in fresh and aged catalysts with Ce. Additionally, the framework Al is better maintained in $CuCe_x$-SSZ-16-800HT, contributing to the excellent NH_3-SCR performance of $CuCe_x$-SSZ-16-800HT. To further investigate the promotional effect of introducing Ce, XPS, EPR, UV-vis, H_2-TPR, and in situ DRIFTS analyses were conducted. The XPS results show that the percentage of Cu^{2+} species is only 32.8% for Cu-SSZ-16-Fresh but rises to 40.4% and 62.6% for $CuCe_{0.77}$-SSZ-16-Fresh and $CuCe_{0.87}$-SSZ-16-Fresh, respectively, after the introduction of Ce. Correspondingly, the Ce and Cu species constitute a redox cycle: $Cu^+ + Ce^{4+} \rightarrow Cu^{2+} + Ce^{3+}$, meaning that the incorporation of Ce is able to increase the Cu^{2+} species in catalysts. After hydrothermal treatment at 800 °C, although the amount of Cu^{2+} species decreases for $CuCe_{0.77}$-SSZ-16-

800HT and CuCe$_{0.87}$-SSZ-16-800HT, it is still higher than that of Cu-SSZ-16-Fresh. The majority of Cu^{2+} is still retained in the CuCe$_x$-SSZ-16-800HT series catalysts, accounting for high NH$_3$-SCR activity. The EPR and UV-vis measurements exhibit similar results to those of XPS, namely that the amounts of Cu^{2+} increase obviously with the addition of Ce. For the H$_2$-TPR analysis, the results suggest that the relative amount of CuO$_x$ reduces from 26.2% to 19.4% and then to 11.5% for the fresh samples. Meanwhile, the proportion of active Cu^{2+} rises after the incorporation of Ce. After hydrothermal aging at 800 °C, even though Cu^{2+}-2Z decreases in the aged catalysts, most active Cu^{2+}-2Z species are retained in the aged catalysts, which is an important reason for the low deactivation of CuCe$_x$-SSZ-16-800HT series catalysts. It is confirmed that the incorporation of Ce can promote the formation of active Cu^{2+} but limit the increase in CuO$_x$. Moreover, the in situ DRIFTS results reveal that adding Ce would significantly increase the relative content of Cu^{2+}-2Z, contributing to remarkable hydrothermal stability. In summary, the introduction of Ce can improve the stability of the Cu-exchanged zeolite skeleton and increase the active Cu^{2+} species in catalysts, thus improving hydrothermal stability. Furthermore, the zeolite skeleton structure and active Cu^{2+} species are better maintained in the samples hydrothermally aged at 800 °C, leading to low deactivation of the CuCe$_x$-SSZ-16-800HT series catalysts.

3. Materials and Methods

3.1. Synthesis and Hydrothermal Treatments

Based on a previous report, the SSZ-16 catalyst was synthesized with a composition of SiO$_2$: 0.045 Al$_2$O$_3$: 0.11 OSDA: 0.8 NaOH: 20 H$_2$O at 150 °C for 9 days [10]. The detailed procedures are available in the Electronic Supplementary Information (ESI). The Cu-SSZ-16 catalyst was obtained by successively exchanging the prepared SSZ-16 with a 0.1 M CH$_3$COONH$_4$ and a 0.01 M Cu(NO$_3$)$_2$ solution. Then, Cu-SSZ-16 was exchanged with a Ce(NO$_3$)$_3$ solution of different concentrations at 80 °C overnight to obtain the CuCe$_x$-SSZ-16 series catalysts. After drying at 100 °C, the catalyst products were calcined in air at 290 °C for 2 h and then at 550 °C for 6 h, thus producing the fresh catalysts. Then the fresh samples were hydrothermally aged at different temperatures in air containing 10 vol% H$_2$O for 10 h to obtain the aged samples.

3.2. Characterization of the Catalysts

The powder X-ray diffraction (PXRD) was applied to analyze the zeolite products with a Bruker D2 Phaser instrument at a scanning speed of 0.2° s^{-1}. The N$_2$ adsorption–desorption analyses were performed on a Micrometrics ASAP 2020 Plus apparatus at the temperature of liquid nitrogen. The scanning electron microscopy (SEM) images were captured with a Hitachi SU8010 microscope at 4 kV. The transmission electron microscopy (TEM) images were recorded on an FEI Tecnai G2 F30, which was operated at an accelerating voltage of 300 kV. The element distributions of the catalysts were detected by energy dispersive spectrometer (EDS) mapping. A PE Avio200 (America) inductively coupled plasma optical emission spectroscope (ICP-OES) was used to determine the elemental compositions. For ICP-OES analysis, a 20 mg sample was mixed with 2 mL concentrated nitric acid, 2 mL HF, and 0.5 mL H$_2$O$_2$. Furthermore, the mixture was treated at 80 °C in a graphite digestion apparatus under sealed conditions. After about 2 h, when the mixture became clear and transparent, it was diluted to the required concentration with H$_2$O. The ^{27}Al solid-state nuclear magnetic resonance (NMR) test was carried out on a Bruker AVANCE III HD 600 MHz spectrometer. The X-ray photoelectron spectra (XPS) were determined with a Thermo Fisher Scientific K-Alpha. The electron paramagnetic resonance (EPR) was analyzed with a JEOL JES-FA200 instrument at −196 °C. The ultraviolet–visible spectra (UV-vis) were determined with a Shimadzu UV 3600 spectrometer, and BaSO$_4$ was used as the reference sample.

The temperature-programmed desorption of NH$_3$ (NH$_3$-TPD) was measured by an MFTP-3060 chemisorption analyzer. First of all, a 100 mg catalyst was pretreated at 400 °C in N₂ atmosphere for 1 h. After cooling to 30 °C and holding for 10 min, 4000 ppm NH_3 was injected into the sample for 30 min. Subsequently, the purging process was conducted at 100 °C with He gas for 1 h to remove physically adsorbed NH_3. After these steps were complete, the TPD profiles of NH_3 were obtained under a He atmosphere from 100 to 700 °C with a temperature ramp rate of 10 °C min^{-1}.

Additionally, to explore the reducibility of zeolite products, temperature-programmed reduction of hydrogen (H_2-TPR) was performed on an MFTP-3060 apparatus. To start with, a 100 mg catalyst was treated using the same purification method as NH_3-TPD, as described above. Then 5% H_2/He was introduced at a 30 mL min^{-1} flow rate to establish a baseline. Eventually, the H_2-TPR profiles were obtained from 100 °C to 850 °C with a 10 °C min^{-1} temperature ramp rate.

The in situ diffuse reflection infrared Fourier spectroscopy (DRIFTS) adsorption analyses under NH_3 were carried out on a Thermo Scientific Nicolet iS20 spectrometer with an in situ diffuse reflection cell equipped with KBr windows. Firstly, the catalysts were purified by N_2 at 400 °C for 1 h with a 50 mL min^{-1} flow rate. Later, the catalysts were cooled to 100 °C to record the background spectra. To complete the NH_3 adsorption process, the catalysts were treated with NH_3 for 30 min. After that, the catalysts were purged for 1 h under N_2 conditions. The spectra were collected by accumulating 32 scans with a resolution of 4 cm^{-1}.

3.3. Catalytic Performance Tests

The NH_3-SCR catalytic experiments were performed in a fixed-bed reactor system with 100 mg (60–100 mesh) catalyst pellets, and the testing range was 150–400 °C. The catalysts were tested at a total flow rate of 400 mL min^{-1}, and the GHSV was 200,000 h^{-1}. Furthermore, the simulated test gases comprised 500 ppm NH_3, 500 ppm NO, 50 ppm SO_2 (when used), 5 vol% O_2, and 5 vol% H_2O, with the balance being N_2. The outlet gases were analyzed with a Thermo Scientific Nicolet Antaris IGS. The equations for calculating NO_x conversion, N_2, NO_2, and N_2O selectivity are as follows:

$$NO_x \text{ conversion} = \frac{[NO]_{in} - [NO]_{out} - [NO_2]_{out} - 2[N_2O]_{out}}{[NO]_{in}} \times 100\% \qquad (1)$$

$$N_2 \text{ selectivity} = \frac{[NH_3]_{in} + [NO]_{in} - [NH_3]_{out} - [NO]_{out} - [NO_2]_{out} - 2[N_2O]_{out}}{[NH_3]_{in} + [NO]_{in} - [NH_3]_{out} - [NO]_{out}} \times 100\% \qquad (2)$$

$$NO_2 \text{ selectivity} = \frac{[NO_2]_{out}}{[NH_3]_{in} + [NO]_{in} - [NH_3]_{out} - [NO]_{out}} \times 100\% \qquad (3)$$

$$N_2O \text{ selectivity} = \frac{2[N_2O]_{out}}{[NH_3]_{in} + [NO]_{in} - [NH_3]_{out} - [NO]_{out}} \times 100\% \qquad (4)$$

where $[NO]_{in}$ and $[NH_3]_{in}$ indicate the concentrations of NO and NH_3 in the inlet gases, respectively, while $[NH_3]_{out}$, $[NO]_{out}$, $[NO_2]_{out}$ and $[N_2O]_{out}$ represent the concentrations of NH_3, NO, NO_2 and N_2O in the outlet gases, respectively.

4. Conclusions

The CuCe$_x$-SSZ-16 series catalysts modified with different amounts of Ce were synthesized via the ion-exchange process and measured under simulated NH_3-SCR conditions. The results reveal that the CuCe$_{0.77}$-SSZ-16 and CuCe$_{0.87}$-SSZ-16 catalysts have better low-temperature (<250 °C) NH_3-SCR performance and outstanding hydrothermal stability compared with Cu-SSZ-16. Meanwhile, the positive effects increase with an increase in the Ce contents, and the CuCe$_{0.87}$-SSZ-16 catalyst shows the best hydrothermal stability. In general, the introduction of 0.87 wt.% Ce effectively promotes the stability of the AFX framework and facilitates the formation of Cu^{2+}-2Z species, which are beneficial to hydrothermal stability. Through a combination of multiple characterization techniques, the effects of adding Ce were investigated. The PXRD results show that the AFX structure is retained in the hydrothermally treated sample with Ce, suggesting that adding Ce can stabilize the skeleton of the catalysts. The ^{27}Al MAS NMR results indicate that adding Ce may have increased the framework aluminum in the catalysts, resulting in better structural stability in the fresh and aged samples. Furthermore, the analyses of Cu species by XPS, EPR, H$_2$-TPR, and in situ DRIFTS demonstrate that Ce ion exchange significantly increases the amount of Cu^{2+}-2Z species in the catalysts and reduces the formation of CuO$_x$, leading to good hydrothermal stability. Even after hydrothermal treatment at 800 °C, most of the framework aluminum and the majority of active Cu^{2+}-2Z species are retained in the aged catalysts with Ce additives, preventing a significant decrease in NH$_3$-SCR activity. In conclusion, CuCe$_{0.87}$-SSZ-16 shows remarkable NH$_3$-SCR performance and outstanding hydrothermal stability and has great application prospects for NO$_x$ removal.

Supplementary Materials: The following supporting information can be downloaded at https://www.mdpi.com/article/10.3390/catal13040742/s1. The synthesis of the catalysts. Figure S1: N$_2$, NO$_2$, and N$_2$O selectivity over Cu-SSZ-16, CuCe$_{0.77}$-SSZ-16, and CuCe$_{0.87}$-SSZ-16 before (a–c) and after hydrothermal treatment at 750 °C (d–f) and 800 °C (g–i). Figure S2: SEM images of Cu-SSZ-16-800HT (a,d), CuCe$_{0.77}$-SSZ-16-800HT (b,e), and CuCe$_{0.87}$-SSZ-16-800HT (c,f). Figure S3: Deconvolution of ^{27}Al NMR spectra over SSZ-16-Fresh, Cu-SSZ-16-Fresh, CuCe$_{0.77}$-SSZ-16-Fresh, and CuCe$_{0.87}$-SSZ-16-Fresh. Figure S4: XPS spectra of Cu 2p (a) and Ce 3d (b) over Cu-SSZ-16-800HT, CuCe$_{0.77}$-SSZ-16-800HT, and CuCe$_{0.87}$-SSZ-16-800HT. Figure S5: UV-vis spectra of Cu-SSZ-16-Fresh, CuCe$_{0.77}$-SSZ-16-Fresh, and CuCe$_{0.87}$-SSZ-16-Fresh. Figure S6: EPR spectra of Cu-SSZ-16-Fresh, CuCe$_{0.77}$-SSZ-16-Fresh, and CuCe$_{0.87}$-SSZ-16-Fresh. Figure S7: Deconvolution of in situ DRIFTS curves over Cu-SSZ-16-Fresh, CuCe$_{0.77}$-SSZ-16-Fresh, and CuCe$_{0.87}$-SSZ-16-Fresh.

Author Contributions: Conceptualization, Y.L. and J.J.; methodology, Z.L. and X.J.; formal analysis, Y.L., R.L. (Ruicong Liang) and Z.L.; investigation, R.L. (Rui Li) and R.L. (Ruicong Liang); data curation, Y.L. and R.L. (Rui Li); writing—original draft preparation, Y.L.; writing—review and editing, R.L. (Rui Li) and J.J.; supervision, J.J.; funding acquisition, J.J. All authors have read and agreed to the published version of the manuscript.

Funding: This research was funded by the National Natural Science Foundation of China (Grant No. 21971259).

Data Availability Statement: Data are available in the main text and the supplementary files.

Conflicts of Interest: The authors declare no conflict of interest.

References

1. Skalska, K.; Miller, J.; Ledakowicz, S. Trends in NO$_x$ abatement: A review. *Sci. Total Environ.* **2010**, *408*, 3976–3989. [CrossRef] [PubMed]
2. Lasek, J.A.; Lajnert, R. On the Issues of NO$_x$ as Greenhouse Gases: An Ongoing Discussion & hellip. *Appl. Sci.* **2022**, *12*, 10429. [CrossRef]
3. Zhang, X.Y.; Dou, T.T.; Wang, Y.; Yang, J.Y.; Wang, X.; Guo, Y.Y.; Shen, Q.; Zhang, X.; Zhang, S.Q. Green synthesis of Cu-SSZ-13 zeolite by seed-assisted route for effective reduction of nitric oxide. *J. Clean. Prod.* **2019**, *236*, 117667. [CrossRef]
4. Borfecchia, E.; Lomachenko, K.A.; Giordanino, F.; Falsig, H.; Beato, P.; Soldatov, A.V.; Bordiga, S.; Lamberti, C. Revisiting the nature of Cu sites in the activated Cu-SSZ-13 catalyst for SCR reaction. *Chem. Sci.* **2015**, *6*, 548–563. [CrossRef]
5. Fu, G.Y.; Yang, R.N.; Liang, Y.Q.; Yi, X.F.; Li, R.; Yan, N.N.; Zheng, A.M.; Yu, L.; Yang, X.B.; Jiang, J.X. Enhanced hydrothermal stability of Cu/SSZ-39 with increasing Cu contents, and the mechanism of selective catalytic reduction of NO$_x$. *Microporous Mesoporous Mater.* **2021**, *320*, 111060. [CrossRef]
6. Li, R.; Zhu, Y.J.; Zhang, Z.P.; Zhang, C.Q.; Fu, G.Y.; Yi, X.F.; Huang, Q.T.; Yang, F.; Liang, W.C.; Zheng, A.M.; et al. Remarkable performance of selective catalytic reduction of NO$_x$ by ammonia over copper-exchanged SSZ-52 catalysts. *Appl. Catal. B Environ.* **2021**, *283*, 119641. [CrossRef]
7. Zhao, Z.C.; Yu, R.; Zhao, R.R.; Shi, C.; Gies, H.; Xiao, F.S.; De Vos, D.; Yokoi, T.; Bao, X.H.; Kolb, U.; et al. Cu-exchanged Al-rich SSZ-13 zeolite from organotemplate-free synthesis as NH$_3$-SCR catalyst: Effects of Na$^+$ ions on the activity and hydrothermal stability. *Appl. Catal. B Environ.* **2017**, *217*, 421–428. [CrossRef]
8. Hernández-Salgado, G.I.; López-Curiel, J.C.; Fuentes, G.A. A Comparative Study of the NH$_3$-SCR Activity of Cu/SSZ-39 and Cu/SSZ-13 with Similar Cu/Al Ratios. *Top. Catal.* **2022**, *65*, 1495–1504. [CrossRef]

9. Chokkalingam, A.; Chaikittisilp, W.; Iyoki, K.; Keoh, S.H.; Yanaba, Y.; Yoshikawa, T.; Kusamoto, T.; Okubo, T.; Wakihara, T. Ultrafast synthesis of AFX-Type zeolite with enhanced activity in the selective catalytic reduction of NO_x and hydrothermal stability. *RSC Adv.* **2019**, *9*, 16790–16796. [CrossRef]
10. Li, R.; Jiang, X.Q.; Lin, J.C.; Zhang, Z.P.; Huang, Q.T.; Fu, G.Y.; Zhu, Y.J.; Jiang, J.X. Understanding the influence of hydrothermal treatment on NH_3-SCR of NO_x activity over Cu_x-SSZ-16. *Chem. Eng. J.* **2022**, *441*, 136021. [CrossRef]
11. Gao, F.; Washton, N.M.; Wang, Y.; Kollár, M.; Szanyi, J.; Peden, C.H.F. Effects of Si/Al ratio on Cu/SSZ-13 NH_3-SCR catalysts: Implications for the active Cu species and the roles of Brønsted acidity. *J. Catal.* **2015**, *331*, 25–38. [CrossRef]
12. Guo, A.Q.; Xie, K.P.; Lei, H.R.; Rizzotto, V.; Chen, L.M.; Fu, M.L.; Chen, P.R.; Peng, Y.; Ye, D.Q.; Simon, U. Inhibition Effect of Phosphorus Poisoning on the Dynamics and Redox of Cu Active Sites in a Cu-SSZ-13 NH_3-SCR Catalyst for NO_x Reduction. *Environ. Sci. Technol.* **2021**, *55*, 12619–12629. [CrossRef]
13. Wu, Q.; Fan, C.; Wang, Y.; Chen, X.P.; Wang, G.M.; Qin, Z.X.; Mintova, S.; Li, J.H.; Chen, J.J. Direct incorporating small amount of Ce (III) in Cu-SAPO-18 catalysts for enhanced low-temperature NH_3-SCR activity: Influence on Cu distribution and Si coordination. *Chem. Eng. J.* **2022**, *435*, 134890. [CrossRef]
14. Chen, Z.; Fan, C.; Pang, L.; Ming, S.J.; Liu, P.; Li, T. The influence of phosphorus on the catalytic properties, durability, sulfur resistance and kinetics of Cu-SSZ-13 for NO_x reduction by NH_3-SCR. *Appl. Catal. B Environ.* **2018**, *237*, 116–127. [CrossRef]
15. Liu, K.; Yan, Z.D.; Shan, W.P.; Shan, Y.L.; Shi, X.Y.; He, H. Quantitative determination of the Cu species, acid sites and NH_3-SCR mechanism on Cu-SSZ-13 and H-SSZ-13 at low temperatures. *Catal. Sci. Technol.* **2020**, *10*, 1135–1150. [CrossRef]
16. Usui, T.; Liu, Z.D.; Ibe, S.; Zhu, J.; Anand, C.; Igarashi, H.; Onaya, N.; Sasaki, Y.; Shiramata, Y.; Kusamoto, T.; et al. Improve the Hydrothermal Stability of Cu-SSZ-13 Zeolite Catalyst by Loading a Small Amount of Ce. *ACS Catal.* **2018**, *8*, 9165–9173. [CrossRef]
17. Luo, J.Y.; Gao, F.; Kamasamudram, K.; Currier, N.; Peden, C.; Yezerets, A. New insights into Cu/SSZ-13 SCR catalyst acidity. Part I: Nature of acidic sites probed by NH_3 titration. *J. Catal.* **2017**, *348*, 291–299. [CrossRef]
18. Martini, A.; Borfecchia, E.; Lomachenko, K.A.; Pankin, I.A.; Negri, C.; Berlier, G.; Beato, P.; Falsig, H.; Bordiga, S.; Lamberti, C. Composition-driven Cu-speciation and reducibility in Cu-CHA zeolite catalysts: A multivariate XAS/FTIR approach to complexity. *Chem. Sci.* **2017**, *8*, 6836–6851. [CrossRef] [PubMed]
19. Zhang, R.Q.; McEwen, J.S.; Kollár, M.; Gao, F.; Wang, Y.L.; Szanyi, J.; Peden, C. NO Chemisorption on Cu/SSZ-13: A Comparative Study from Infrared Spectroscopy and DFT Calculations. *ACS Catal.* **2014**, *4*, 4093–4105. [CrossRef]
20. Paolucci, C.; Khurana, I.; Parekh, A.A.; Li, S.C.; Shih, A.J.; Li, H.; Di, I.; John, R.; Albarracin-Caballero, J.D.; Yezerets, A.; et al. Dynamic multinuclear sites formed by mobilized copper ions in NO_x selective catalytic reduction. *Science* **2017**, *357*, 898–903. [CrossRef]
21. Song, J.; Wang, Y.L.; Walter, E.D.; Washton, N.M.; Mei, D.H.; Kovarik, L.; Engelhard, M.H.; Prodinger, S.; Wang, Y.L.; Peden, C.; et al. Toward Rational Design of Cu/SSZ-13 Selective Catalytic Reduction Catalysts: Implications from Atomic-Level Understanding of Hydrothermal Stability. *ACS Catal.* **2017**, *7*, 8214–8227. [CrossRef]
22. Gao, F.; Szanyi, J. On the hydrothermal stability of Cu/SSZ-13 SCR catalysts. *Appl. Catal. A Gen.* **2018**, *560*, 185–194. [CrossRef]
23. Chen, J.L.; Peng, G.; Liang, T.Y.; Zhang, W.B.; Zheng, W.; Zhao, H.R.; Guo, L.; Wu, X.Q. Catalytic Performances of Cu/MCM-22 Zeolites with Different Cu Loadings in NH_3-SCR. *Nanomaterials* **2020**, *10*, 2170. [CrossRef] [PubMed]
24. Kim, Y.J.; Lee, J.K.; Min, K.M.; Hong, S.B.; Nam, I.-S.; Cho, B.K. Hydrothermal stability of CuSSZ13 for reducing NO_x by NH_3. *J. Catal.* **2014**, *311*, 447–457. [CrossRef]
25. Fickel, D.W.; D'Addio, E.; Lauterbach, J.; Lobo, R.F. The ammonia selective catalytic reduction activity of copper-exchanged small-pore zeolites. *Appl. Catal. B Environ.* **2011**, *102*, 441–448. [CrossRef]
26. Xie, J.L.; Jin, Q.Q.; Fang, D.; Ye, Y.L.; Hou, S.S.; Wang, X.H.; He, F. Effect of La/Ce modification over Cu based Y zeolite catalysts on high temperature selectivity for selective catalytic reduction with ammonia. *J. Clean. Prod.* **2022**, *362*, 132255. [CrossRef]
27. Wang, Y.; Li, G.G.; Zhang, S.Q.; Zhang, X.Y.; Zhang, X.; Hao, Z.P. Promoting effect of Ce and Mn addition on Cu-SSZ-39 zeolites for NH_3-SCR reaction: Activity, hydrothermal stability, and mechanism study. *Chem. Eng. J.* **2020**, *393*, 124782. [CrossRef]
28. Mao, J.W.; Xu, B.; Hu, Y.K.; Zhang, C.Y.; Meng, H.M. Effect of Ce metal modification on the hydrothermal stability of Cu-SAPO-34 catalyst. *J. Fuel Chem. Technol.* **2020**, *48*, 1208–1216. [CrossRef]
29. Deng, D.; Deng, S.J.; He, D.D.; Wang, Z.H.; Chen, Z.P.; Ji, Y.; Yan, G.P.; Hou, G.J.; Liu, L.C.; He, H. A comparative study of hydrothermal aging effect on cerium and lanthanum doped Cu/SSZ-13 catalysts for NH_3-SCR. *J. Rare Earth.* **2021**, *39*, 969–978. [CrossRef]
30. Li, R.; Liang, Y.Q.; Zhang, Z.P.; Huang, Q.T.; Jiang, X.Q.; Yang, R.N.; Yu, L.; Jiang, J.X. Understanding roles of Ce on hydrothermal stability of Cu-SSZ-52 catalyst for selective catalytic reduction of NO_x with NH_3. *Catal. Today* **2022**, *405-406*, 125–134. [CrossRef]
31. Fickel, D.W.; Lobo, R.F. Copper Coordination in Cu-SSZ-13 and Cu-SSZ-16 Investigated by Variable-Temperature XRD. *J. Phys. Chem. C* **2010**, *114*, 1633–1640. [CrossRef]
32. Shi, Y.J.; Li, Z.M.; Wang, J.L.; Zhou, R.X. Synergistic effect of Pt/Ce and USY zeolite in Pt-based catalysts with high activity for VOCs degradation. *Appl. Catal. B Environ.* **2021**, *286*, 119936. [CrossRef]
33. Chen, B.H.; Xu, R.N.; Zhang, R.D.; Liu, N. Economical Way to Synthesize SSZ-13 with Abundant Ion-Exchanged Cu^+ for an Extraordinary Performance in Selective Catalytic Reduction (SCR) of NO_x by Ammonia. *Environ. Sci. Technol.* **2014**, *48*, 13909–13916. [CrossRef] [PubMed]

34. Prodinger, S.; Derewinski, M.A.; Wang, Y.L.; Washton, N.M.; Walter, E.D.; Szanyi, J.; Gao, F.; Wang, Y.L.; Peden, C. Sub-micron Cu/SSZ-13: Synthesis and application as selective catalytic reduction (SCR) catalysts. *Appl. Catal. B Environ.* **2017**, *201*, 461–469. [CrossRef]
35. Klinowski, J. Solid-state NMR studies of molecular sieve catalysts. *Chem. Rev.* **1991**, *91*, 1459–1479. [CrossRef]
36. Zhao, Y.Y.; Choi, B.C.; Kim, D. Effects of Ce and Nb additives on the de-NO_x performance of SCR/CDPF system based on Cu-beta zeolite for diesel vehicles. *Chem. Eng. Sci.* **2017**, *164*, 258–269. [CrossRef]
37. Liu, W.J.; Long, Y.F.; Liu, S.N.; Zhou, Y.Y.; Tong, X.; Yin, Y.J.; Li, X.Y.; Hu, K.; Hu, J.J. Promotional effect of Ce in NH_3-SCO and NH_3-SCR reactions over Cu-Ce/SCR catalysts. *J. Ind. Eng. Chem.* **2022**, *107*, 197–206. [CrossRef]
38. Zhao, S.; Huang, L.M.; Jiang, B.Q.; Cheng, M.; Zhang, J.W.; Hu, Y.J. Stability of Cu–Mn bimetal catalysts based on different zeolites for NO_x removal from diesel engine exhaust. *Chinese J. Catal.* **2018**, *39*, 800–809. [CrossRef]
39. Shan, Y.L.; Shi, X.Y.; Yan, Z.D.; Liu, J.J.; Yu, Y.B.; He, H. Deactivation of Cu-SSZ-13 in the presence of SO_2 during hydrothermal aging. *Catal. Today* **2019**, *320*, 84–90. [CrossRef]
40. Chen, L.; Li, J.H.; Ablikim, W.; Wang, J.; Chang, H.Z.; Ma, L.; Xu, J.Y.; Ge, M.F.; Arandiyan, H. CeO_2–WO_3 Mixed Oxides for the Selective Catalytic Reduction of NO_x by NH_3 Over a Wide Temperature Range. *Catal. Lett.* **2011**, *141*, 1859–1864. [CrossRef]
41. Zhang, Z.P.; Li, R.M.; Wang, M.J.; Li, Y.S.; Tong, Y.M.; Yang, P.P.; Zhu, Y.J. Two steps synthesis of $CeTiO_x$ oxides nanotube catalyst: Enhanced activity, resistance of SO_2 and H_2O for low temperature NH_3-SCR of NO_x. *Appl. Catal. B Environ.* **2021**, *282*, 119542. [CrossRef]
42. Li, H.R.; Yi, X.F.; Miao, J.F.; Chen, Y.T.; Chen, J.S.; Wang, J.X. Improved Sulfur Resistance of COMMERCIAl V_2O_5-WO_3/TiO_2 SCR Catalyst Modified by Ce and Cu. *Catalysts* **2021**, *11*, 906. [CrossRef]
43. Chen, L.; Ren, S.; Jiang, Y.H.; Liu, L.; Wang, M.M.; Yang, J.; Chen, Z.C.; Liu, W.Z.; Liu, Q.C. Effect of Mn and Ce oxides on low-temperature NH_3-SCR performance over blast furnace slag-derived X supported catalysts. *Fuel* **2022**, *320*, 123969. [CrossRef]
44. Bie, X.; Jiao, K.; Gong, C.; Qu, B.; Liu, D.; Ma, S. The Role of Medium Acid Sites Tuned by Ce Adding in Moderate-Temperature NH_3-SCR. *Catal. Lett.* **2022**, *152*, 2270–2279. [CrossRef]
45. Vennestrøm, P.; Katerinopoulou, A.; Tiruvalam, R.R.; Kustov, A.; Moses, P.G.; Concepcion, P.; Corma, A. Migration of Cu Ions in SAPO-34 and Its Impact on Selective Catalytic Reduction of NO_x with NH_3. *ACS Catal.* **2013**, *3*, 2158–2161. [CrossRef]
46. Guan, B.; Jiang, H.; Peng, X.S.; Wei, Y.F.; Liu, Z.Q.; Chen, T.; Lin, H.; Huang, Z. Promotional effect and mechanism of the modification of Ce on the enhanced NH_3-SCR efficiency and the low temperature hydrothermal stability over Cu/SAPO-34 catalysts. *Appl. Catal. A Gen.* **2021**, *617*, 118110. [CrossRef]
47. Xiang, X.; Cao, Y.; Sun, L.; Wu, P.F.; Cao, L.; Xu, S.T.; Tian, P.; Liu, Z.M. Improving the low-temperature hydrothermal stability of Cu-SAPO-34 by the addition of Ag for ammonia selective catalytic reduction of NO_x. *Appl. Catal. A Gen.* **2018**, *551*, 79–87. [CrossRef]
48. Chen, Z.Q.; Liu, L.; Qu, H.X.; Zhou, B.J.; Xie, H.F.; Zhong, Q. Migration of cations and shell functionalization for Cu-Ce-La/SSZ-13@ZSM-5: The contribution to activity and hydrothermal stability in the selective catalytic reduction reaction. *J. Catal.* **2020**, *392*, 217–230. [CrossRef]
49. Nanba, T.; Masukawa, S.; Ogata, A.; Uchisawa, J.; Obuchi, A. Active sites of Cu-ZSM-5 for the decomposition of acrylonitrile. *Appl. Catal. B Environ.* **2005**, *61*, 288–296. [CrossRef]
50. Ma, Y.Y.; Li, Z.F.; Zhao, N.; Teng, Y.L. One-pot synthesis of Cu-Ce co-doped SAPO-5/34 hybrid crystal structure catalysts for NH_3-SCR reaction with SO_2 resistance. *J. Rare Earth.* **2021**, *39*, 1217–1223. [CrossRef]
51. Zhang, T.; Liu, J.; Wang, D.X.; Zhao, Z.; Wei, Y.C.; Cheng, K.; Jiang, G.Y.; Duan, A.J. Selective catalytic reduction of NO with NH_3 over HZSM-5-supported Fe–Cu nanocomposite catalysts: The Fe–Cu bimetallic effect. *Appl. Catal. B Environ.* **2014**, *148–149*, 520–531. [CrossRef]
52. Su, W.K.; Li, Z.G.; Peng, Y.; Li, J.H. Correlation of the changes in the framework and active Cu sites for typical Cu/CHA zeolites (SSZ-13 and SAPO-34) during hydrothermal aging. *PCCP* **2015**, *17*, 29142–29149. [CrossRef] [PubMed]
53. Ma, L.; Cheng, Y.S.; Cavataio, G.; McCabe, R.W.; Fu, L.X.; Li, J.H. Characterization of commercial Cu-SSZ-13 and Cu-SAPO-34 catalysts with hydrothermal treatment for NH_3-SCR of NO_x in diesel exhaust. *Chem. Eng. J.* **2013**, *225*, 323–330. [CrossRef]
54. Luo, J.Y.; Wang, D.; Kumar, A.; Li, J.H.; Kamasamudram, K.; Currier, N.; Yezerets, A. Identification of two types of Cu sites in Cu/SSZ-13 and their unique responses to hydrothermal aging and sulfur poisoning. *Catal. Today* **2016**, *267*, 3–9. [CrossRef]
55. Wang, A.; Arora, P.; Bernin, D.; Kumar, A.; Kamasamudram, K.; Olsson, L. Investigation of the robust hydrothermal stability of Cu/LTA for NH_3-SCR reaction. *Appl. Catal. B Environ.* **2019**, *246*, 242–253. [CrossRef]

Disclaimer/Publisher's Note: The statements, opinions and data contained in all publications are solely those of the individual author(s) and contributor(s) and not of MDPI and/or the editor(s). MDPI and/or the editor(s) disclaim responsibility for any injury to people or property resulting from any ideas, methods, instructions or products referred to in the content.

Article

CO_2-Assisted Dehydrogenation of Propane to Propene over Zn-BEA Zeolites: Impact of Acid–Base Characteristics on Catalytic Performance

Svitlana Orlyk [1], Pavlo Kyriienko [1], Andriy Kapran [1], Valeriy Chedryk [1], Dmytro Balakin [2], Jacek Gurgul [3], Malgorzata Zimowska [3], Yannick Millot [4] and Stanislaw Dzwigaj [4,*]

[1] L.V. Pisarzhevskii Institute of Physical Chemistry, National Academy of Sciences of Ukraine, 31 Prosp. Nauky, 03028 Kyiv, Ukraine
[2] Institute of Physics, National Academy of Sciences of Ukraine, 46 Prosp. Nauky, 03028 Kyiv, Ukraine
[3] Jerzy Haber Institute of Catalysis and Surface Chemistry, Polish Academy of Sciences, Niezapominajek 8, PL-30239 Kraków, Poland
[4] Laboratoire de Réactivité de Surface, Sorbonne Université-CNRS, UMR 7197, 4 Place Jussieu, F-75005 Paris, France
* Correspondence: stanislaw.dzwigaj@sorbonne-universite.fr

Abstract: Research results about the influence of BEA zeolite preliminary dealumination on the acid–base characteristics and catalytic performance of 1% Zn-BEA compositions in propane dehydrogenation with CO_2 are presented. The catalyst samples, prepared through a two-step post-synthesis procedure involving partial or complete dealumination of the BEA specimen followed by the introduction of Zn^{2+} cations into the T-positions of the zeolite framework, were characterized using XRD, XPS, MAS NMR, SEM/EDS, low-temperature N_2 ad/desorption, C_3H_8/C_3H_6 (CO_2, NH_3)-TPD, TPO-O_2, and FTIR-Py techniques. Full dealumination resulted in the development of a mesoporous structure and specific surface area (BET) with a twofold decrease in the total acidity and basicity of Zn-BEA, and the formation of Lewis acid sites and basic sites of predominantly medium strength, as well as the removal of Brønsted acid sites from the surface. In the presence of the ZnSiBEA catalyst, which had the lowest total acidity and basicity, the obtained selectivity of 86–94% and yield of 30–33% for propene (at 923 K) exceeded the values for ZnAlSiBEA and ZnAlBEA. The results of propane dehydrogenation with/without carbon dioxide showed the advantages of producing the target olefin in the presence of CO_2 using Zn-BEA catalysts.

Keywords: Zn-BEA zeolites; dealumination; acid–base characteristics; propane dehydrogenation with CO_2; propene

1. Introduction

The world production of propene—the raw material for the synthesis of polypropylene and many important organic compounds (propene oxide, acrylic acid, propylene glycol, etc.)—exceeds 100 million tons per year. Conventional propene production through the steam cracking or catalytic cracking of petroleum does not meet the growing market needs. The catalytic dehydrogenation of propane, especially direct (PDH) and oxidative dehydrogenation using O_2 or N_2O, and CO_2-mediated dehydrogenation (CO_2-PDH) are considered a promising alternative to the oil-based cracking process [1,2]. The participation of CO_2 in the dehydrogenation of alkanes is of interest as a potential approach to utilizing carbon dioxide [3–5]. An important task for the realization of these processes is the development of active and selective catalysts that are not rapidly deactivated (especially in direct dehydrogenation) by coking.

Besides metal oxide catalysts for PDH or CO_2-PDH, zeolites containing cations or oxide nanoparticles of active components (mainly Cr, Ga, and Pt-Sn as components of known metal oxide catalysts) are of great interest for the dehydrogenation of alkanes [1,4,6–12].

It is known that the catalytic properties of zeolite catalysts are largely determined by their acid–base characteristics, which, in turn, depend on the Si/Al ratio, the nature of the active component, its form (nanoparticles, clusters, isolated cations), and quantity/density. In particular, partial or complete dealumination of the zeolite has been shown to increase the catalyst selectivity for the target product, especially for the dehydrogenation of propane in the presence of CO_2. This is due to changes in the acid–base properties of the system, in particular, a reduction in the concentration of Brønsted acid sites (BAS) until they are removed from the surface—as a result of complete dealumination—and changes in the form of the active component, in particular, the formation of isolated cations or the stabilization of oxidized subnanoclusters of the active component in/near vacant positions in the dealuminated zeolite [7,13–16].

A number of recent works have shown the potential of eco-benign Zn-containing catalysts for the dehydrogenation of alkanes, in which ZnO nanoparticles applied to the zeolite act as the main component [13–15,17–19] or zinc species as isolated cations and ZnO clusters are cocatalysts in bimetallic systems (Pt-Zn, Cr-Zn, Ni-Zn) based on high-silica zeolites [20–24]. Thus, studies of Zn-containing zeolites have used samples containing zinc oxide in amounts of ZnO $\geq$ 3–20 wt % [13–15,18]; in research on bimetallic systems with isolated Zn(II) forming Lewis acid sites (LAS), the main focus is on the effect of zinc as a transition metal cocatalyst. At the same time, the promoting effect of Zn^{2+} cations in Zn^{2+}/H-BEA and the synergistic effect of Zn sites and BAS on the activation of the C—H bonds of methane is stronger than that of ZnO species in ZnO/H-BEA [25]. Therefore, it can be expected that Zn-containing zeolites with isolated Zn (II) may also be of interest as catalysts for propane dehydrogenation.

Despite the progress in the study of Zn-containing zeolite catalysts for DH or CO_2-DH processes of lower alkanes, it is not yet known how the location of zinc as isolated atoms will affect their catalytic properties. The effect of the Si/Al ratio on the catalytic properties of Zn-containing zeolites with BEA structure, which, according to Zhao et al. [18], dominate over catalysts based on zeolites with other structural types in terms of activity and selectivity to propene, has also not been clarified.

The CO_2-PDH process on the catalysts, which do not undergo redox transformations under reaction conditions (including Ga-, Zn-containing), is considered to be mainly direct propane dehydrogenation (1), the equilibrium of which shifts in the direction of propene production due to the consumption of hydrogen in the reverse water–gas shift reaction (RWGSR) (2) [1–4]:

$$C_3H_8 \leftrightarrow C_3H_6 + H_2 \qquad \Delta H_{298K} = + 124 \text{ kJ mol}^{-1} \qquad (1)$$

$$CO_2 + H_2 \leftrightarrow CO + H_2O \qquad \Delta H_{298K} = + 41 \text{ kJ mol}^{-1} \qquad (2)$$

Carbon dioxide may also participate in coke gasification through the reverse Boudouard reaction $CO_2 + C \leftrightarrow 2CO$ ($\Delta H_{298K} = \Delta H_{298K} = + 172 \text{ k Jmol}^{-1}$) that enhances the catalyst stability.

In this paper, we report on the influence of the preliminary dealumination of BEA zeolite on the acid–base characteristics of synthesized Zn-BEA samples and their catalytic properties in the CO_2-mediated dehydrogenation of propane to propene. Zn-BEA specimens with different Si/Al ratios and a zinc loading of 1 wt % were prepared using a two-step post-synthesis procedure including preliminary partial and full dealumination of initial BEA zeolite followed by the incorporation of zinc cations into the vacant T-atom sites of the zeolite framework. The catalytic behavior of the Zn-BEA zeolites in propane dehydrogenation was tested both in the presence and absence of carbon dioxide in the initial reaction mixture.

2. Results and Discussion

2.1. Structure, Texture, and Acid–Base Characteristics of Zn-BEA Zeolites with Different Si/Al Ratios

The presence of diffraction peaks typical of BEA zeolites in the corresponding XRD patterns indicates that the dealumination of the initial TEABEA sample (Si/Al = 17) with nitric acid and subsequent incorporation of zinc cations into the SiBEA framework does not affect the crystallinity of the structure, as shown in Figure 1. The increase in the unit cell parameter d_{302} to 3.976 Å (ZnSiBEA; $2\theta = 22.36°$) compared to 3.920 Å (SiBEA; $2\theta = 22.68°$) is due to the expansion of the BEA zeolite matrix as a result of the interaction of zinc ions with OH groups of vacant T-atom sites and, as a consequence, their incorporation into the zeolite structure, resulting in an increase in the Zn-O bond length compared to Si-O or Al-O.

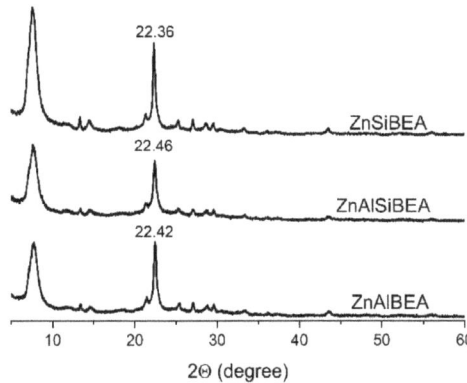

Figure 1. X-ray diffraction patterns of Zn-BEA samples after calcination at 923 K.

Each studied Zn-BEA sample is characterized by the presence of micropores with a total volume of ~0.2 cm^3/g and an average diameter of ~1 nm, as given in Table 1. The volume of mesopores of the samples is 0.32–0.37 cm^3/g. As can be seen from the data presented, the complete dealumination led to the formation of mesopores with the largest diameter/surface of ~60 nm/80 m^2/g and specific surface area (BET) of 605 m^2/g.

Table 1. Texture characteristics of zeolite Zn-BEA samples.

Sample	Micropores		Mesopores			S_{BET}, m^2/g	Adsorption Volume at $p/p_0 = 1$, cm^3/g
	Volume V_{mi}, cm^3/g	Diameter d_{mi} *, nm	Volume V_{me}, cm^3/g	Diameter d_{me}, nm	$(S_{me} + S_{outer})$, m^2/g		
ZnAlBEA	0.19	1.00	0.37	32 ± 5	70	535	0.58
ZnAlSiBEA	0.18	1.01	0.32	50 ± 15	60	505	0.52
ZnSiBEA	0.21	1.05	0.33	~60 *	80	605	0.56

* Determined through the Saito–Foley method.

In Figure S1, SEM images with a magnification of 25,000 (left) and 50,000 (right) of ZnSiBEA (Figure S1a,b,) ZnAlSiBEA, (Figure S1c,d) and ZnAlBEA (Figure S1e,f) are presented. They illustrate the morphology of the examined samples. SEM/EDS analysis reveals that they are composed of Si, O, Zn, and Al, and the amount of Al increases from 0.4 wt % in ZnSiBEA to 1.2 wt % in ZnAlSiBEA and 2.6 wt % in ZnAlBEA. The amount of Zn is close to 1.2 wt %; however, ZnSiBEA exhibits the highest quantity of zinc (1.4 wt %). It is worth noting that the ZnSiBEA sample exhibits the presence of the largest 160–100 nm particles, while ZnAlSiBEA led to crystallites not exceeding 120 nm. The morphology of the ZnAlBEA particles is very similar and in the range of 150–100 nm.

The relative abundance of elements on the surface of the Zn-BEA catalysts obtained from the XPS survey scans in the depth of max. 11.2 nm are presented in Table 2.

Table 2. Surface elemental composition of Zn-BEA zeolites (at. %).

Sample	Zn	Si	Al	O	C
ZnAlBEA	0.23	33.24	2.07	58.53	5.92
ZnAlSiBEA	0.25	34.53	0.59	59.68	4.95
ZnSiBEA	0.24	36.86	0.11	57.95	4.84

These were computed with the assumption that the samples are made of pure and uniform SiO_2 with a density equal to 2.18 g cm^{-3} [26]. The Si/Al ratios calculated for ZnAlBEA (16), ZnAlSiBEA (58), and ZnSiBEA (335) prove strong dealumination of BEA zeolite through the two-step post-synthesis procedure. High-resolution spectra of Zn 2p, Si 2p, Al 2p, O 1s, and C 1s were used to investigate the chemical states of the active phase in the catalysts.

The C 1s core lines (Figure 2A) of the Zn-BEA catalysts are composed of three characteristic peaks at 285.0 eV (organic contaminants), 285.9–286.3 eV (C–O groups), and 289.9–290.5 eV (O–C=O groups). ZnAlBEA has an additional fourth component with a BE of 288.1 eV (13%) related to C=O groups. The dealuminated catalyst (ZnSiBEA) shows significantly lower content of C–O groups (28%) compared to the others (42–44%), whereas the amount of O–C=O groups does not exceed 8% in any sample (Table S1 in Supplementary Materials). The hydrocarbon contamination was used as an internal calibration for XPS spectra, as mentioned below in Section 3.1.

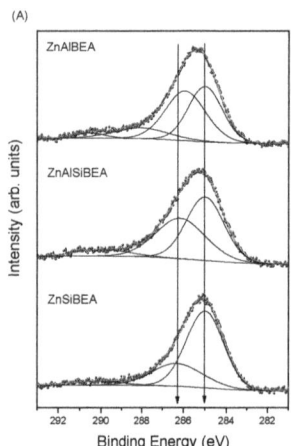 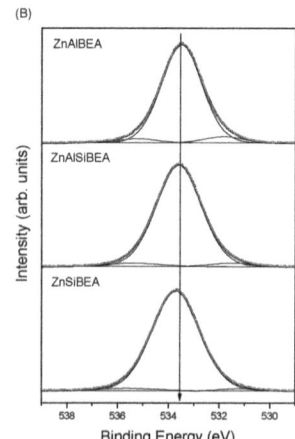 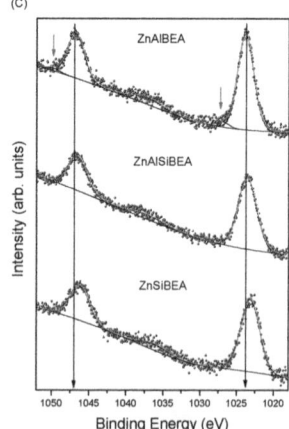

Figure 2. XPS spectra of Zn-BEA zeolites in the BE regions of C 1s (**A**), O 1s (**B**), and Zn 2p (**C**). Lines marking the most intense components. Blue arrows show an additional component in the Zn 2p spectrum of ZnAlBEA.

The O 1s spectra (Figure 2B) of the catalysts show three components: (i) a mainline (over 90% of total spectrum area) at BE of 533.5–533.7 eV related to oxygen located in the BEA zeolite lattice [27–30], (ii) oxygen from defective sites of the zeolite matrix

(BE = 531.2–531.8 eV) [31–33], and (iii) a peak at BE > 535 eV assigned to physisorbed water, and/or the oxygen of organic contaminants (Table S1 in Supplementary Materials). The relative decrease in aluminum in the catalysts causes an increase in lattice components at the cost of reducing the area of the low-BE component. This strongly suggests that as well as Si-O-Zn bonds, Al-O bonds also contribute to this component. However, a slightly lower BE range of the O 1s line (530.5–531.3 eV) was attributed to the thin films of Al_2O_3 [34,35]. The peak corresponding to the Zn–O bond (529.9 eV) [31,36] was not observed.

A single silicon component was detected with a Si $2p_{3/2}$ BE value close to 104.0 eV in ZnAlBEA, 104.2 eV in the partially dealuminated catalyst, and 104.3 eV in ZnSiBEA (Figure S2), and a single aluminum component was detected with an Al $2p_{3/2}$ BE value close to 75.6 eV in ZnAlBEA, 75.5 eV in the partially dealuminated catalyst, and 75.0 eV in ZnSiBEA (Table 3) (Figure S3). Such a contribution can be associated with the Al^{3+} in tetrahedral positions similar to the case of Faujasites [37]. The spin-orbit splitting of the Al 2p doublet was constrained to Δ_{SO} = 0.41 eV.

Table 3. XPS data of Zn-BEA zeolites.

Core Excitation	ZnAlBEA		ZnAlSiBEA		ZnSiBEA		
	BE (eV)	Area (%)	BE (eV)	Area (%)	BE (eV)	Area (%)	
Zn 2p3/2	1023.8	95.9	1023.5	100	1023.1	100	A
	1027.0	4.1					B
Si 2p3/2	104.0	100	104.2 2	100	104.3	100	
Al 2p3/2	75.6	100	75.55	100	75.0	100	

Figure 2C presents Zn 2p core-level spectra obtained for the ZnAlBEA, ZnAlSiBEA, and ZnSiBEA samples. They can be well fitted by a single symmetric doublet with fairly high Zn $2p_{3/2}$ BE values of 1023.1–1023.8 eV. It should be noted that the Zn $2p_{3/2}$ line for ZnO was quoted at 1022.0 eV [36], whereas metallic Zn was reported at a BE of 1021.6 eV [38]. It is clear that the BE shift observed between bulk ZnO and our Zn-BEA zeolites indicates different electronic states of Zn. Such an increase in BE could be a result of the incorporation of Zn into the zeolite lattice. In this case, Zn species are localized at the vacant T-atom sites, where the oxygen from the zeolite matrix exhibits higher electronegativity than the O^{2-} ligand in bulk ZnO. This results in a reduction in the valence electron density of Zn in the Zn–O–Si bond and an increased binding energy [31,39]. One can note that a similar effect was observed in several zeolites doped with Zn, e.g., MFI [31], ZSM-5 [25,40], FAU [37], BEA [41,42], and Y zeolites [43]. All these papers attribute the Zn $2p_{3/2}$ peak at about 1023.2 eV to the isolated $[Zn(OH)]^+$ species, which are formed from the tight interactions of zinc species with BEA zeolite. Such $[Zn(OH)]^+$ species can decompose to form water and $[Zn–O–Zn]^{2+}$ [25,42,44]. Therefore, the Zn 2p doublets can be reasonably assigned to the Zn(II) species located in the framework of the BEA zeolite with tetrahedral symmetry. This is also confirmed by the spin-orbit splitting of 23.0 eV characteristic of divalent Zn species.

Moreover, one can identify a very small component (4%) in ZnAlBEA with a BE of 1027.0 eV related to unknown Zn(II) species. However, it has been stated that wet chemistry-based techniques can lead to the incorporation of various Zn species into the zeolite, including isolated Zn^{2+} or $[Zn(OH)]^+$ cations localized at the exchange positions, as well as binuclear $[Zn–O–Zn]^{2+}$ or multinuclear $[Zn–(O–Zn)_n]^{2+}$ clusters [39,44,45]. Perhaps cluster formation is the reason for the appearance of this additional component.

All Si 2p spectra are well fitted by a single doublet with a spin-orbit splitting of 0.61 eV. The high binding energies of Si $2p_{3/2}$ shown in Table 3 prove that only Si(IV) species are present in our BEA catalysts. It is worth mentioning that these values are slightly larger than those reported for MFI and MOR zeolites elsewhere. The dealumination process does not cause the appearance of an additional component in the Si 2p spectra, but only a shift of the main peak towards higher binding energies. At the same time, the Zn 2p lines are shifted towards lower BE, which is additional evidence that zinc is built into the framework positions and interacts with the zeolite matrix.

Figure 3 shows the MAS NMR spectra of ZnAlBEA, ZnAlSiBEA, and ZnSiBEA. In all samples, we observe signals around -115 ppm, which correspond to silicon atoms in a $Si(OSi)_4$ environment (named Q^4) located in different crystallographic sites [46]. While the resonances are broad for the ZnAlBEA and ZnAlSiBEA samples, for ZnSiBEA, the resonances are narrow. This increase in resolution may be related to the dealumination and, thus, the departure of aluminum atoms, and the incorporation of zinc atoms could also have an effect on the resolution of the different contributions of the $Si(OSi)_4$ species. For ZnAlBEA, the DP MAS NMR spectrum shows a broad signal around -100 ppm composed of two contributions. The first one, at -103.5 ppm, corresponds to $Si(OSi)_3(OAl)$ species, and the second one, at -101.5 ppm, to $Si(OSi)_3(OH)$ species [47,48]. This last contribution is highlighted by the CPMAS experiments since it is strongly exalted in a non-quantitative way. We observe a decrease in this large signal for ZnAlSiBEA and an even larger decrease for ZnSiBEA. This decrease is due to both the departure of aluminum ions and the reaction between the zinc ions and the silanols of the vacant T-atom sites. In the CP spectra, a small fraction of Si atoms in a $Si(OH)_2(OSi)_2$ environment is also highlighted by the peak at 92.0 ppm.

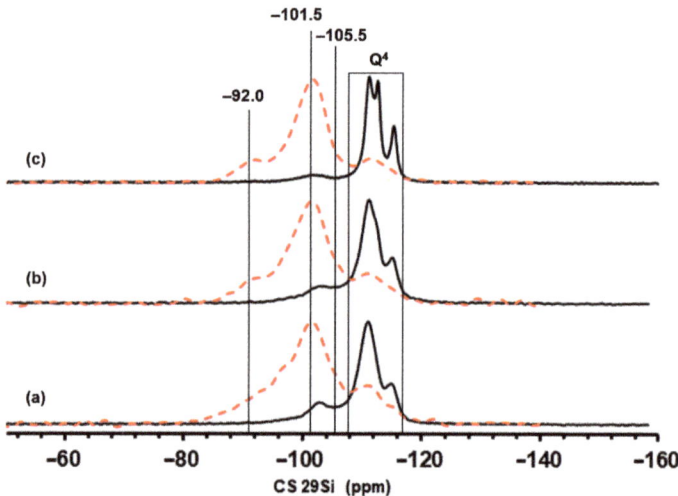

Figure 3. ^{29}Si MAS NMR spectra of ZnAlBEA (a), ZnAlSiBEA (b), and ZnSiBEA (c). Black curves are direct polarization (DP) spectra and red dotted curves are cross polarization (CP) spectra.

Figure 4 shows the data (profiles) for studying Zn-BEA zeolites through the $NH_3(CO_2)$-TPD technique, and corresponding acid–base characteristics are presented in Table 4.

Table 4. Acid–base characteristics of zeolite samples of Zn-BEA according to TPD-$NH_3(CO_2)$ profiles.

Sample	Concentration of Acidic Sites, rel. un. [1]				Concentration of Basic Sites, rel. un. [1]		
	Weak (293–423 K) [2]	Medium Strength (423–673 K) [2]	Strong (>673 K) [2]	Total	Weak (293–423 K) [2]	Medium Strength (423–673 K) [2]	Total
ZnAlBEA	0.26	0.36	0.38	1.00	0.78	0.10	0.88
ZnAlSiBEA	0.14	0.38	0.08	0.60	0.89	0.11	1.00
ZnSiBEA	0.04	0.31	0.08	0.43	0.22	0.26	0.48

[1] Rel. un. (Related unit)—ratio of the peak area over a certain temperature range to the peak area under the curve corresponding to the sample with maximum acidity/basicity; [2] desorption temperatures of NH_3 and CO_2.

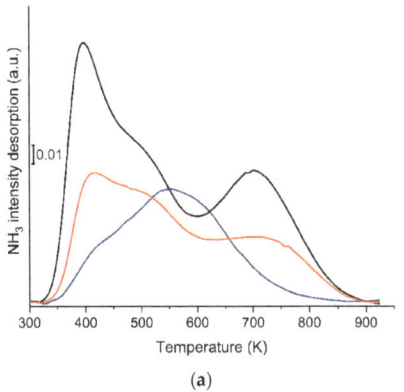
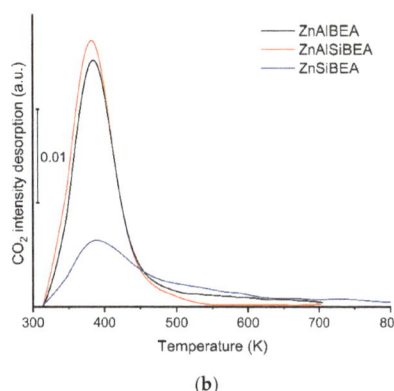

Figure 4. Normalized NH3-TPD (**a**) and CO_2-TPD (**b**) profiles for the ZnAlBEA, ZnAlSiBEA, and ZnSiBEA compositions.

The ZnAlBEA sample is characterized by the highest total acidity among the zeolites tested. The shape of the TPD-NH_3 profile for the ZnAlSiBEA sample is similar to ZnAlBEA, but its total acidity is lower (0.60), which is likely due to a reduction in the number of acid sites formed with the participation of Al(III), which is consistent with [7,49]. In the case of ZnSiBEA, the lowest total acidity is observed (0.43). In addition, the NH_3-TPD profile for ZnSiBEA has a significantly different shape compared to that of the samples containing Al. Considering the weak acidity of the SiBEA surface [7,50], the acid sites present on ZnSiBEA may be formed predominantly by Zn(II), incorporated into vacant T-atom sites of the dealuminated BEA zeolite.

Analyzing the CO_2-TPD data for the as-prepared Zn-BEA compositions, it should be noted that the corresponding profiles of carbon dioxide desorption for all studied samples have a single maximum in the temperature range of 383–388 K (Figure 4). The surface of both ZnAlBEA and ZnAlSiBEA exhibits predominantly weak basic sites, while the surface of the completely dealuminated ZnSiBEA specimen is characterized by a larger fraction of basic sites of medium strength (Table 4). A more intense shoulder is observed in the CO_2-TPD profile for the ZnSiBEA sample at temperatures above 423 K compared to the profiles for ZnAlBEA and ZnAlSiBEA. The function of the medium basic sites of medium strength for the ZnSiBEA specimen is likely performed by oxygen anions/vacancies of [Zn-O-Si] structures at the T-positions of the fully dealuminated zeolite or ZnO particles, dispersed on the SiBEA surface.

The FTIR spectra of pyridine adsorbed on zeolite samples are presented in Figure 5.

The spectra contain the absorption bands (a.b.) of the skeletal vibrations of the heteroaromatic ring (1446 (shoulder), 1453, 1454, 1456, 1490, 1495 (shoulder), 1578, 1600 (shoulder), 1612, 1614, 1616, and 1622 (shoulder) cm^{-1}) [14,51]. At the same time, a. b. 1446, 1578, 1600, and 1622 cm^{-1} disappear after evacuation at 523 and 623 K, which allows us to attribute them to weakly bound pyridine, likely through hydrogen bonding with OH groups on the surface of the zeolite samples. The a.b. at 1453–1456, 1490, 1495, and 1612–1616 cm^{-1} refer to pyridine coordinated to Lewis acid sites of the surface [14,15]. Taking into account previous results on the zeolites SiBEA, AlBEA [50,52], AlSiBEA [49], Zn/H-BEA [53], and ZnZr-SiBEA [54], the bands at 1456 cm^{-1} can be attributed to LAS formed with Al^{3+} cations, whereas the bands at 1454–1453 cm^{-1} can be attributed to LAS formed with Zn^{2+}. The FTIR-Py spectra of ZnAlBEA and ZnAlSiBEA samples show a. b. at 1547 and 1638 cm^{-1}, relating to the pyridinium ion (PyH^+) [14,49,51], which indicates the presence of BAS, due to the presence of bridging OH groups bound to aluminum cations at the T-positions of the zeolite framework. The decrease in the intensity of these bands after the heat treatment of these samples in a vacuum at 523 and 623 K is due to the desorption of pyridine associated with BAS of weak/medium strength. The higher intensity of a. b.

at 1453–1456, 1490, and 1616 cm^{-1} in the FTIR spectra of pyridine adsorbed at 423 K on ZnAlBEA and ZnAlSiBEA compared to the corresponding bands for ZnSiBEA is caused by the higher total concentration of acid centers on the surface of the samples based on the initial and partially dealuminated zeolite (Table 4).

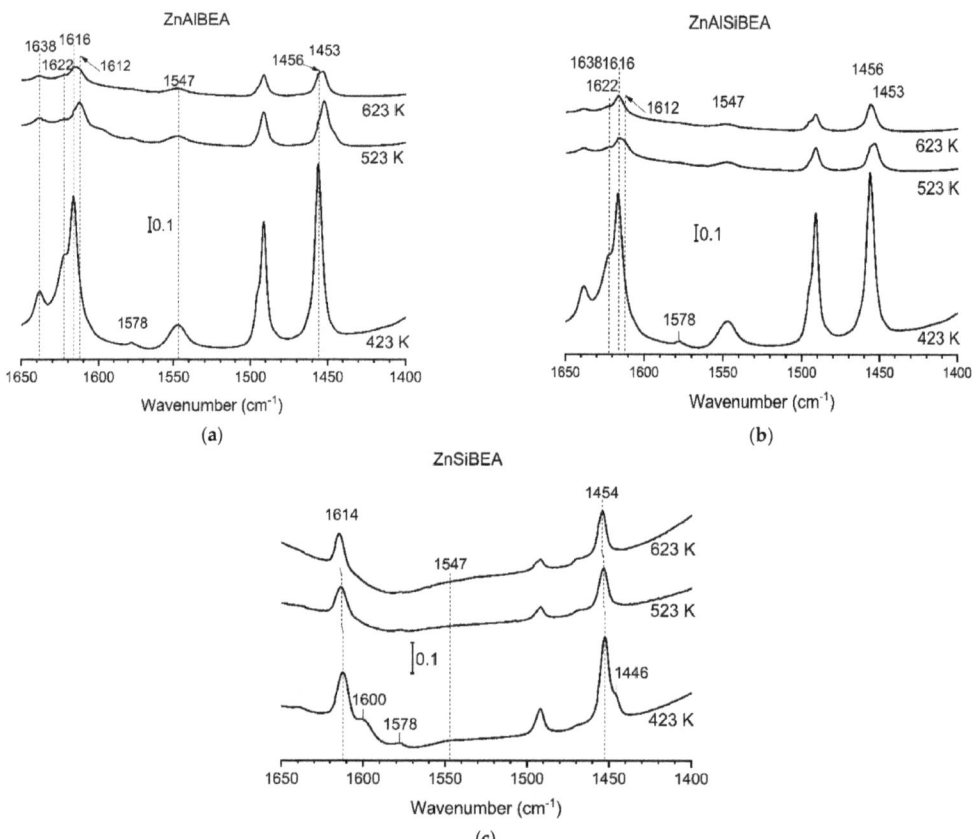

Figure 5. Normalized FTIR absorption spectra of pyridine adsorbed at 423 K on ZnAlBEA (a), ZnAlSiBEA (b), and ZnSiBEA (c) after heat treatment at 873 K and subsequent evacuation at different temperatures.

Thus, according to the analysis of FTIR-Py data, the preliminary complete dealumination of BEA zeolite followed by the incorporation of zinc atoms ensures the formation of the ZnSiBEA composition, on the surface of which there are LAS (Zn(II) (likely bound to silanol groups) and BAS are absent, whose function on the surface of ZnAlBEA and ZnAlSiBEA is realized by bridging OH groups bound to aluminum cations at the T-positions of the zeolite framework.

2.2. Catalytic Properties of Zn-BEA Zeolites in Propane Dehydrogenation

According to the results obtained, the initial propane conversions for the ZnAlBEA and ZnAlSiBEA catalyst samples exceed those in the presence of the fully dealuminated ZnSiBEA composition (Table 5 and Figure 6). At the same time, for ZnAlSiBEA and ZnAlBEA specimens at 873 and 923 K, propane conversion decreases quite rapidly with increasing TOS, whereas for ZnSiBEA, some decrease in X_{C3H8} with increasing TOS is observed only at 923 K. At temperatures of 873–923 K, the highest formation selectivities

(86–94%) and propene yields (16–18% at 873 K and 30–33% at 923 K) are obtained in the presence of the ZnSiBEA catalyst. The change in selectivity on propene in the ZnSiBEA, ZnAlSiBEA, and ZnAlBEA series occurs symbatically with a change in the Si/Al ratio (1000, 100, and 17, respectively). The highest propene yield is achieved in the presence of the ZnSiBEA catalyst.

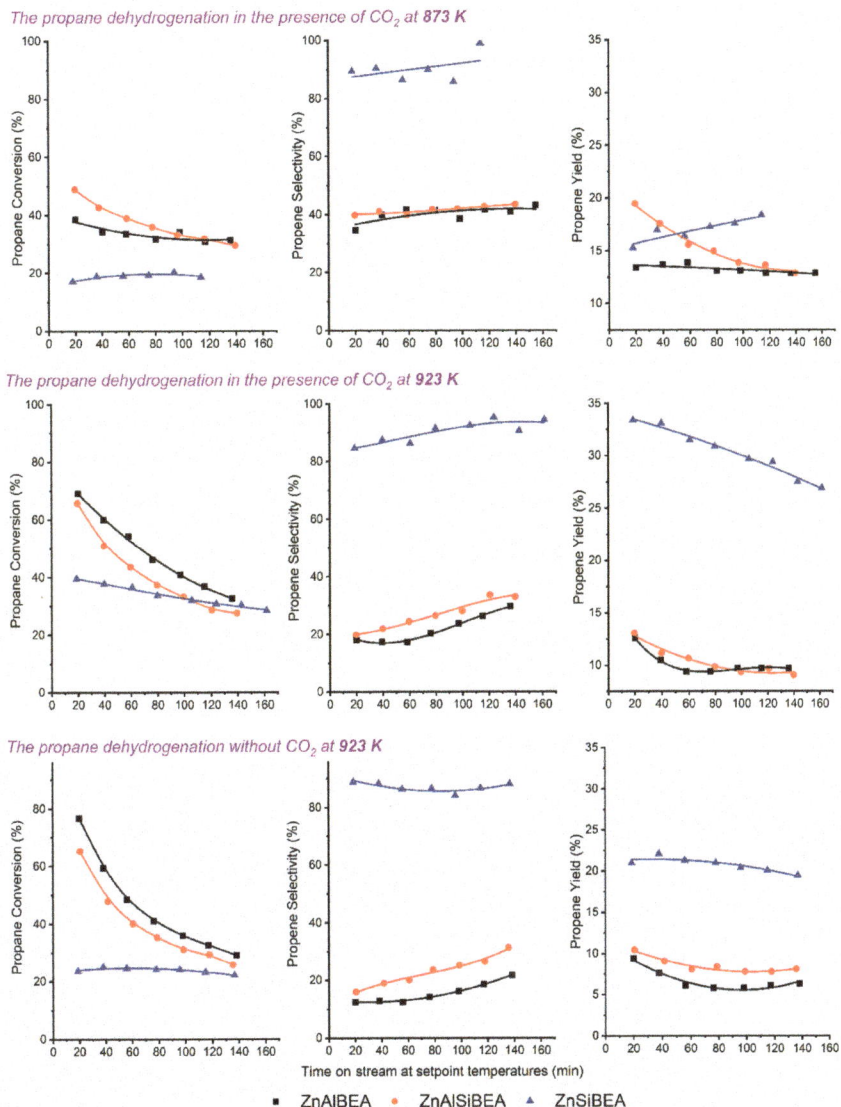

Figure 6. Indices of the propane dehydrogenation in the presence/absence of CO_2 on the Zn-BEA catalysts.

Table 5. Indices of the propane dehydrogenation in the presence of CO_2 on Zn-BEA catalysts for different TOS.

Catalyst	TOS, min	823 K			873 K			923 K		
		X_{C3H8}	S_{C3H6}	Y_{C3H6}	X_{C3H8}	S_{C3H6}	Y_{C3H6}	X_{C3H8}	S_{C3H6}	Y_{C3H6}
ZnAlBEA	30	38	25	9.5	36	37	13.3	64	17	10.9
	60	34	29	9.9	33	41	13.5	52	18	9.4
	120	–	–	–	31	42	13.0	35	27	9.5
ZnAlSiBEA	30	29	36	10.4	45	40	18.0	58	21	12.2
	60	23	42	9.7	39	41	16.0	44	24	10.6
	120	–	–	–	31	43	13.3	29	33	9.6
ZnSiBEA	30	8	57	4.6	18	90	16.2	38	86	32.7
	60	9	57	5.1	20	90	18.0	36	88	31.7
	120	–	–	–	19	94	17.9	32	94	30.1

The by-products of propane conversion are mainly methane and others of propane and propene cracking; coking of the catalyst surface was also observed. The analysis of the obtained data regarding the by-products of the CO_2-PDH process indicates that a greater amount of by-products (CH_4, C_2H_6, C_2H_4) is fixed on the Al-containing catalysts. According to the number of by-products formed, the catalysts are arranged in the following order: ZnAlBEA > ZnAlSiBEA > ZnSiBEA.

In the context of the above, it should be noted that Brønsted acid sites, as is well known, intensify the cracking and oligomerization of the olefins including propene, followed by carbonization and, accordingly, in this case, blocking of the active centers of the ZnAlBEA and ZnAlSiBEA samples. Thus, the better catalytic performance of the zeolite composition ZnSiBEA is caused by the absence of BAS on its surface. We note, however, that the absence of Brønsted acid sites does not exclude the possibility of side reactions on the ZnSiBEA catalyst, which are less intense anyway compared to the ZnAlBEA and ZnAlSiBEA samples.

To determine the effect of BEA dealumination on Zn-BEA coking in the CO_2-PDH process, O_2-TPO profiles of spent catalysts were obtained (120 min at 923 K, at which propane cracking is more intense). The results are shown in Figure 7. Assuming that the intensity of the O_2-TPO curves is proportional to the calcined coking products, the area under the O_2-TPO curve and the maximum temperature are comparable for the ZnAlSiBEA and ZnAlBEA samples. This is consistent with similar propane conversion/propene yield values at TOS = 120 min (Figure 7). For ZnSiBEA, the area under the O_2-TPO curve is much smaller compared to that for ZnAlSiBEA and ZnAlBEA, which is consistent with the greater stability of the ZnSiBEA catalyst compared to Al-containing samples.

The more intense coking of the ZnAlSiBEA and ZnAlBEA samples (as opposed to ZnSiBEA) may be due to the greater number of adsorption centers and their holding capacity. The correlation between the ability to retain propene (according to C_3H_6-TPD) and the coking of the catalyst was shown in [24].

To evaluate the number and strength of the propane and propene adsorption centers under reaction conditions, a TPD study of propane and propene (after their adsorption on the catalyst surface from the mixture of propane and propene) was performed.

The results obtained (Figure 8) indicate that the surface of ZnAlBEA has the highest number of centers capable of retaining propane and propene.

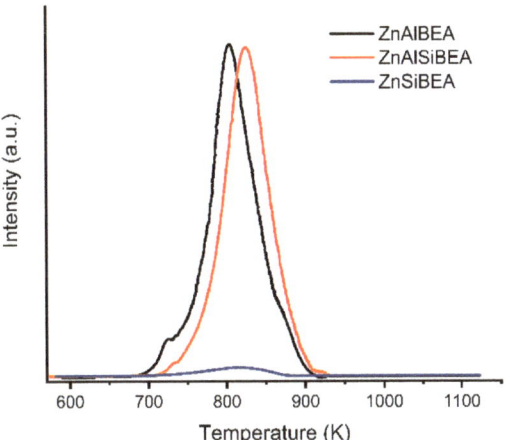

Figure 7. Normalized O$_2$-TPO profiles of Zn-BEA catalysts after 2.5 h in the CO$_2$-PDH reaction mixture at 923 K.

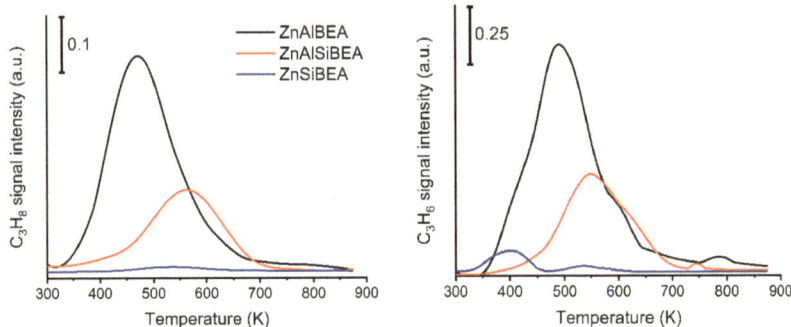

Figure 8. Normalized TPD profiles of propane and propene for the Zn-BEA catalysts.

On the ZnAlSiBEA surface, the number of such centers is smaller, although quite significant, especially the strong centers, as indicated by the temperature of the desorption maximum. In the case of ZnSiBEA, the number of such centers is the smallest (compared to ZnAlSiBEA and ZnAlBEA). Therefore, it can be assumed that the surface coking of Zn-BEA samples occurs primarily as a result of propane/propene cracking on Al-containing sites (especially BAS). Brønsted acid sites are capable of protonating (in the case of ZnAlBEA and ZnAlSiBEA) the produced olefin in the π-bond with the formation of carbenium cations C$_3$H$_7^+$, thus intensifying the course of the side reactions of cracking, oligomerization, and carbonization of the catalyst surface [45,55,56]. The observed result is consistent with the fact that high acidity can be detrimental to the selectivity of olefins due to the difficulty in desorption of the resulting intermediates [5]. The presence of only strong adsorption centers on the ZnAlSiBEA surface (Figure 8) may explain the shift of the maximum on the O$_2$-TPO profile compared to ZnAlBEA (Figure 7).

In general, the initial propane conversion on Zn-BEA catalysts decreases as the Si/Al ratio increases, while propene selectivity and catalyst stability improve. Similar trends of decreasing initial propane conversion and changing selectivity of the CO$_2$-PDH process were found on ZnO/HZSM-5 [14] and Ga$_2$O$_3$/HZSM-48 [57] catalysts, as well as in the PDH process on ZnO/HZSM-5 [13]. Given the significant amount of BAS on the surface of ZnAlSiBEA and ZnAlBEA, it can be assumed that the formation of propane cracking products is catalyzed primarily by BAS, which, according to Dzwigaj et al. [23], contribute to the side reactions of oligomerization, alkylation and cracking in the dehydrogenation of alkanes on zeolites.

As noted above, the ZnSiBEA sample is characterized by a more developed mesoporous structure and specific surface area (BET) in comparison with other compositions (Table 1), which contributes to the target CO_2-PDH process. However, most important factor determining the catalytic performance of a fully dealuminated specimen is its acid–base characteristics. The ZnSiBEA sample is characterized by the presence of LAS (Zn^{2+}) and basic sites (O^{2-} anions and oxygen vacancies) of mostly medium strength on the surface, as well as the absence of Brønsted acid sites, intensifying side reactions with subsequent carbonization of the catalyst surface. The higher concentration of medium-strength basic sites for ZnSiBEA may be a favorable precondition for the most likely route of alkane molecule activation at acid–base paired sites—through dissociation of the $C^{\delta-}$–$H^{\delta+}$ bond (as a rule, the limiting stage of the propane DH [1–4,58–61]) via deprotonation by nucleophilic O^{2-} anions (Brønsted base sites) and subsequent coordination of the formed carbanions $C_3H_7^-$ with Zn^{2+} cations (LAS), also predominantly of medium strength. In this regard, it should be emphasized that Brønsted base sites play an important role in propane transformations, as they are the ones that ensure the heterolysis of the $C^{\delta-}$–$H^{\delta+}$ bond and, thus, "release" the electron pair for the coordination of propyl anions with LAS. The basic properties of the most propene-selective ZnSiBEA sample also facilitate carbon dioxide activation, involving the formation of $CO_2^{\cdot-}$ radical anions [62,63] with dissociative adsorption on oxygen vacancies of the zeolite framework lattice [64]. In order to determine the effect of CO_2 on X_{C3H8}, S_{C3H6}, and Y_{C3H6} targets for all catalysts, a propane dehydrogenation reaction was performed in the absence of CO_2 in the reaction medium at 923 K (the temperature at which the highest rate of decrease in targets with increasing TOS is observed). The results shown in Figure 6 indicate a positive effect of CO_2 on propane conversion, formation selectivity, and propene yield for all Zn-BEA catalyst samples. In the case of ZnAlBEA at 923 K, the presence of CO_2 also contributes to a decrease in the catalyst deactivation; thus, the reduction of X_{C3H8} at TOS = 20 min → 140 min in the CO_2-PDH reaction is 70 → 32, and in the PDH reaction is 77 → 30.

Thus, the results obtained demonstrate the benefits of producing propene from propane in the presence of carbon dioxide. It is noteworthy that the direct dehydrogenation of propane to propene $C_3H_8 \leftrightarrow C_3H_6 + H_2$ is the reaction with volume increasing, whereas the reverse water–gas shift $CO_2 + H_2 \leftrightarrow CO + H_2O$ is a molecularity-invariant reaction; both are endothermic. As a result, a higher temperature and higher CO_2 concentration are thermodynamically favorable for a higher propane conversion to propene and greater assistance of CO_2 through the RWGSR. The equilibrium yield of olefin approaches 100% at 973 K and an initial molar ratio of CO_2/C_3H_8 = 5/1 according to thermodynamic calculations [65]. However, the selectivity with reference to propene is significantly reduced by sintering, agglomeration, and surface carbonization, thus causing catalyst deactivation. Therefore, the optimal temperature range for propene formation is 873–923 K.

A more detailed characterization of the effect of the Si/Al ratio in Zn-BEA zeolites on the propane conversion and average propene yield in CO_2–PDH and PDH reactions over Zn-BEA at TOS = 20 min → 140 min (T = 923 K) is shown in Figure S4. According to the results of C_3H_8 dehydrogenation and cracking studies on Zn/H-MFI catalysts [66], their higher activity and selectivity for C_3H_6 formation in the absence of cofeed H_2 (or H_2 removal by $CO_2 + H_2$ = CO + H_2O reaction) may be a consequence of the conversion of $[ZnH]^+$ cations into bridging Zn^{2+} cations.

Figure 9 and Figure S5 show the temperature dependence of propane conversion and the selectivity of propene formation and yield in the propane dehydrogenation process in the presence/absence of CO_2 in the reaction medium for the ZnSiBEA catalyst, which provides better propene yields with stable operation over time.

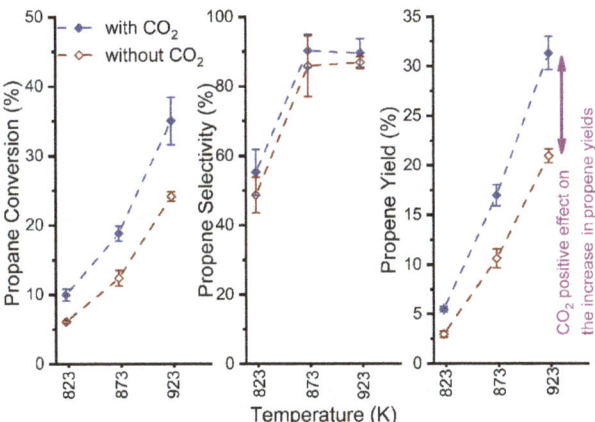

Figure 9. Mean values (for TOS = 20–120 min) of propane conversion, propene selectivity, and yield of the propane dehydrogenation in the presence/absence of CO_2 on the ZnSiBEA catalyst.

These results clearly indicate a positive effect of CO_2 on the production of the target product, propene. At the same time, the presence of CO_2 in the reaction medium does not counteract the coking of the catalyst, as evidenced by the slopes of the curves of changes in propane conversion and propene yield over time (dependence of X_{C3H8} and Y_{C3H6} on TOS).

When CO_2 is introduced into the reaction mixture, the apparent activation energy of propene formation, determined from Arrhenius plots (Figure S6), decreases from 124 ± 9 kJ·mol^{-1} to 110 ± 7 kJ·mol^{-1} (it is worth noting that the calculated E_a values are close to those obtained in [67], but for ZrO_2-based catalysts).

3. Materials and Methods

3.1. Zeolite Sample Preparation and Characterization

Zeolite samples were prepared based on templated tetraethylammonium BEA zeolite (TEABEA, Si/Al = 17) manufactured by Research Institute of Petroleum Processing (RIPP), Haidian District, Beijing, China.

To obtain the exemplary starting AlBEA zeolite, the TEABEA was calcined in air at 550 °C for 15 h. To obtain partially and fully dealuminated zeolite, the starting AlBEA zeolite was treated with HNO_3 solution (6 or 13 mol·L^{-1}) for 4 h at 353 K according to the method described in [49,68]. Partially and fully dealuminated zeolites with Si/Al = 100 (AlSiBEA) and 1000 (SiBEA) compositions were separated through centrifugation, washed with distilled water, and dried for 24 h at 353 K.

In order to introduce 1 wt % of Zn into the AlBEA, AlSiBEA, and SiBEA zeolites, these samples were treated with excess aqueous zinc(II) nitrate solutions at pH 3.0 to obtain the ZnAlBEA, ZnAlSiBEA, and ZnSiBEA series.

X-ray diffraction patterns of the prepared zeolite specimens were recorded on a Bruker AXS GmbH D8 Advance diffractometer (series II) (nickel filter, CuK$_\alpha$ radiation, λ = 0.154184 nm).

The texture characteristics (S_{BET}, pore volume and size, mesopore surface) of the studied samples were determined using N_2 ad/desorption at a low temperature (77 K) on a Sorptomatic 1990 porous materials analyzer with preliminary evacuation (573 K, 0.001 Pa/7.50 Torr). The pore size distribution was calculated using the Saito–Foley (micropores) and Barrett–Joyner–Halenda (mesopores) methods, and the volume of micropores and specific surface of mesopores were determined using the t-plot method.

The morphology of the samples was determined out using a JEOL JSM–7500F Field Emission Scanning Electron Microscope (JEOL, Akishima, Japan) equipped with a retractable backscattered-electron detector (RBEI) and energy-dispersive spectra (EDS) detection system of a characteristic X-ray radiation Ztec Live for EDS system (Oxford Instruments, Abingdon, London, UK).

The X-ray photoelectron spectroscopy (XPS) investigations were carried out in a multichamber ultra-high-vacuum system equipped with a hemispherical analyzer (SES R4000, Gammadata Scienta). A Mg K_α X-ray source (1253.6 eV) was used for photoelectron generation. The anode was operated at 180 W (12 kV, 15 mA). The spectrometer was calibrated according to ISO 15472:2001. The energy resolution of the system (pass energy 100 eV) determined for the Ag $3d_{5/2}$ excitation line was 0.9 eV. The base pressure in the analytical chamber was 1×10^{-10} mbar and approximately 6×10^{-9} mbar during the experiment. The powder samples were examined after being pressed in indium foil and mounted on a special holder. The analysis area of the samples was about 4 mm^2 (5 × 0.8 mm). High-resolution spectra were collected at a pass energy of 100 eV (with a 25 meV step), while survey scans were collected at a pass energy of 200 eV (with a 0.25 eV step). The experimental curves were fitted in CasaXPS 2.3.23 using a combination of Gaussian and Lorentzian lines with variable ratios (70:30) after subtracting a Shirley-type background. The relative ratio of the intensities of the $2p_{3/2}$ and $2p_{1/2}$ lines in the doublets was set to 2:1. All binding energies were charge-corrected to the carbon C 1s excitation, which was set to 285.0 eV.

Solid-state magic angle spinning nuclear magnetic resonance (MAS NMR) experiments were performed on a Bruker (Billerica, MA, USA) AVANCE500 spectrometer at 11.7 T in 4 mm zirconia rotors spinning at 14 kHz. ^{29}Si direct polarization (DP) MAS NMR and ^{1}H-^{29}Si cross polarization (CP) MAS NMR were performed with a 5 mm zirconia rotor with a 5 kHz spinning speed, 2 μs excitation pulse, and 10 s recycle delay. 3-(trimethylsilyl)-1-propanosulfonic sodium salt was used for setting the Hartmann–Hahn conditions. The proton π/2 pulse duration, the contact time, and recycle delay were 3 μs, 5 ms, and 5 s, respectively.

One-pass temperature-programmed desorption of propane/propene, carbon dioxide, and/or ammonia (C_3H_8/C_3H_6 (CO_2, NH_3)-TPD) was carried out in an ultra-high-vacuum (UHV) black chamber-type system, controlling the desorbed molecules using a time-of-flight MSX-3PC mass spectrometer (Electron, Iviv, Ukraine). Zeolite samples of 0.02 g each were preheated at 873 K for 2 h under a pressure of 10^{-9} Torr and then cooled to room temperature in a vacuum. The adsorption of propane and propene (gas mixture 10% C_3H_8 + 10% C_3H_6 in He), ammonia (99.99%), and carbon dioxide (99.99%) were carried out with the respective molecular probe gases for 12 h. The programmed temperature rise was carried out at a rate of 9 K·min^{-1}. The application of the TPD technique is described in detail in [69].

The acidity and basicity of the samples were evaluated based on signal intensity proportional to the amount of NH_3 and CO_2 adsorbed on the sample surface at a given temperature, normalized to the sample mass. The areas under the curves for the TPD profiles, corresponding to acidic or basic sites of a given strength on the surface of the samples, were calculated after the deconvolution of the spectrum into a minimum number of components (using a Gaussian distribution) so that the total curve of the deconvoluted spectrum coincided with the experimental curve.

The nature of acid sites on the surfaces of the samples was investigated through Fourier-transform infrared spectroscopy of adsorbed pyridine as a probing molecule (FTIR-Py) using a Spectrum One FTIR spectrometer (Perkin Elmer, Waltham, MA, USA). Samples in the form of thin wafers, pressed from fine powders with suitable catalytic compositions, were pre-heated at 693 K for 1 h under a 10^{-3} Torr vacuum in a quartz cuvette reactor. Adsorption of gaseous pyridine was carried out at 423 K and then evacuated at 423, 523, and 623 K for 0.5 h. FTIR spectra of adsorbed Py were recorded at room temperature (spectrometer beam temperature) with a resolution of 1 cm^{-1} and 24 scans.

The temperature-programmed oxidation of the catalysts with O_2 (O_2-TPO) was performed on an AMI-300Lite Catalyst Characterization Instrument (Altamira Instruments,

Pittsburgh, PA, USA). Prior to testing, samples (0.1 g after 2.5 h of work at 923 K in CO_2-PDH) were treated at 573 K for 20 min in He gas at a flow rate of 25 mL·min^{-1}. After cooling to 323 K, the samples were immediately reheated in 20 vol% O_2 in He from 323 K to 1073 K (a flow rate of 25 mL·min^{-1}) with a linear temperature ramp of 5 K·min^{-1}. The signal changes of the effluent gases were analyzed using a thermal conductivity detector.

3.2. Catalytic Activity Measurements

Catalytic experiments were carried out in a flow-type quartz reactor at atmospheric pressure, at a temperature of 823–923 K, with a gas hour space velocity (GHSV) of 6000 h^{-1} (catalyst loading 0.2 g, grain size of 0.25–0.5 mm, and reaction mixture (RM) flow rate of 30 cm^3·min^{-1}). The RM composition for CO_2-PDH was 2.5 vol. % of C_3H_8 and 15 vol. % CO_2 in He; for PDH, it was −2.5 vol. % of C_3H_8 in He. The weight hour space velocity (WHSV) was 0.4 g$_{C3H8}$·g$_{cat}^{-1}$·h^{-1}. The reagents and reaction products (C_3H_8, CO_2, C_3H_6, CH_4, C_2H_4, C_2H_6) were analyzed using gas chromatography (Krystallux 4000M, MetaChrom, Yoshkar-Ola, Russian Federation) equipped with a thermal conductivity detector and a column packed with Porapak Q. The gas sample was preliminarily dried by passing it through a calcium chloride trap.

Before evaluation, the catalysts were pretreated in He flow at the required temperatures for 30 min. It should be noted that the study of catalytic properties was performed with a gradual temperature rise in the range of 823–923 K with a step of 50 K, and with interstage regeneration of the catalyst sample in situ before the RM was introduced into the reactor. Regeneration of the sample was carried out through calcination in air at 873 K for 2 h to remove coke particles.

The catalytic properties of samples in the CO_2-PDH and PDH processes were characterized on the basis of propane conversion (X_{C3H8}), selectivity (S_{C3H6}), and yield (Y_{C3H6}) with respect to propene. The indices of the catalytic process were calculated using the following formulas:

$$X_{C3H8} = (C_{C3H8\ inlet} - C_{C3H8\ outlet})/C_{C3H8\ inlet} \cdot 100\%,$$

$$S_{C3H6} = C_{C3H6}/(C_{C3H8\ inlet} - C_{C3H8\ outlet})\ 100\%,$$

$$Y_{C3H6} = X_{C3H8} \cdot S_{C3H6}/100\%,$$

where $C_{C3H8\ inlet(outlet)}$ is the mole concentration of propane at the inlet (outlet) of the reactor and C_{C3H6} is the mole concentration of produced propene.

The propene formation rate was calculated per unit mass of catalyst (mol$_{C3H6}$·kg$_{cat}^{-1}$·s^{-1}) as follows:

$$r_{C3H6} = F_{C3H8} \cdot (Y_{C3H6}/100\%)/m_{cat},$$

where F_{C3H8} is the molar flow rate of propane (mol/s) and m_{cat} is the mass of catalyst (kg).

Based on these calculations, the Arrhenius plots for propene formation in the PDH and CO_2-PDH processes were drawn, and the associated activation energies were determined from the slopes of the corresponding plots.

4. Conclusions

The effect of the preliminary dealumination of BEA zeolite on the acid–base characteristics and catalytic performance of the 1%Zn-BEA compositions in the dehydrogenation of propane in the presence/absence of CO_2 was determined.

The post-synthesis procedure of preparing Zn-BEA catalyst samples including partial and full preliminary dealumination of the TEABEA initial specimen (Si/Al = 17) followed by introducing Zn^{2+} cations into vacant T-atom sites of the zeolite framework leads to a reduction in the total acidity of ZnAlSiBEA (0.60, rel. un.) and ZnSiBEA (0.43) compared to ZnAlBEA (1.0).

Full dealumination also results in the development of mesoporous structure and specific surface area (BET) while halving the total basicity of Zn-BEA, creating acid sites (Lewis) and basic sites of predominantly medium strength, and removing Brønsted acid sites from the surface.

In the presence of the ZnSiBEA sample, which has the lowest total acidity and basicity, the achieved selectivity of 86–94% and yield of 30–33% related to propene in the CO_2-PDH process (at 923 K) exceed those for the ZnAlBEA and ZnAlSiBEA compositions.

The positive effect of full dealumination on selectivity and lower deactivation with increasing TOS are mainly attributed to the lack of BAS on the surface of ZnSiBEA, which are capable of protonating (in the case of ZnAlBEA and ZnAlSiBEA) the produced olefin in the π-bond and, thus, intensifying the course of the side reactions of oligomerization and cracking with subsequent carbonization of the catalyst surface.

A comparison of the achieved selectivity and yields for propene in the dehydrogenation of propane with/without carbon dioxide demonstrates the advantages of target olefin production in the presence of CO_2 using Zn-BEA zeolite catalysts.

Supplementary Materials: The following supporting information can be downloaded at: https://www.mdpi.com/article/10.3390/catal13040681/s1, Table S1: XPS data obtained from C 1s and O 1s regions of Zn-BEA zeolites. SEM images with magnification of 25000 (left) and 50000 (right) of (a,b) ZnSiBEA, (c,d) ZnAlSiBEA, and (e,f) ZnAlBEA; Figure S2: XPS spectra of Zn-BEA zeolites in the BE region of Si2p; Figure S3: XPS spectra of Zn-BEA zeolites in the BE region of Al 2p; Figure S4: Impact of Si/Al in Zn-BEA zeolites on the propane conversion and the average propene yield in CO_2-PDH and PDH reactions on Zn-BEA (TOS = 20 min → 140 min, T = 923 K). Figure S5: Propane conversion and propene yield versus TOS in the propane dehydrogenation with (full symbols) and without (empty symbols) CO_2 on the ZnSiBEA catalyst. Figure S6: Arrhenius plots of propene formation in PDH and CO_2-PDH processes on the ZnSiBEA catalyst.

Author Contributions: Conceptualization: S.O. and S.D.; Methodology: S.D., P.K., M.Z. and A.K.; Software: V.C.; Validation: V.C.; Formal Analysis: S.O., S.D. and A.K.; Investigation: P.K., V.C., D.B., J.G., Y.M. and S.D.; Resources: S.O. and S.D.; Data Curation: S.O. and S.D.; Writing—Original Draft Preparation: A.K. and P.K.; Writing—Review and Editing: S.O. and S.D.; Visualization: P.K., V.C., D.B., J.G., Y.M. and A.K.; Supervision: S.O. and S.D. All authors have read and agreed to the published version of the manuscript.

Funding: This research received no external funding.

Data Availability Statement: Suggested Data Availability Statements are available in section "MDPI Research Data Policies" at https://www.mdpi.com/ethics (accessed on 26 March 2023).

Conflicts of Interest: The authors declare no conflict of interest.

References

1. Otroshchenko, T.; Jiang, G.; Kondratenko, V.A.; Rodemerck, U.; Kondratenko, E.V. Current status and perspectives in oxidative, non-oxidative and CO_2-mediated dehydrogenation of propane and isobutane over metal oxide catalysts. *Chem. Soc. Rev.* **2021**, *50*, 473–527. [CrossRef] [PubMed]
2. Hu, Z.P.; Yang, D.; Wang, Z.; Yuan, Z.Y. State-of-the-art catalysts for direct dehydrogenation of propane to propylene. *Chinese J. Catal.* **2019**, *40*, 1233–1254. [CrossRef]
3. Gomez, E.; Yan, B.; Kattel, S.; Chen, J.G. Carbon dioxide reduction in tandem with light-alkane dehydrogenation. *Nat. Rev. Chem.* **2019**, *3*, 638–649. [CrossRef]
4. Atanga, M.A.; Rezaei, F.; Jawad, A.; Fitch, M.; Rownaghi, A.A. Oxidative dehydrogenation of propane to propylene with carbon dioxide. *Appl. Catal. B Environ.* **2018**, *220*, 429–445. [CrossRef]
5. Gambo, Y.; Adamu, S.; Tanimu, G.; Abdullahi, I.M.; Lucky, R.A.; Ba-Shammakh, M.S.; Hossain, M.M. CO_2-mediated oxidative dehydrogenation of light alkanes to olefins: Advances and perspectives in catalyst design and process improvement. *Appl. Catal. A Gen.* **2021**, *623*, 118273. [CrossRef]
6. Schreiber, M.W.; Plaisance, C.P.; Baumgärtl, M.; Reuter, K.; Jentys, A.; Bermejo-Deval, R.; Lercher, J.A. Lewis-Brønsted acid pairs in Ga/H-ZSM-5 to catalyze dehydrogenation of light alkanes. *J. Am. Chem. Soc.* **2018**, *140*, 4849–4859. [CrossRef]
7. Michorczyk, P.; Zeńczak-Tomera, K.; Michorczyk, B.; Węgrzyniak, A.; Basta, M.; Millot, Y.; Valentin, L.; Dzwigaj, S. Effect of dealumination on the catalytic performance of Cr-containing Beta zeolite in carbon dioxide assisted propane dehydrogenation. *J. CO2 Util.* **2020**, *36*, 54–63. [CrossRef]

8. Ni, L.; Khare, R.; Bermejo-Deval, R.; Zhao, R.; Tao, L.; Liu, Y.; Lercher, J.A. Highly active and selective sites for propane dehydrogenation in zeolite Ga-BEA. *J. Am. Chem. Soc.* **2022**, *144*, 12347–12356. [CrossRef]
9. Wang, Z.-Y.; He, Z.-H.; Li, L.-Y.; Yang, S.-Y.; He, M.-X.; Sun, Y.-C.; Wang, K.; Chen, J.-G.; Liu, Z.-T. Research progress of CO_2 oxidative dehydrogenation of propane to propylene over Cr-free metal catalysts. *Rare Met.* **2022**, *41*, 2129–2152. [CrossRef]
10. Castro-Fernández, P.; Mance, D.; Liu, C.; Abdala, P.M.; Willinger, E.; Rossinelli, A.A.; Serykh, A.I.; Pidko, E.A.; Copéret, C.; Fedorov, A.; et al. Bulk and surface transformations of Ga_2O_3 nanoparticle catalysts for propane dehydrogenation induced by a H_2 treatment. *J. Catal.* **2022**, *408*, 155–164. [CrossRef]
11. Ye, T.; Carter, J.H.; Chen, B.; Li, X.; Ye, Y.; Taylor, S.H.; Hutchings, G.J. Iron-chromium mixed metal oxides catalyse the oxidative dehydrogenation of propane using carbon dioxide. *Catal. Commun.* **2022**, *162*, 106383. [CrossRef]
12. Li, L.-Y.; Wang, Z.-Y.; Yang, S.-Y.; Chen, J.-G.; He, Z.-H.; Wang, K.; Luo, Q.-X.; Liu, Z.-W.; Liu, Z.-T. Understanding the role of Fe doping in tuning the size and dispersion of GaN nanocrystallites for CO_2-assisted oxidative dehydrogenation of propane. *ACS Catal.* **2022**, *12*, 8527–8543. [CrossRef]
13. Chen, C.; Hu, Z.P.; Ren, J.T.; Zhang, S.; Wang, Z.; Yuan, Z.Y. ZnO supported on high-silica HZSM-5 as efficient catalysts for direct dehydrogenation of propane to propylene. *Mol. Catal.* **2019**, *476*, 110508. [CrossRef]
14. Ren, Y.; Zhang, F.; Hua, W.; Yue, Y.; Gao, Z. ZnO supported on high silica HZSM-5 as new catalysts for dehydrogenation of propane to propene in the presence of CO_2. *Catal. Today* **2009**, *148*, 316–322. [CrossRef]
15. Chen, C.; Hu, Z.; Ren, J.; Zhang, S.; Wang, Z.; Yuan, Z.Y. ZnO nanoclusters supported on dealuminated zeolite β as a novel catalyst for direct dehydrogenation of propane to propylene. *ChemCatChem* **2019**, *11*, 868–877. [CrossRef]
16. Orlyk, S.M.; Kantserova, M.R.; Chedryk, V.I.; Kyriienko, P.I.; Balakin, D.Y.; Millot, Y.; Dzwigaj, S. Ga(Nb,Ta)SiBEA zeolites prepared by two-step postsynthesis method: Acid–base characteristics and catalytic performance in the dehydrogenation of propane to propylene with CO_2. *J. Porous Mater.* **2021**, *28*, 1511–1522. [CrossRef]
17. Liu, J.; He, N.; Zhang, Z.; Yang, J.; Jiang, X.; Zhang, Z.; Su, J.; Shu, M.; Si, R.; Xiong, G.; et al. Highly-dispersed zinc species on zeolites for the continuous and selective dehydrogenation of ethane with CO_2 as a soft oxidant. *ACS Catal.* **2021**, *11*, 2819–2830. [CrossRef]
18. Zhao, D.; Tian, X.; Doronkin, D.E.; Han, S.; Kondratenko, V.A.; Grunwaldt, J.D.; Perechodjuk, A.; Vuong, T.H.; Rabeah, J.; Eckelt, R.; et al. In situ formation of ZnO_x species for efficient propane dehydrogenation. *Nature* **2021**, *599*, 234–238. [CrossRef]
19. Zhao, D.; Li, Y.; Han, S.; Zhang, Y.; Jiang, G.; Wang, Y.; Guo, K.; Zhao, Z.; Xu, C.; Li, R.; et al. ZnO nanoparticles encapsulated in nitrogen-doped carbon material and silicalite-1 composites for efficient propane dehydrogenation. *iScience* **2019**, *13*, 269–276. [CrossRef]
20. Zhang, Y.; Zhou, Y.; Huang, L.; Zhou, S.; Sheng, X.; Wang, Q.; Zhang, C. Structure and catalytic properties of the Zn-modified ZSM-5 supported platinum catalyst for propane dehydrogenation. *Chem. Eng. J.* **2015**, *270*, 352–361. [CrossRef]
21. Xie, L.; Chai, Y.; Sun, L.; Dai, W.; Wu, G.; Guan, N.; Li, L. Optimizing zeolite stabilized Pt-Zn catalysts for propane dehydrogenation. *J. Energy Chem.* **2021**, *57*, 92–98. [CrossRef]
22. Qi, L.; Babucci, M.; Zhang, Y.; Lund, A.; Liu, L.; Li, J.; Chen, Y.; Hoffman, A.S.; Bare, S.R.; Han, Y.; et al. Propane dehydrogenation catalyzed by isolated Pt atoms in ≡SiOZn–OH nests in dealuminated zeolite Beta. *J. Am. Chem. Soc.* **2021**, *143*, 21364–21378. [CrossRef] [PubMed]
23. Huang, C.; Han, D.; Guan, L.; Zhu, L.; Mei, Y.; He, D.; Zu, Y. Bimetallic Ni-Zn site anchored in siliceous zeolite framework for synergistically boosting propane dehydrogenation. *Fuel* **2022**, *307*, 121790. [CrossRef]
24. Sun, Q.; Wang, N.; Fan, Q.; Zeng, L.; Mayoral, A.; Miao, S.; Yang, R.; Jiang, Z.; Zhou, W.; Zhang, J.; et al. Subnanometer bimetallic platinum–zinc clusters in zeolites for propane dehydrogenation. *Angew. Chemie Int. Ed.* **2020**, *59*, 19450–19459. [CrossRef]
25. Gabrienko, A.A.; Arzumanov, S.S.; Toktarev, A.V.; Danilova, I.G.; Prosvirin, I.P.; Kriventsov, V.V.; Zaikovskii, V.I.; Freude, D.; Stepanov, A.G. Different efficiency of Zn^{2+} and ZnO species for methane activation on Zn-modified zeolite. *ACS Catal.* **2017**, *7*, 1818–1830. [CrossRef]
26. Tanuma, S.; Powell, C.J.; Penn, D.R. Calculations of electron inelastic mean free paths. V. Data for 14 organic compounds over the 50–2000 eV range. *Surf. Interf. Anal.* **1994**, *21*, 165–176. [CrossRef]
27. Bandala, E.R.; Sadek, R.; Gurgul, J.; Łątka, K.; Zimowska, M.; Valentin, L.; Rodriguez-Narvaez, O.M.; Dzwigaj, S. Assessment of the capability of Fe and Al modified BEA zeolites to promote advanced oxidation processes in aqueous phase. *Chem. Eng. J.* **2021**, *409*, 127379. [CrossRef]
28. Chalupka, K.A.; Sadek, R.; Szkudlarek, L.; Mierczynski, P.; Maniukiewicz, W.; Rynkowski, J.; Gurgul, J.; Casale, S.; Brouri, D.; Dzwigaj, S. The catalytic activity of microporous and mesoporous NiCoBeta zeolite catalysts in Fischer–Tropsch synthesis. *Res. Chem. Intermed.* **2021**, *47*, 397–418. [CrossRef]
29. Pamin, K.; Gurgul, J.; Mordarski, G.; Millot, Y.; Nogier, J.-P.; Valentin, L.; Dzwigaj, S. Efficient transformation of cyclohexanone to ε-caprolactone in the oxygen-aldehyde system over single-site titanium BEA zeolite. *Microporous Mesoporous Mater.* **2021**, *322*, 111159. [CrossRef]
30. Kocemba, I.; Rynkowski, J.; Gurgul, J.; Socha, R.P.; Łątka, K.; Krafft, J.-M.; Dzwigaj, S. Nature of the active sites in CO oxidation on FeSiBEA zeolites. *Appl. Catal. A Gen.* **2016**, *519*, 16–26. [CrossRef]
31. Hu, P.; Iyoki, K.; Yamada, H.; Yanaba, Y.; Ohara, K.; Katada, N.; Wakihara, T. Synthesis and characterization of MFI-type zincosilicate zeolites with high zinc content using mechanochemically treated Si–Zn oxide composite. *Microporous Mesoporous Mater.* **2019**, *288*, 109594. [CrossRef]

32. Liu, Y.; Shen, J.; Chen, Z.; Yang, L.; Liu, Y.; Han, Y. Effects of amorphous-zinc-silicate-catalyzed ozonation on the degradation of p-chloronitrobenzene in drinking water. *Appl. Catal. A Gen.* **2011**, *403*, 112–118. [CrossRef]
33. Hastir, A.; Kohli, N.; Singh, R.C. Comparative study on gas sensing properties of rare earth (Tb, Dy and Er) doped ZnO sensor. *Phys. Chem. Solids* **2017**, *105*, 23–34. [CrossRef]
34. Iatsunskyi, I.; Kempiński, M.; Jancelewicz, M.; Załęski, K.; Jurga, S.; Smyntyna, V. Structural and XPS characterization of ALD Al_2O_3 coated porous silicon. *Vacuum* **2015**, *113*, 52–58. [CrossRef]
35. Sygellou, L.; Gianneta, V.; Xanthopoulos, N.; Skarlatos, D.; Georga, S.; Krontiras, C.; Ladas, S.; Kennou, S. ZrO_2 and Al_2O_3 thin films on Ge(100) grown by ALD: An XPS investigation. *Surf. Sci. Spectra.* **2011**, *18*, 58–67. [CrossRef]
36. Biesinger, M.C.; Lau, L.W.M.; Gerson, A.R.; Smart, R.S.C. Resolving surface chemical states in XPS analysis of first row transition metals, oxides and hydroxides: Sc, Ti, V, Cu and Zn. *Appl. Surf. Sci.* **2010**, *257*, 887–898. [CrossRef]
37. Hunsicker, R.A.; Klier, K.; Gaffney, T.S.; Kirner, J.G. Framework zinc-substituted zeolites: synthesis, and core-level and valence-band XPS. *Chem. Mater.* **2002**, *14*, 4807–4811. [CrossRef]
38. Wöll, C. The chemistry and physics of zinc oxide surfaces. *Progr. Surf. Sci.* **2007**, *82*, 55–120. [CrossRef]
39. Tamiyakul, S.; Ubolcharoen, W.; Tungasmita, D.N.; Jongpatiwut, S. Conversion of glycerol to aromatic hydrocarbons over Zn-promoted HZSM-5 catalysts. *Catal. Today* **2015**, *256*, 325–335. [CrossRef]
40. Niu, X.; Gao, J.; Miao, Q.; Dong, M.; Wang, G.; Fan, W.; Qin, Z.J. Wang. Influence of preparation method on the performance of Zn-containing HZSM-5 catalysts in methanol-to-aromatics. *Microporous Mesoporous Mater.* **2014**, *197*, 252–261. [CrossRef]
41. Santos, R.C.; Almeida, D.F.; de Aguiar Pontes, D.; Lau, L.Y.; Magalhães Pontes, L.A. Thiophene cracking on zinc modified beta zeolite. *Mol. Catal.* **2019**, *470*, 112–119. [CrossRef]
42. Zhang, N.; Li, R.; Zhang, G.; Dong, L.; Zhang, D.; Wang, G.; Li, T. Zn-modified Hβ zeolites used in the adsorptive removal of organic chloride from model naphtha. *ACS Omega* **2020**, *5*, 11987–11997. [CrossRef] [PubMed]
43. Gong, T.; Qin, L.; Lu, J.; Feng, H. ZnO modified ZSM-5 and Y zeolites fabricated by atomic layer deposition for propane conversion. *Phys. Chem. Chem. Phys.* **2016**, *18*, 601–614. [CrossRef] [PubMed]
44. Su, X.; Zan, W.; Bai, X.; Wang, G.; Wu, W. Synthesis of microscale and nanoscale ZSM-5 zeolites: Effect of particle size and acidity of Zn modified ZSM-5 zeolites on aromatization performance. *Catal. Sci. Technol.* **2017**, *7*, 1943–1952. [CrossRef]
45. Almutairi, S.M.T.; Mezari, B.; Magusin, P.C.M.M.; Pidko, E.A.; Hensen, E.J.M. Structure and reactivity of Zn-Modified ZSM-5 zeolites: The importance of clustered cationic Zn complexes. *ACS Catal.* **2012**, *2*, 71–83. [CrossRef]
46. Fyfe, C.A.; Strobl, H.; Kokotailo, G.T.; Pasztor, C.T.; Barlow, G.E.; Bradley, S. Correlations between lattice structures of zeolites and their ^{29}Si MAS n.m.r. spectra: Zeolites KZ-2, ZSM-12, and Beta. *Zeolites* **1988**, *8*, 132–136. [CrossRef]
47. Pérez-Pariente, J.; Sanz, J.; Fornés, V.; Corma, A. ^{29}Si and ^{27}Al MAS NMR study of zeolite β with different Si/Al ratios. *J. Catal.* **1990**, *124*, 217–223. [CrossRef]
48. Dzwigaj, S.; Millot, Y.; Méthivier, C.; Che, M. Incorporation of Nb(V) into BEA zeolite investigated by XRD, NMR, IR, DR UV–vis, and XPS. *Microporous Mesoporous Mater.* **2010**, *130*, 162–166. [CrossRef]
49. Popovych, N.O.; Kyriienko, P.I.; Soloviev, S.O.; Orlyk, S.M.; Dzwigaj, S. Influence of partial dealumination of BEA zeolites on physicochemical and catalytic properties of AgAlSiBEA in H_2-promoted SCR of NO with ethanol. *Microporous Mesoporous Mater.* **2016**, *226*, 10–18. [CrossRef]
50. Dzwigaj, S.; Popovych, N.; Kyriienko, P.; Krafft, J.M.; Soloviev, S. The similarities and differences in structural characteristics and physico-chemical properties of AgAlBEA and AgSiBEA zeolites. *Microporous Mesoporous Mater.* **2013**, *182*, 16–24. [CrossRef]
51. Parry, E.P. An infrared study of pyridine adsorbed on acidic solids. Characterization of surface acidity. *J. Catal.* **1963**, *2*, 371–379. [CrossRef]
52. Popovych, N.; Kyriienko, P.; Soloviev, S.; Orlyk, S.; Dzwigaj, S. Catalytic properties of AgAlBEA and AgSiBEA zeolites in H2-promoted selective reduction of NO with ethanol. *Microporous Mesoporous Mater.* **2015**, *203*, 163–169. [CrossRef]
53. Penzien, J.A.L.J.; Abraham, A.; van Bokhoven, J.A.; Jentys, A.; Müller, T.E.; Sievers, C. Generation and characterization of well-defined Zn^{2+} Lewis acid sites in ion exchanged zeolite BEA. *J. Phys. Chem. B* **2004**, *108*, 4116–4126. [CrossRef]
54. Larina, O.V.; Shcherban, N.D.; Kyriienko, P.I.; Remezovskyi, I.M.; Yaremov, P.S.; Khalakhan, I.; Mali, G.; Soloviev, S.O.; Orlyk, S.M.; Dzwigaj, S. Design of effective catalysts based on ZnLaZrSi oxide systems for obtaining 1,3-butadiene from aqueous ethanol. *ACS Sustain. Chem. Eng.* **2020**, *8*, 16600–16611. [CrossRef]
55. Phadke, N.M.; Mansoor, E.; Bondil, M.; Head-Gordon, M.; Bell, A.T. Mechanism and kinetics of propane dehydrogenation and cracking over Ga/H-MFI prepared via vapor-phase exchange of H-MFI with $GaCl_3$. *J. Am. Chem. Soc.* **2019**, *141*, 1614–1627. [CrossRef] [PubMed]
56. Gounder, R.; Iglesia, E. Catalytic hydrogenation of alkenes on acidic zeolites: Mechanistic connections to monomolecular alkane dehydrogenation reactions. *J. Catal.* **2011**, *277*, 36–45. [CrossRef]
57. Ren, Y.; Wang, J.; Hua, W.; Yue, Y.; Gao, Z. Ga_2O_3/HZSM-48 for dehydrogenation of propane: Effect of acidity and pore geometry of support. *J. Ind. Eng. Chem.* **2012**, *18*, 731–736. [CrossRef]
58. Sattler, J.J.H.B.; Ruiz-Martinez, J.; Santillan-Jimenez, E.; Weckhuysen, B.M. Catalytic dehydrogenation of light alkanes on metals and metal oxides. *Chem. Rev.* **2014**, *114*, 10613–10653. [CrossRef]
59. Gambo, Y.; Adamu, S.; Abdulrasheed, A.A.; Lucky, R.A.; Ba-Shammakh, M.S.; Hossain, M.M. Catalyst design and tuning for oxidative dehydrogenation of propane–A review. *Appl. Catal. A Gen.* **2021**, *8*, 117914. [CrossRef]

60. Liu, Y.; Li, Z.H.; Lu, J.; Fan, K. Periodic density functional theory study of propane dehydrogenation over perfect Ga_2O_3(100) surface. *J. Phys. Chem. C* **2008**, *112*, 20382–20392. [CrossRef]
61. Copéret, C. C-H bond activation and organometallic intermediates on isolated metal centers on oxide surfaces. *Chem. Rev.* **2010**, *110*, 656–680. [CrossRef] [PubMed]
62. Ansari, M.B.; Park, S.-E. Carbon dioxide utilization as a soft oxidant and promoter in catalysis. *Energy Environ. Sci.* **2012**, *5*, 9419–9437. [CrossRef]
63. Mukherjee, D.; Park, S.-E.; Reddy, B.M. CO_2 as a soft oxidant for oxidative dehydrogenation reaction: An eco benign process for industry. *J. CO2 Util.* **2016**, *16*, 301–312. [CrossRef]
64. Nowicka, E.; Reece, C.; Althahban, S.; Mohammed, K.; Kondrat, S.; John, D. Elucidating the role of CO_2 in the soft oxidative dehydrogenation of propane over ceria-based catalysts. *ACS Catal.* **2018**, *8*, 3454–3468. [CrossRef]
65. Michorczyk, P.; Zeńczak, K.; Niekurzak, R.; Ogonowski, J. Dehydrogenation of propane with CO_2–a new green process for propene and synthesis gas production. *Polish J. Chem. Technol.* **2012**, *14*, 77–82. [CrossRef]
66. Nozik, D.; Tinga, F.M.P.; Bell, A.T. Propane dehydrogenation and cracking over Zn/H-MFI prepared by solid-state ion exchange of $ZnCl_2$. *ACS Catal.* **2021**, *11*, 14489–14506. [CrossRef]
67. Otroshchenko, T.P.; Kondratenko, V.A.; Rodemerck, U.; Linke, D.; Kondratenko, E.V. Non-oxidative dehydrogenation of propane, *n*-butane, and isobutane over bulk ZrO_2-based catalysts: Effect of dopant on the active site and pathways of product formation. *Catal. Sci. Technol.* **2017**, *7*, 4499–4510. [CrossRef]
68. Dzwigaj, S.; Peltre, M.J.; Massiani, P.; Davidson, A.; Che, M.; Sen, T.; Sivasanker, S. Incorporation of vanadium species in a dealuminated β zeolite. *Chem. Commun.* **1998**, *1*, 87–88. [CrossRef]
69. Kyriienko, P.I.; Larina, O.V.; Balakin, D.Y.; Stetsuk, A.O.; Nychiporuk, Y.M.; Soloviev, S.O.; Orlyk, S.M. 1,3-Butadiene production from aqueous ethanol over $ZnO/MgO-SiO_2$ catalysts: Insight into H_2O effect on catalytic performance. *Appl. Catal. A Gen.* **2021**, *616*, 118081. [CrossRef]

Disclaimer/Publisher's Note: The statements, opinions and data contained in all publications are solely those of the individual author(s) and contributor(s) and not of MDPI and/or the editor(s). MDPI and/or the editor(s) disclaim responsibility for any injury to people or property resulting from any ideas, methods, instructions or products referred to in the content.

Article

Effects of Synthesis Variables on SAPO-34 Crystallization Templated Using Pyridinium Supramolecule and Its Catalytic Activity in Microwave Esterification Synthesis of Propyl Levulinate

Yik-Ken Ma [1], Taghrid S. Alomar [2], Najla AlMasoud [2], Zeinhom M. El-Bahy [3], Stephen Chia [4], T. Jean Daou [5,6], Fitri Khoerunnisa [7], Tau Chuan Ling [8,*] and Eng-Poh Ng [1,*]

1. School of Chemical Sciences, Universiti Sains Malaysia, USM, Penang 11800, Malaysia
2. Department of Chemistry, College of Science, Princess Nourah bint Abdulrahman University, P.O. Box 84428, Riyadh 11671, Saudi Arabia
3. Department of Chemistry, Faculty of Science, Al-Azhar University, Nasr City, Cairo 11884, Egypt
4. Centre for Global Archaeological Research, Universiti Sains Malaysia, USM, Penang 11800, Malaysia
5. Axe Matériaux à Porosités Contrôlées, Institut de Science de Matériaux de Mulhouse UMR 7361, ENSCMu, Université de Haute-Alsace, 3b Rue Alfred Werner, 68093 Mulhouse, France
6. Université de Strasbourg, 67000 Strasbourg, France
7. Chemistry Education Department, Universitas Pendidikan Indonesia, Jl. Setiabudhi 258, Bandung 40514, Indonesia
8. Institute of Biological Sciences, Faculty of Science, University of Malaya, Kuala Lumpur 50603, Malaysia
* Correspondence: tcling@um.edu.my (T.C.L.); epng@usm.my (E.-P.N.)

Abstract: A detailed investigation of the hydrothermal crystallization of SAPO-34 in the presence of the novel 1-propylpyridinium hydroxide ([PrPy]OH) organic structural directing agent is presented. The synthesis conditions are systematically tuned to investigate the effects of various parameters (viz. concentrations of each reactant, crystallization time, and temperature) on the nucleation and crystallization of SAPO-34. The results show that a careful variation in each of the synthesis parameters results in the formation of competing phases such as SAPO-5, SAPO-35, and SAPO-36. Pure and fully crystalline SAPO-34 can be crystallized using a precursor hydrogel of a molar ratio of 2.0 Al: 4.7 P: 0.9 Si: 6.7 [PrPy]OH: 148 H_2O at 200 °C for only 19 h, which is a shorter time than that found in previous studies. The prepared SAPO-34 is also very active in the esterification of levulinic acid and 1-propanol. By using microwave heating, 91.5% conversion with 100% selectivity toward propyl levulinate is achieved within 20 min at 190 °C. Hence, the present study may open a new insight into the optimum synthesis study of other zeolites using novel pyridinium organic moieties and the opportunity of replacing conventional harmful and non-recyclable homogeneous catalysts in levulinate biofuel synthesis.

Keywords: SAPO-34; zeolites; crystallization; esterification; propyl levulinate

1. Introduction

Aluminophosphates (AlPO-*n*) and silicoaluminophosphates (SAPO-*n*) are zeolite-like microporous solids that have shown numerous promising industrial applications in adsorption, ion exchange, and catalysis [1–3]. Among them, SAPO-34 (CHA topology), which has a three-dimensional pore system with a diameter size of 3.80 × 3.80 Å^2 and a large CHA cage (9.4 Å in diameter), is one of the most important zeolites and has widely been used in the methanol-to-olefins (MTO), gas separation, and hydroisomerization reactions [4,5].

In general, several strategies can be used to crystallize SAPO-34, namely the hydrothermal technique [6], interzeolite conversion [7], and dry gel conversion [8]. During the crystallization process, aliphatic or aromatic amines, such as diethylamine [9], triethylamine [10], tetraethylammonium hydroxide [11], piperidine [12], or morpholine [13], are added and served as structure-directing agents (SDAs) to direct the crystallization of SAPO-34 zeolite. However, the classical time-consuming hydrothermal treatment may severely impact the scale-up production and cost, where the shortest crystallization time for the formation of SAPO-34 so far requires at least 24 h at 200 °C [9,12,13]. In addition, the use of various types and amounts of SDAs with different electronic and hydrophilic/hydrophobic properties in SAPO-34 synthesis also leads to an alteration in the colloidal properties, and thermodynamic stability and activity of the precursor hydrogel, which in turn affects the overall nucleation and crystallization kinetics, final phase purity, chemical composition, crystal size, and morphology of the synthesized zeolites [14,15]. Therefore, a detailed study of synthesis variables via hydrothermal condition improvement is of utmost importance for enabling comprehensive control of the zeolite crystallization process.

Pyridinium-based molecules are a new type of SDA that, thus far, have seldom been synthesized and used in the preparation of zeolites. Due to their unique aromatic structure, delocalization of π electrons, and presence of electronegative N atom in the cyclic ring, pyridinium-based SDAs show different polarity, surface charge, and electron density compared to aliphatic aminic SDAs, which may lead to different crystallization profiles of zeolites (particularly in SAPO-34) [16]. In addition, other synthesis parameters, such as the concentrations of various reactants, crystallization time, and temperature, also directly affect the thermodynamic and crystallization process of zeolites [17,18]. Nevertheless, knowledge about the effects of these parameters on the crystallization of SAPO-34 in the presence of pyridinium-based SDA still remains limited.

In order to better understand the crystallization phenomenon of SAPO-34 and the roles of synthesis variables in the formation of SAPO-34 in the presence of heterocyclic pyridinium SDA, a systematic set of experiments is performed. Specifically, 1-propylpyridinium is first prepared prior to being applied in the hydrothermal crystallization of SAPO-34. Concurrently, a study of the influence of crystallization conditions by altering the synthesis parameters is also carried out. Finally, the acidity and surface properties of SAPO-34 crystals are studied before being tested in the production of propyl levulinate biofuel additive via the esterification of levulinic acid and propanol using a microwave heating method.

2. Results and Discussion
2.1. Single-Parameter Tuning Synthesis of SAPO-34
2.1.1. Time-Dependent Formation Study of SAPO-34

Ostwald's Law of successive reaction is frequently invoked in the formation of zeolitic materials where successive phase transformations into more thermodynamically stable zeolite phases occur due to their metastability. As such, the effect of crystallization time was studied by heating the hydrogel of a molar composition of 2.0 Al: 4.7 P: 0.9 Si: 6.7 [PrPy]OH: 148 H_2O for 0, 6, 12, 16, 19, and 30 h at 200 °C. Initially, at 0 h, the hydrogel is amorphous indicating no crystalline phase is formed upon completion of hydrogel preparation (Table 1). The white hydrogel is subjected to hydrothermal treatment to allow chemical reactions (dissolution, polymerization, induction, nucleation, crystallization, etc.) to occur. The white suspension solid dissolves entirely in the mother liquor after 6 h of heating, indicating the formation of monomeric and oligomeric Si, P, and Al oxides in the mother liquor [19].

The precursor hydrogel is further heated for 12 h where a small amount of soft solid is recovered upon centrifugation. The solid displays an irregular shape according to the FESEM analysis, proving its XRD amorphous characteristics (Figure 1a). Hence, it reveals that polymerization of Si, Al, and P oligomers has occurred, leading to the sedimentation of a dense amorphous solid. Nucleation is witnessed at 16 h where several XRD peaks corresponding to SAPO-34 slowly appear (Figure 1b). The solid comprises cubic SAPO-34 crystals with rough surfaces covered by amorphous entities.

Table 1. The porous and acidity properties of SAPO-34 (S-3) crystallized using [PrPy]OH.

Sample	Si/(P + Al) Ratio	S_{BET} (m^2 g^{-1}) [a]	S_{Micro} (m^2 g^{-1}) [b]	V_{Total} (cm^3 g^{-1}) [c]	NH$_3$-TPD (mmol g^{-1})		
					Weak-to-Medium [d]	Mild-to-Strong [e]	Total
S-3	0.24	673	661	0.27	1.27	1.25	2.52

[a] Specific surface area; [b] Micropore surface area; [c] Total pore volume; [d] Calculated based on TPD desorption curves at 168 °C and 235 °C; [e] Calculated based on desorption curves at 404 °C and 479 °C.

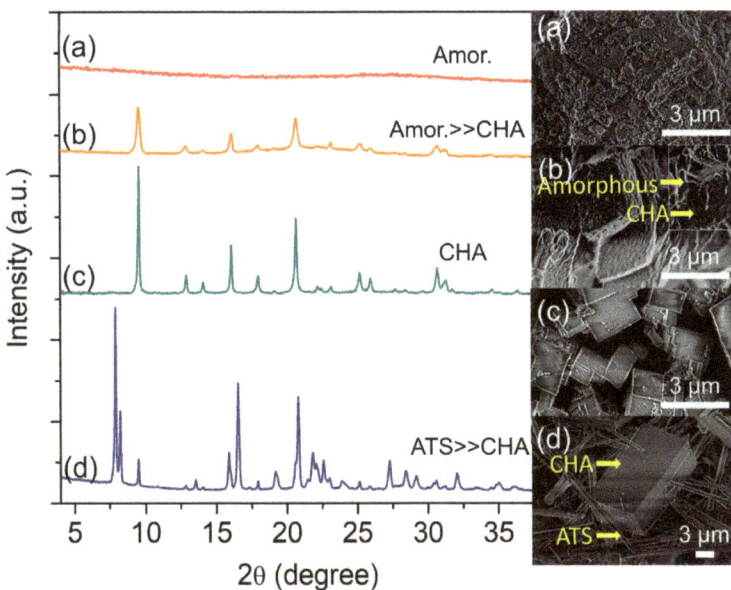

Figure 1. XRD patterns and FESEM micrographs of (a) S-1, (b) S-2, (c) S-3, and (d) S-4 solids heated at 150 °C for 12 h, 16 h, 19 h, and 30 h, respectively.

Further extending the crystallization time to 19 h confirms the complete formation of SAPO-34 where the amorphous solid is entirely consumed for the crystal growth, leading to well-defined cubic crystals with sharp edges (ca. 1.86 µm) (Figure 1c). As shown in Figure 1, the XRD diffraction peaks corresponding to SAPO-34 crystallites (9.48° [100], 16.06° [11-1], 17.93° [111], 18.86° [200], 20.62° [20-1]) are intense and no additional peaks are observed, reflecting its high crystallinity and purity.

The crystallization time is further extended to 30 h to study the metastable phase formation. As seen in Figure 1, the main XRD peaks of SAPO-34 at 2θ = 9.49° "100", 18.86° "200", and 17.93° "111" are becoming weaker, and they are at the expense of the peaks of SAPO-36 (2θ = 7.90° "110", 8.24° "020", 15.86° "220", 16.52° "040", and 20.78° "310") (Figure 1d). The intrazeolite transformation process is also detected by FESEM analysis whereby the cuboid-shaped crystals (SAPO-34) are co-crystallized together with needle-like crystals (SAPO-36), hence confirming the metastability feature of SAPO-34.

2.1.2. Effect of P/Al Molar Ratio

Phosphoric acid plays a significant role in the crystallization of SAPO-34 since it is one of the basic building blocks of the zeolite. It also alters the pH of the hydrogel that governs the entire crystallization process [20]. Hence, the effect of phosphoric acid amount (presented in the form of P/Al molar ratio) is investigated by heating the precursor hydrogel of 2.0 Al: w P: 0.9 Si: 6.7 [PrPy]OH: 148 H$_2$O (w = 4.1, 4.7, 5.3 and 5.9) at 200 °C for 19 h. Amorphous particles are obtained when w = 4.1 (P/Al = 2.05) (Figure 2a) due to weakly alkaline hydrogel (pH = 7.2) that inhibits the nucleation of SAPO-34 [21]. When the P/Al ratio increases to 2.35 (w = 4.7), the pH of the precursor gel drops to 6.7. Under this weakly acidic condition, SAPO-34 crystals with conventional cubic morphology (ca. 1.86 μm) are formed (Figure 2b).

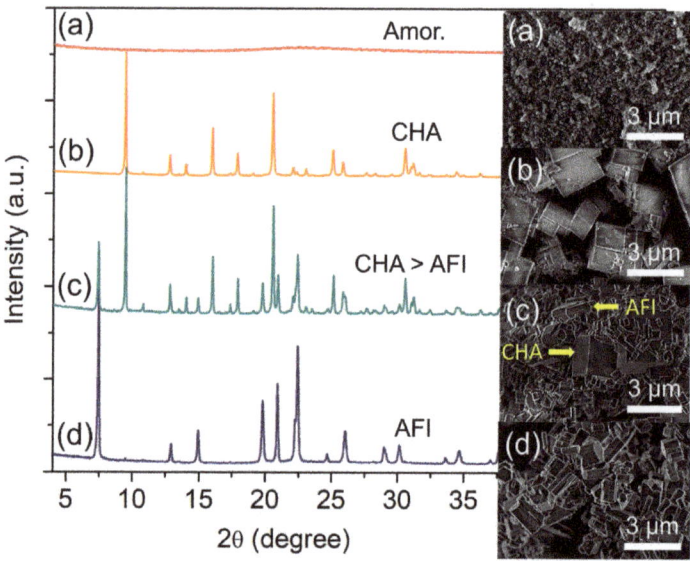

Figure 2. XRD patterns and FESEM micrographs of (**a**) S-5 (w = 4.1), (**b**) S-3 (w = 4.7), (**c**) S-6 (w = 5.3), and (**d**) S-7 (w = 5.9) samples. The samples were prepared using a hydrogel of 2.0 Al: w P: 0.9 Si: 6.7 [PrPy]OH: 148 H_2O heated at 200 °C for 19 h.

Further adding phosphoric acid increases the acidity of the precursor hydrogel (pH 6.2 when w = 5.3, and pH 5.6 when w = 5.9) which also witnesses the interzeolite transformation of SAPO-34 into SAPO-5 whereby the latter zeolite phase is preferentially formed under acidic conditions (Figure 2c,d) [22]. As shown, SAPO-5 (2θ = 7.47° "100", 19.85° "210", 20.97° "002", 22.46° "300") is formed at the expense of SAPO-34 where the metastable SAPO-34 particles are partially disintegrated in the mother liquor, releasing double 6-ring (D6R) secondary building units—the seeding sites—for the crystallization of SAPO-5 [23]. Furthermore, the mild acidic hydrogel (pH = 5.0–6.0) at w = 5.9 also favors the formation of SAPO-5, producing a hexagonal prism of crystals with sharp edges and smooth surfaces (ca. 1.16 × 1.25 μm^2).

2.1.3. Effect of Si Content

Silicon (Si) is another important parameter since it is the origin of the acid sites of SAPO-34 besides altering the entire crystallization process [24,25]. Hence, the effects of Si content on the formation of SAPO-34 were studied by heating the hydrogel of molar composition 2.0 Al: 4.7 P: x Si: 6.7 [PrPy]OH: 148 H_2O (x = 0, 0.4, 0.9 and 1.3) at 200 °C for 19 h. The XRD analysis reveals that the sample without the addition of Si is identified as an aluminophosphate (AlPO) clay solid with a thin layered morphology (ca. 1.78 × 0.04 μm^2) (Figure 3a). At x = 0.4, the clay material completely transforms into pure SAPO-36 (ATS topology)—a 12-membered ring large pore zeolite with needle-like shape (ca. 9.18 μm)—based on the major diffraction peaks at 2θ = 7.90° "110", 8.29° "020", 15.81° "220", 16.41° "040", and 20.67° "310" (Figure 3b) [23].

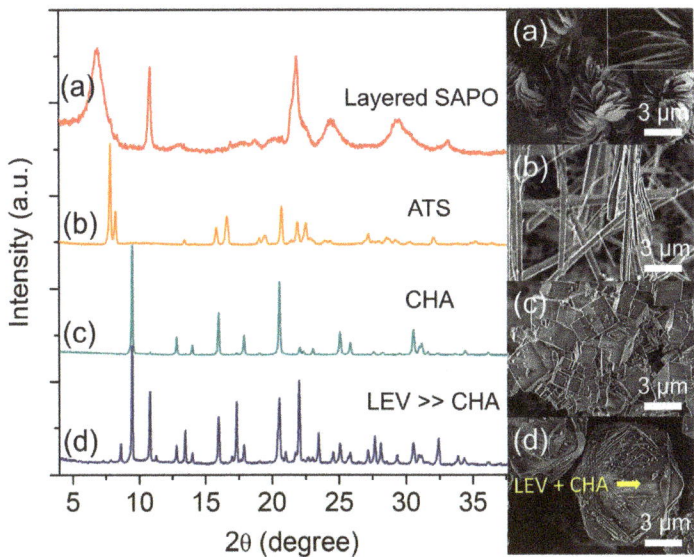

Figure 3. XRD patterns and FESEM micrographs of (**a**) S-8 ($x = 0$), (**b**) S-9 ($x = 0.4$), (**c**) S-3 ($x = 0.9$), and (**d**) S-10 ($x = 1.3$) samples. The samples were prepared using the hydrogel of 2.0 Al: 4.7 P: x Si: 6.7 [PrPy]OH: 148 H$_2$O heated at 200 °C for 19 h.

When the Si content is slightly increased to $x = 0.9$, the phase transformation from needle-like SAPO-36 into cuboid SAPO-34 crystallites (1.86 µm) is observed, thus showing the active participation of Si in the framework rearrangement process (Figure 3c). With further slightly increasing the Si content to $x = 1.3$, a partial transformation of cuboid SAPO-34 (7.5 µm) into rhombohedral SAPO-35 (7.9 µm, LEV topology) occurs. This is proven by the several SAPO-35 XRD major peaks (2θ = 8.63° "101", 10.82° "012", 13.49° "110", 17.33° "202", and 21.98° "122") that co-exist with those of SAPO-34 (Figure 3d) [23]. Thus, the results suggest that besides the active involvement of Si atoms as building blocks for self-organization, the mother liquor of SAPO-34 might contain double 6-rings (D6R) secondary building units too which can act as secondary seeds for further promoting SAPO-35 transformation (note: ATS, CHA, and LEV-type microporous solids have the same D6R secondary building units) [26].

2.1.4. Effect of [PrPy]OH Content

Four precursor mixtures (2.0 Al: 4.7 P: 0.9 Si: y [PrPy]OH: 148 H$_2$O) with different amounts of [PrPy]OH ($y = 5.3, 6.0, 6.7$, and 7.4) were hydrothermally heated at 200 °C for 19 h and the initial pH of the mixture was also measured. When $y = 5.3$, a mildly acidic hydrogel is formed (pH 5.76), which leads to the crystallization of pure SAPO-5 crystals with a hexagonal prism rod shape (ca. 0.70×1.18 µm^2) (Figure 4a) [27]. Partial transformation of SAPO-5 into SAPO-34 is detected when the amount of [PrPy]OH organic template increases to $y = 6.0$ (Figure 4b). At this point, the pH becomes nearly neutral (pH = 6.31). As shown in the XRD pattern, SAPO-5 (2θ = 7.45° "100", 12.92° "110", 14.97° "200", 19.85° "210", 20.97° "002", and 22.44° "300") remains as the major crystalline phase over SAPO-34 (2θ = 9.51° "100", 16.06° "11-1", 17.94° "111", 20.62° "20-1") in line with the FESEM observation; SAPO-5 hexagonal prismatic crystals are intergrown on the cuboid SAPO-34 crystals. Hence, increasing the hydrogel pH (by increasing the template concentration) is beneficial for inducing the crystallization of SAPO-34.

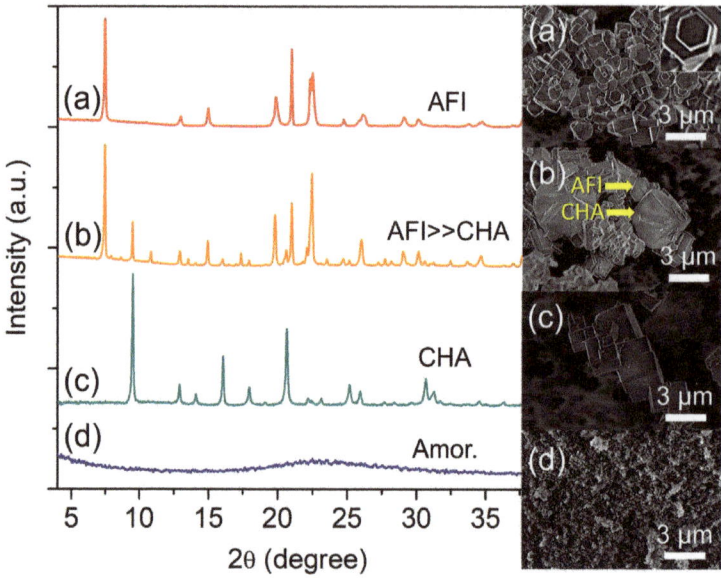

Figure 4. XRD patterns and FESEM micrographs of (**a**) S-11 (y = 5.3), (**b**) S-12 (y = 6.0), (**c**) S-3 (y = 6.7), and (**d**) S-13 (y = 7.4) samples. The samples were prepared using the hydrogel of 2.0 Al: 4.7 P: 0.9 Si: y [PrPy]OH: 148 H_2O heated at 200 °C for 19 h.

The above speculation is proven when the pH of the hydrogel is adjusted to be very close to neutral (pH = 6.67, y = 6.7) where fully crystalline cuboid SAPO-34 crystals (ca. 1.33 µm) are formed (Figure 4c). However, further increasing the template content to x = 7.4 (pH = 7.12) leads to the amorphization of the sample. Thus, the results demonstrate that [PrPy]$^+$ supramolecule (dimensional size of 6.8 × 3.2 Å^2) helps in the formation of pore framework structures of SAPO-5 and SAPO-34 via stabilizing their respective 12-membered rings (7.3 × 7.3 × 7.3 Å^3) and 8-membered rings (3.8 × 3.8 × 3.8 Å^3) through side-on and head-on orientations [24,28]. Furthermore, the addition of [PrPy]$^+$ template in OH$^-$ form also controls the pH of the hydrogel by providing a suitable environment for promoting nucleation for crystallizing SAPO-5 and SAPO-34 materials [29].

2.1.5. Effect of H_2O Content

Water content is one of the important components in precursor hydrogel as it controls the concentration of reactants besides serving as the solvent in the hydrothermal synthesis of zeolite [30]. As such, the effect of water content on the crystallization of SAPO-34 was studied at 200 °C for 19 h by varying the amount of water in the hydrogel of 2.0 Al: 4.7 P: 0.9 Si: 6.7 [PrPy]OH: z H_2O (z = 116, 148, 180, 240). From the experimental observation, the crystallization condition is very sensitive to the water content. By varying the amount of water, several competing phases, such as SAPO-36 (ATS), SAPO-35 (LEV), SAPO-34 (CHA), and SAPO-5 (AFI), are detected. For instance, at z = 116, SAPO-36 forms as the competing phase to SAPO-34 (Figure 5a) where the large cuboid SAPO-34 crystals (ca. 6.7 µm) are grown with their surface covered by short needle-shaped crystals of SAPO-36 (ca. 1.1 µm). The SAPO-36 phase, however, disappears at z = 148, forming only pure SAPO-34 as the final product (Figure 5b). As seen, the crystallites are almost homogeneous and have uniform size distribution (ca. 1.86 µm) due to the effects of water in altering the intermolecular forces and supersaturation condition of the precursor [31].

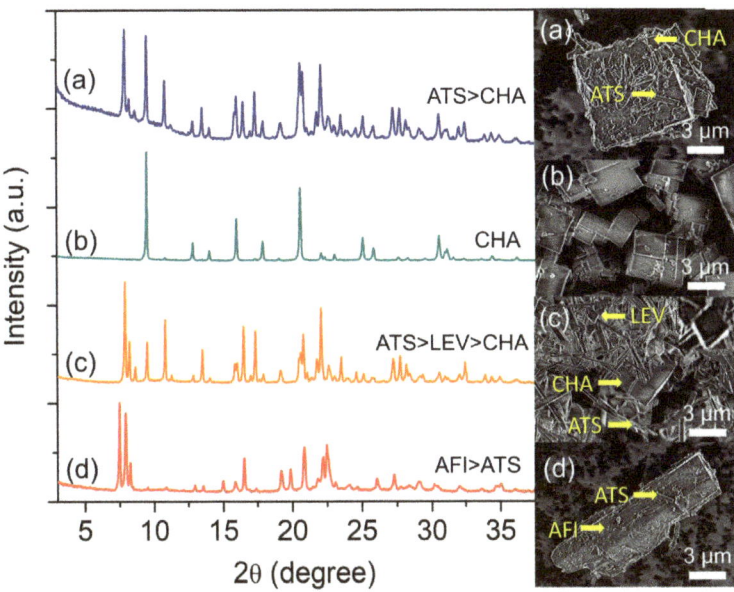

Figure 5. XRD patterns and FESEM micrographs of (**a**) S-14 (z = 116), (**b**) S-3 (z = 148), (**c**) S-15 (z = 180), and (**d**) S-16 (z = 240) samples. The samples were prepared using the hydrogel of 2.0 Al: 4.7 P: 0.9 Si: 6.7 [PrPy]OH: z H$_2$O heated at 200 °C for 19 h.

Increasing the water content to z = 180 witnesses the re-appearance of SAPO-36 (needle-like shape) where SAPO-35 (rhombohedral shape) together with SAPO-34 (cubic shape) exist as the minor phase (Figure 5c). The main diffraction peaks related to SAPO-36 are observed at 2θ = 7.90° "110", 8.18° "020", and 16.43° "040", whereas those of SAPO-35 and SAPO-34 are respectively found at 2θ = 10.84° "012" and 9.49° "100". However, SAPO-5 (2θ = 7.49° "100", 20.82° "002", and 22.38° "211") appears as the major phase over SAPO-36 when more water is added (z = 240) (Figure 5d) because, at high water content, the hydrogel becomes more acidic which favors the crystallization of SAPO-5 [32]. A similar effect is also detected by the FESEM analysis whereby needle-like SAPO-36 crystals are grown on the surface of large rod-like SAPO-5 crystals.

Compared to other synthesis parameters, water content is less selective in controlling the zeolite phases because water is used as a solvent for dissolving reactants and at the same time, transporting nutrients to the inorganic matrices for nucleation and crystal growth during hydrothermal synthesis [33]. The presence of different amounts of water and autogenic pressure generated from water vapor at high temperatures also tend to alter the nutrient concentration and solubility, which indirectly affect the supersaturation profile of the zeolites. As a result, the resulting final products with multiple phases are observed [34].

2.1.6. Effect of Crystallization Temperature

The heating temperature is the most important synthesis parameter since energy is essential for chemical reactions (e.g., polycondensation, nucleation, crystallization, etc.) to occur [35]. Hence, the effect of heating temperature on the crystallization of SAPO-34 was studied by heating the hydrogels of 2.0 Al: 4.7 P: 0.9 Si: 6.7 [PrPy]OH: 148 H$_2$O at 160, 180, 200, and 220 °C for 19 h. At 160 °C, the sample is amorphous according to XRD analysis, but few cuboid crystallites (ca. 0.8 μm) can be seen amid spherical amorphous entities indicating that nucleation of SAPO-34 has occurred at 160 °C (Figure 6a). By increasing the temperature to 180 °C, it is possible to see more SAPO-34 crystals are formed which leads to the emergence of weak X-ray diffraction peaks at 2θ = 9.51° "100", 16.05° "11-1", and 20.65° "20-1" (Figure 6b). As shown in Figure 6, the XRD peaks are broad, indicating small crystallites, which is confirmed by the FESEM data (ca. 1.1 μm) [24].

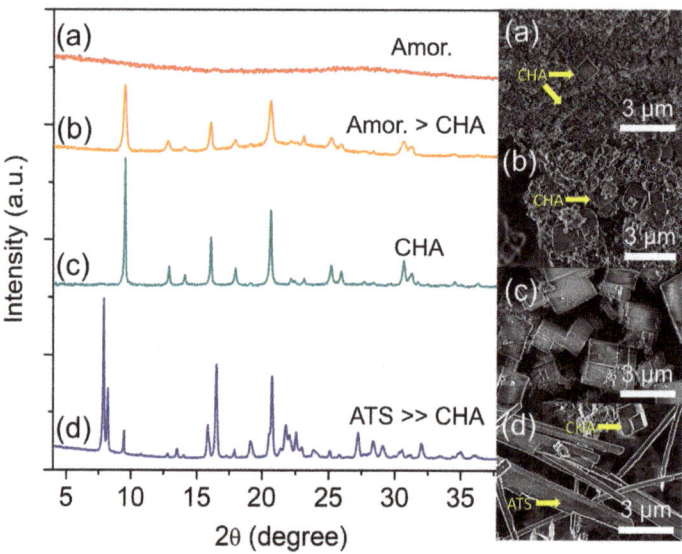

Figure 6. XRD patterns and FESEM micrographs of (**a**) S-17, (**b**) S-18, (**c**) S-3, and (**d**) S-19 samples prepared using the hydrogel of 2.0 Al: 4.7 P: 0.9 Si: 6.7 [PrPy]OH: 148 H$_2$O and heated at 160 °C, 180 °C, 200 °C and 220 °C for 19 h, respectively.

The crystallization and crystal growth rates are accelerated when increasing the heating temperature to 200 °C. As shown, only strong and narrow XRD diffraction peaks corresponding to SAPO-34 are seen. Furthermore, no amorphous hump is detected revealing that all amorphous particles have been consumed serving as nutrients for SAPO-34 formation [36]. The same phenomenon is also shown by FESEM analysis where large cuboid crystals (ca. 1.86 μm) without amorphous spherical particles are captured. However, a partial transformation of SAPO-34 (CHA topology, FD = 15.1 T/1000 Å^3) into SAPO-36 of a denser framework (ATS topology, FD = 16.1 T/1000 Å^3) is observed with heating temperature increases to 220 °C. As shown, the XRD results indicate SAPO-36 is formed at the expense of SAPO-34 where the needle-like structures (ca. 7.5 μm, SAPO-36) dominate the FESEM image over the cuboid SAPO-34 particles (Figure 6d). Hence, the results indicate that the crystallization of SAPO-34 zeolite is a thermally activated process. However, its metastable nature leads it to the transformation into another denser zeolite phase, especially at higher temperatures [37].

The TGA/DTG technique is used to prove the role of [PrPy]$^+$ as a supramolecular structure-directing agent on SAPO-34. Figure 7 shows the TGA/DTG thermogram of uncalcined SAPO-34 (S-3). The first stage of weight loss (7.3%) at below 200 °C is due to physisorbed water, the second weight loss (7.9%) at 200–566 °C is due to the decomposition of [PrPy]$^+$ molecules, while the third weight loss (1.0%) at 566–795 °C is due to carbon coke deposition from incomplete combustion of [PrPy]$^+$. As seen, the decomposition of [PrPy]$^+$ is delayed at higher temperatures due to the occlusion of [PrPy]$^+$ in the pores of SAPO-34. From the weight loss, it is found that one molecule of [PrPy]$^+$ is packed and enfolded by ca. 18 units of TO$_2$ primary building units (T = Si, Al, or P). It is believed that the occlusion of [PrPy]$^+$ adapts the "ship in a bottle" pathway where the TO$_2$ units polymerize around the [PrPy]$^+$ (3.2 Å × 6.8 Å × 2.5 Å) before the supramolecule is trapped inside the pores (3.8 Å × 3.8 Å) [28].

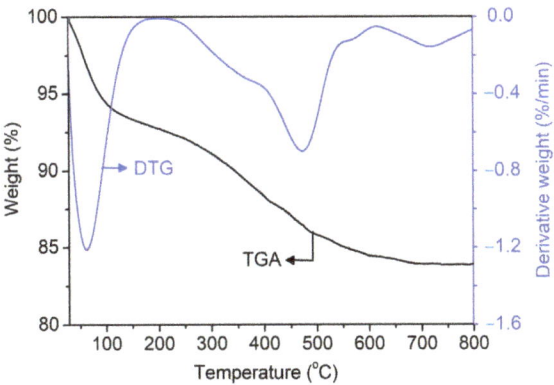

Figure 7. TGA/DTG thermograms of uncalcined SAPO-34 (S-3) sample.

2.2. Surface Characterizations of SAPO-34

The porosity of SAPO-34 crystallized using PrPy]OH supramolecule was studied using nitrogen adsorption-desorption analysis at −196 °C and the data is shown in Figure 8 and Table 1. The solid displays a type I curve shape without steep N_2 uptake at high P/P_o, indicating it merely has microporosity [38]. The solid has high porosity showing a high specific surface area (S_{BET} = 673 m^2 g^{-1}), micropore surface area (661 m^2 g^{-1}), and total pore volume (V_{Total} = 0.27 cm^3 g^{-1}). In addition, the SAPO-34 solid also exhibits moderate insertion of Si atoms (Si/(Al + P) ratio = 0.24) into the CHA zeolite framework resulting in the formation of surface acidity of different strengths (Table 1). As shown, the microporous solid possesses weak-to-medium (desorption curves at 168 °C and 235 °C) and medium-to-strong (desorption curves at 404 °C and 479 °C) acid sites arising from T–OH (T = Al, P, or Si) defect sites and framework tetrahedral Si sites, respectively [39]. The total number of acid sites of [PrPy]OH-synthesized SAPO-34 is found to be 2.52 mmol g^{-1} which is higher than that of the classical TEAOH-synthesized SAPO-34 (1.79 mmol g^{-1}) [11].

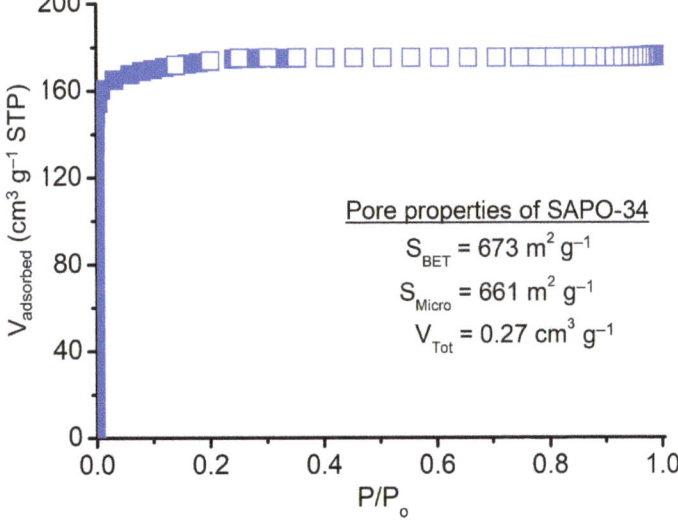

Figure 8. Nitrogen adsorption-desorption isotherm of SAPO-34 (S-3).

2.3. Catalytic Reaction Study

SAPO-34 zeolite is an important solid catalyst for chemical production and hence the catalytic behavior of SAPO-34 crystallized using [PrPy]OH was investigated in the microwave-assisted esterification of levulinic acid with 1-propanol. The reaction was first run at 130 °C without catalyst and no reaction conversion is observed after 60 min. This shows that microwave heating is inactive in the esterification reaction. When SAPO-34 zeolite is added, a significant increase in reaction conversion (60.6%) with 100% selectivity towards propyl levulinate is observed after 20 min of heating (Figure 9). The esterification conversion keeps increasing with the extension of the heating time and 82.2% of conversion is achieved at 60 min. Thus, the results show that SAPO-34 catalyst is active in this reaction whereby the catalytic activity of SAPO-34 comes from the (P-O-Si-O-Al)$^-$H$^+$ active sites located at the surface of the solid.

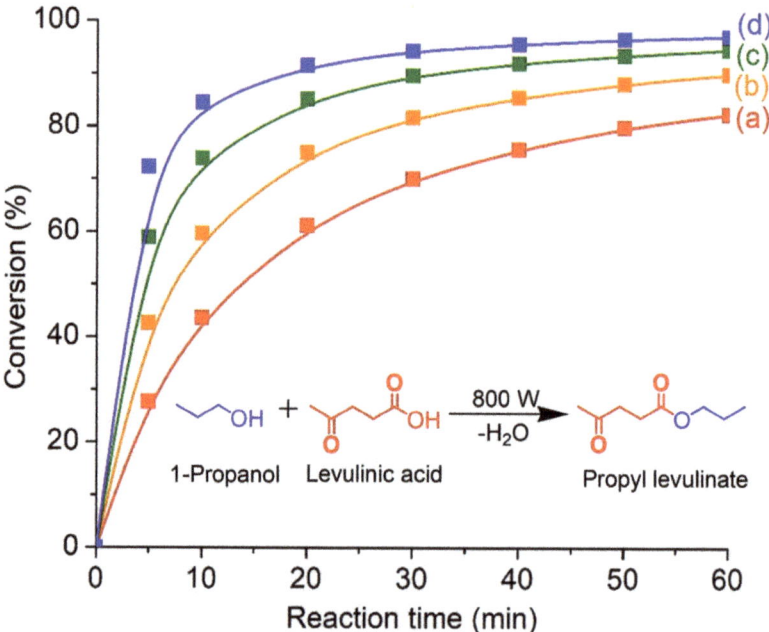

Figure 9. Microwave-assisted esterification of levulinic acid with 1-propanol in the presence of SAPO-34 (S-3) catalyst at (**a**) 130 °C, (**b**) 150 °C, (**c**) 170 °C, and (**d**) 190 °C for 60 min. Reaction conditions: Catalyst loading = 0.100 g; 1-propanol:levulinic acid molar ratio = 11:1; microwave power = 800 W; solvent-free.

The esterification reaction is also carried out at elevated temperatures (130–190 °C) over various times (0–60 min). As shown, a high reaction temperature is beneficial for the esterification of levulinic acid since the reaction kinetics are exponentially proportional to the temperature as shown by the Arrhenius theory [40]. For instance, the conversion is significantly enhanced from 60.6% to 91.2% when the temperature increases from 130 °C to 190 °C after 20 min of reaction. Thus, it is speculated that higher temperatures provide higher kinetic energy to the reactants (levulinic acid and 1-propanol) and the organic guest molecules are hence able to diffuse rapidly into the micropores of SAPO-34. The confinement environment of the zeolite then facilitates the orientation and chemisorption of reactants molecules, enabling the esterification to take place [41].

A catalytic comparative study between SAPO-34 and common homogeneous (e.g., sulfuric acid, hydrochloric acid, acetic acid) and heterogeneous (H-LTL, SAPO-18, H-FAU-Y) catalysts was also carried out (Figure 10, see Supplementary Materials: Table S1). All heterogeneous catalysts are 100% selective to propyl levulinate, while medium pore SAPO-34 and SAPO-18 zeolites (>90%) show better performance than large pore H-LTL and H-FAU-Y (ca. 75.0%) zeolites. The better catalytic activity of SAPO-34 (3.8 × 3.8 × 3.8 Å^3) and SAPO-18 (3.8 × 3.8 × 3.6 Å^3) could be due to their higher acidity and small pores that can exhibit proximity effect, viz. allowing oriented molecular diffusion inside the narrow pores and organize the levulinic acid (2.49 × 5.98 × 1.78 Å^3) and 1-propanol (3.11 × 5.35 × 2.54 Å^3) molecules on the acid sites so that the reactants molecules are much closer together than they would be in large pore FAU-Y (7.4 × 7.4 × 7.4 Å^3) and LTL (7.5 × 2.1 × 2.1 Å^3) zeolites [42]. Conversely, strong homogeneous catalysts like sulfuric acid and hydrochloric acid experience full conversion of levulinic acid into propyl levulinate. Nevertheless, they are not reusable and the system has to undergo laborious neutralization, separation, and purification processes to isolate the reaction product [43]. For acetic acid, the conversion is the lowest (60.1%) due to its weak acidity.

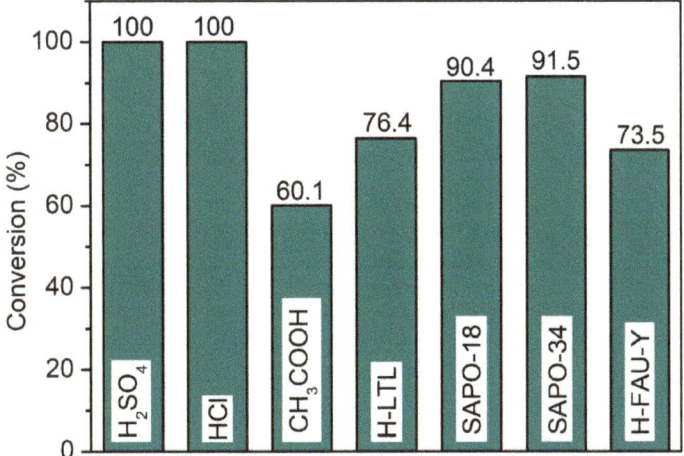

Figure 10. Catalytic performance of various homogeneous and heterogeneous acid catalysts in the esterification of levulinic acid with 1-propanol. Reaction conditions: Catalyst loading = 7.70 µmol equivalent to 0.100 g SAPO-34; 1-propanol:levulinic acid molar ratio = 11:1; reaction temperature = 190 °C; heating time = 20 min; microwave power = 800 W; solvent-free.

The major concern with heterogeneous catalysts is their stability and recyclability [44]. Hence, the catalyst recyclability of SAPO-34 was tested (Figure 11). The results reveal that the catalyst reactivity and stability are preserved even after five reaction cycles where an insignificant loss in conversion and selectivity are observed; a slight decrease in reaction conversion might be due to the physical loss of the catalyst during the recovery operation. Hence, SAPO-34 can be a promising eco-friendly and recyclable solid catalyst for the replacement of conventional homogeneous catalysts in biofuel synthesis in addition to being an efficient molecular sieve for gas separation.

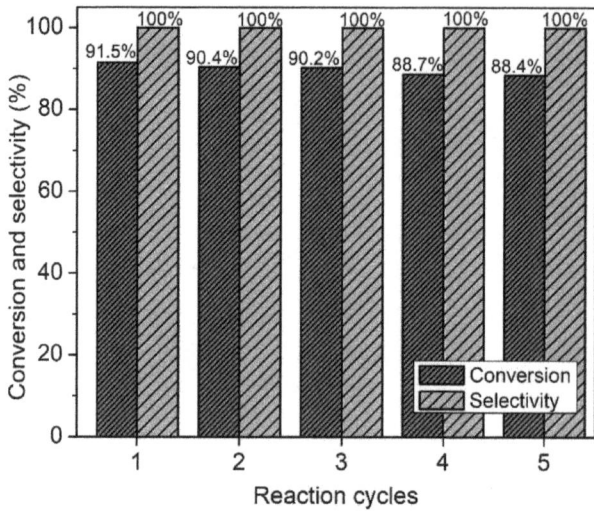

Figure 11. SAPO-34 catalyst reusability test after five reaction cycles.

3. Materials and Methods

3.1. Preparation of 1-Propylpyridinium Hydroxide, [PrPy]OH

Scheme 1 displays the synthesis procedure of the novel [PrPy]OH organic template. Typically, pyridine (82.20 g, 99%, Acros Organics, Geel, Belgium) and 1-bromopropane (191.71 g, 99%, Acros Organics, Geel, Belgium) were first mixed and heated under reflux for 18 h at 70 °C. The mixture was rotary evaporated under reduced pressure to remove excessive and unreacted 1-bromopropane, producing 1-propylpyridinium bromide ([PrPy]Br). The yellowish solid was soaked and purified with acetone (90 mL) several times prior to drying at 100 °C overnight to give an 87% yield of [PrPy]Br. The purity of [PrPy]Br was confirmed by ^{1}H and ^{13}C NMR, IR, and CHN analyses: ^{1}H NMR: δ (ppm) = 1.02 (triplet), 2.12 (multiplet), 4.70 (triplet), 8.18 (triplet), 8.65 (triplet), 8.97 (doublet). ^{13}C NMR: δ (ppm) = 10.25, 24.08, 63.73, 127.86, 144.02, 145.58. FTIR: ν (cm^{-1}) = 1169 (C–N), 1488 (C=N), 1503 & 1632 (C=C), 2878 (C_{sp^3}–H), 3054 (C_{sp^2}–H), 3405 (O–H). CHN analysis: $C_8H_{12}NBr$ theoretical = C, 47.54%; H, 5.95%; N, 6.92%; Experimental = C, 46.80%; H, 5.84%; N, 6.84%.

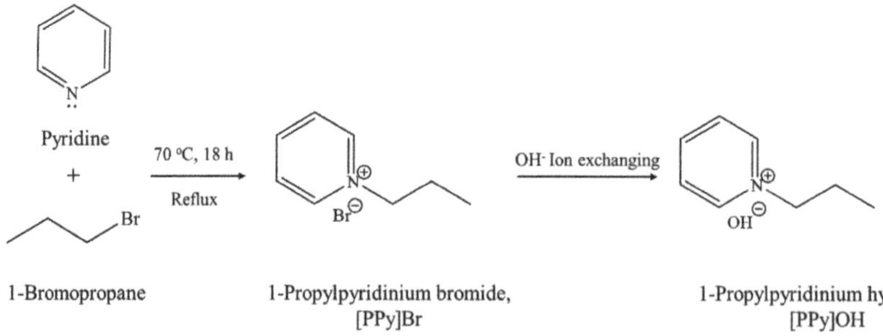

Scheme 1. The synthesis pathway of [PrPy]OH organic structural directing agent.

Subsequently, 50.00 g of both [PrPy]Br ionic liquid and Amberlite® IRN-78 OH$^-$ resins (Acros Organics, Geel, Belgium) were mixed with 50.00 g of deionized water. The ion-exchange treatment was then performed at room temperature (17 h, 250 rpm). The template solution and the resins are isolated using vacuum filtration where the percentage of OH⁻ ion exchange was confirmed by using 1.0 mL of the solution titrated with HCl (0.1 M). A similar ion-exchange procedure was performed until 87.0% OH⁻ exchange was achieved before allowing the sample to slowly concentrate to 32.8 wt.% in an electric oven (55 °C).

3.2. Single-Parameter Tuning Synthesis and Crystallization of SAPO-34 Microporous Solid

The synthesis of SAPO-34 microporous solid was conducted as follows: Aluminum triisopropoxide (0.6245 g, 98%, Sigma-Aldrich, Darmstadt, Germany) was first mixed with the [PrPy]OH template solution (4.5002 g) and stirred (600 rpm) for 30 min at room temperature. Afterward, deionized water (0.8350 g) was introduced. Then, orthophosphoric acid (0.8065 g, 85%, Acros Organics, Geel, Belgium) was slowly added whereby the entire addition process required 30 min. Lastly, tetraorthosilicate (0.2767 g, 98%. Sigma-Aldrich, Darmstadt, Germany) was introduced and stirred, forming a hydrogel mixture composed of 2.0 Al: 4.7 P: 6.7 [PrPy]OH: 0.9 Si: 148 H_2O molar composition. The mixture was loaded into an autoclave for crystallization at 200 °C for 19 h. The resulting solid product was washed with deionized water via centrifugation (9500 rpm, 8 min) until pH = 7 before being dried in an oven (70 °C, 16 h).

The effects of various parameters in the crystallization of SAPO-34 were also studied by varying one parameter of the synthesis conditions each time as summarized in Table 2 using similar preparation protocols. The samples were designated as S-n whereby n corresponded to the number of the sample.

Table 2. The hydrothermal crystallization conditions of SAPO-34 by varying the synthesis variables and the phase products obtained.

Entry	Variable Parameters	Hydrogel Composition in Molar Ratio					T (°C)	t (h)	pH before Synthesis	pH after Synthesis	Products *
		Al	P	Si	[PPy]	H_2O					
S-0								6	6.68	6.75	No solid
S-1								12	6.72	6.80	Am
S-2	Crystallization time	2.0	4.7	0.9	6.7	148	200	16	6.81	6.92	Am ≫ CHA
S-3								19	6.67	6.88	CHA
S-4								30	6.70	6.90	ATS ≫ CHA
S-5			4.1						7.18	7.23	Am
S-3	P_2O_5	2.0	4.7	0.9	6.7	148	200	19	6.67	6.88	CHA
S-6			5.3						6.18	6.77	CHA > AFI
S-7			5.9						5.61	6.34	AFI
S-8				0					6.76	7.00	Layered SAPO
S-9				0.5					7.06	7.15	ATS
S-3	SiO_2	2.0	4.7	0.9	6.7	148	200	19	6.67	6.88	CHA
S-10				1.4					6.86	7.00	LEV ≫ CHA
S-11					5.3				5.76	6.79	AFI
S-12	[PPy]₂O	2.0	4.7	0.9	6.0	148	200	19	6.31	6.64	AFI ≫ CHA
S-3					6.7				6.67	6.88	CHA
S-13					7.4				7.12	7.21	Am
S-14						116			7.03	7.09	ATS ≫ CHA
S-3						148			6.67	6.88	CHA
S-15	H_2O	2.0	4.7	0.9	6.7	180	200	19	6.55	6.86	ATS ≫ LEV > CHA
S-16						240			6.49	6.85	AFI ≫ ATS
S-17							160		6.79	7.09	Am ≫ CHA
S-18	Heating temperature	2.0	4.7	0.9	6.7	148	180	19	6.73	7.09	Am > CHA
S-3							200		6.67	6.88	CHA
S-19							220		6.80	6.44	ATS ≫ CHA

* Am = Amorphous, AFI = SAPO-5, ATS = SAPO-36, CHA = SAPO-34, LEV = SAPO-35.

3.3. Characterization

The elemental analysis of [PrPy]Br was performed using a Perkin Elmer 2400 Series II CHNS/O instrument (Waltham, MA, USA). The liquid ^{1}H and ^{13}C NMR spectra were acquired via a Bruker Advance 500 MHz spectrometer (Waltham, MA, USA) using a single pulse excitation with $\pi/2$ (3 μs) pulses, operating at 500 and 126 MHz, respectively. Deuterated oxide (D_2O) and tetramethylsilane (TMS) were used as the solvent and reference, respectively. The identity of the functional groups of [PrPy]Br was confirmed using a Perkin Elmer's System 2000 spectrometer (resolution 4 cm^{-1}, 50 scans, Waltham, MA, USA). The KBr pellet was prepared according to a KBr: sample weight ratio of 50:1. The crystallinity and zeolite phase composition were investigated using a Bruker D8 Advance diffractometer (Waltham, MA, USA) with copper Kα as the radiation source (λ = 0.154 nm at 40 kV and 10 mA, scanning rate 0.2° min^{-1}). The morphology and surface properties of solids were studied using a Hitachi Regulus 8220 FESEM microscope (Tokyo, Japan) at a voltage of 5 kV. The elemental composition of SAPO-34 (S-3 sample) was determined through a Phillips X'Unique XRF spectrometer (Cambridge, UK). The organic and inorganic moieties in the solid were studied using a Mettler TGA SDTA851 instrument (Columbus, OH, USA, heating rate of 20 °C min^{-1}) under a flow of air. The porous properties of SAPO-34 were investigated using a Micromeritics ASAP 2010 instrument (Norcross, GA, USA) at −196 °C. Before the analysis, ca. 0.080 g of calcined SAPO-34 powder was first degassed under vacuum at 250 °C for 6 h (to remove adsorbed impurities). The number and acid strength of synthesized SAPO-34 were investigated via BELCAT-B analyzer (York, PA, USA) using the NH$_3$-TPD technique. First, ca. 80 mg of calcined SAPO-34 sample was heated at 400 °C for 4 h under a helium flow of 30 cm^3 min^{-1}. Then, the solid was allowed to adsorb with NH$_3$. The excess NH$_3$ was removed by purging with helium for 20 min prior to initiating NH$_3$ desorption from 50 °C to 700 °C using a heating rate of 10 °C min^{-1}.

3.4. Catalytic Reaction Study

Esterification of levulinic acid and 1-propanol was used as a model reaction to study the catalytic performance of SAPO-34 (S-3 sample) under microwave heating (Monowave 200, Anton Paar, Graz, Austria). The resulting reaction product—propyl levulinate—is a very useful compound for biofuel applications [45]. First, the calcined SAPO-34 (0.100 g, 250 °C, 2 h) was mixed with levulinic acid (0.160 g, 1.34 mmol, 98%, Merck, Darmstadt, Germany) and 1-propanol (0.893 g, 14.7 mmol, 99%, Merck, Darmstadt, Germany) in a glass vessel. The resulting reaction mixture was stirred vigorously (500 rpm, 1 min) to ensure its homogeneity before heating to 160–190 °C for 0–60 min. The liquid product was separated using centrifugation (9000 rpm, 3 min) and subjected to qualitative and quantitative analyses using GC-MS (Agilent 7000 Series Triple Quad, Santa Clara, CA, USA) and GC-FID (Agilent's HP6890 GC, Santa Clara, CA, USA), respectively. For quantitative analysis, toluene was used as the internal standard. For the recyclability study, the spent catalyst after each cycle was soaked and washed with diethyl ether five times (10 mL) before re-activation. A similar catalytic reaction procedure was repeated for the next four cycles.

4. Conclusions

In conclusion, the effects of synthesis variables on the crystallization of SAPO-34 in the presence of a 1-propylpyridinium hydroxide ([PrPy]OH) template have been systematically investigated. The findings show that the nucleation, crystallization, structural purity and metastability, crystallite size, and morphology are strongly affected by the synthesis conditions of SAPO-34, namely, the chemical composition of the initial precursor gel, heating temperature, and heating duration. During the course of the slight altering of the synthesis parameters, SAPO-5 (AFI), SAPO-35 (LEV), and SAPO-36 (ATS) as the common competing phases are observed. More specifically, the pH of the precursor gel influences the final zeolite phase where SAPO-34 favors a nearly neutral precursor gel (pH = 6.6–6.8) while SAPO-5 tends to predominate at lower pH values (pH = 5.6–5.8). Meanwhile, SAPO-36 forms at lower silicon content, higher crystallization temperatures, and longer heating times due to its higher metastability and denser framework nature whereas SAPO-35 can only form at high silicon content as revealed in the parameter effects study. In addition, the [PrPy]OH SDA not only helps in forming a highly crystalline SAPO-34 zeolite structure but is also an SDA for SAPO-5 in a weakly acidic environment. Nevertheless, a precise water content ($H_2O/Al = 74$) is very important to ensure the crystallization of pure SAPO-34. Lastly, the SAPO-34 zeolite synthesized using [PrPy]OH is also catalytically active in the synthesis of propyl levulinate via esterification reaction where it shows 91.5% of conversion in only 20 min when heated at 190 °C under microwave heating conditions. Hence, this study provides insights into the structural-directing ability and selectivity of the novel [PrPy]OH organic template under various synthesis parameters. In addition to being reusable, the prepared SAPO-34 hence provides a promising green alternative for the economical and sustainable production of renewable fuel blending levulinate compound.

Supplementary Materials: The following supporting information can be downloaded at: https://www.mdpi.com/article/10.3390/catal13040680/s1, Table S1: Physicochemical properties of various zeolites.

Author Contributions: Conceptualization, T.S.A., N.A. and E.-P.N.; methodology, Y.-K.M.; software, T.S.A. and N.A.; formal analysis, Y.-K.M., T.J.D. and E.-P.N.; investigation, Y.-K.M., T.J.D. and F.K.; resources, S.C.; data curation, T.J.D.; writing—original draft preparation, Y.-K.M.; writing—review and editing, Z.M.E.-B. and E.-P.N.; supervision, T.C.L. and E.-P.N.; project administration, T.S.A., N.A. and E.-P.N.; funding acquisition, T.S.A., N.A. and E.-P.N. All authors have read and agreed to the published version of the manuscript.

Funding: Fundamental Research Grant Scheme (FRGS/1/2022/STG05/USM/02/12, Ministry of Higher Education of Malaysia), and Princess Nourah bint Abdularahman University Researchers Supporting Project number (PNURSP2023R47), Princess Nourah bint Abdularahman University, Riyadh, Saudi Arabia.

Acknowledgments: The authors would like to acknowledge the Fundamental Research Grant Scheme (FRGS/1/2022/STG05/USM/02/12, Ministry of Higher Education of Malaysia) for financial support. The authors also acknowledge the financial support from Princess Nourah bint Abdularahman University Researchers Supporting Project number (PNURSP2023R47), Princess Nourah bint Abdularahman University, Riyadh, Saudi Arabia.

Conflicts of Interest: The authors declare no conflict of interest.

References

1. Ng, E.-P.; Awala, H.; Komaty, S.; Mintova, S. Microwave-green synthesis of AlPO-n and SAPO-n (n = 5 and 18) nanosized crystals and their assembly in layers. *Microporous Mesoporous Mater.* **2019**, *280*, 256–263. [CrossRef]
2. Tosheva, L.; Ng, E.-P.; Mintova, S.; Hölzl, M.; Metzger, T.H.; Doyle, A.M. AlPO$_4$-18 Seed Layers and Films by Secondary Growth. *Chem. Mater.* **2008**, *20*, 5721–5726. [CrossRef]
3. Ng, E.-P.; Delmotte, L.; Mintova, S. Selective capture of water using microporous adsorbents to increase the lifetime of lubricants. *ChemSusChem* **2009**, *2*, 255–260. [CrossRef] [PubMed]
4. Usman, M. Recent Progress of SAPO-34 Zeolite Membranes for CO_2 Separation: A Review. *Membranes* **2022**, *12*, 507. [CrossRef]
5. Park, J.W.; Lee, J.Y.; Kim, K.S.; Hong, S.B.; Seo, G. Effects of cage shape and size of 8-membered ring molecular sieves on their deactivation in methanol-to-olefin (MTO) reactions. *Appl. Catal. A Gen.* **2008**, *339*, 36–44. [CrossRef]
6. Zhao, L.; Yang, G.; Hu, S.; Sun, Y.; Ma, Z.; Peng, P.; Ng, E.-P.; Tian, P.; Guo, H.; Mintova, S. SAPO-34 crystals with nanosheet morphology synthesized by pyrophosphoric acid as new phosphorus source. *Microporous Mesoporous Mater.* **2022**, *333*, 111753. [CrossRef]
7. Nishitoba, T.; Nozaki, T.; Park, S.; Wang, Y.; Kondo, J.N.; Gies, H.; Yokoi, T. CHA-type zeolite prepared by interzeolite conversion method using FAU and LTL-type zeolite: Effect of the raw materials on the crystallization mechanism, and physico-chemical and catalytic properties. *Catalysts* **2020**, *10*, 1204. [CrossRef]
8. Li, J.; Li, Z.; Han, D.; Wu, J. Facile synthesis of SAPO-34 with small crystal size for conversion of methanol to olefins. *Powder Technol.* **2014**, *262*, 177–182. [CrossRef]
9. Liu, G.; Tian, P.; Liu, Z. Chin. Synthesis of SAPO-34 molecular sieves templated with diethylamine and their properties compared with other templates. *J. Catal.* **2012**, *33*, 174–182.

10. Zhou, H.; Wang, Y.; Wei, F.; Wang, D.; Wang, Z. Kinetics of the reactions of the light alkenes over SAPO-34. *Appl. Catal. A Gen.* **2008**, *348*, 135–141. [CrossRef]
11. Askari, S.; Halladj, R.; Sohrabi, M. Methanol conversion to light olefins over sonochemically prepared SAPO-34 nanocatalyst. *Microporous Mesoporous Mater.* **2012**, *163*, 334–342. [CrossRef]
12. Dumitriu, E.; Azzouz, A.; Hulea, V.; Lutic, D.; Kessler, H. Synthesis, characterization and catalytic activity of SAPO-34 obtained with piperidine as templating agent. *Microporous Mesoporous Mater.* **1997**, *10*, 1–12. [CrossRef]
13. Prakash, A.M.; Unnikrishnan, S. Synthesis of SAPO-34: High silicon incorporation in the presence of morpholine as template. *J. Chem. Soc. Faraday Trans.* **1994**, *90*, 2291–2296. [CrossRef]
14. Ng, E.-P.; Itani, L.; Sekhon, S.S.; Mintova, S. Micro- to macroscopic observations of MnAlPO-5 nanocrystal growth in ionic-liquid media. *Chem. Eur. J.* **2010**, *16*, 12890–12897. [CrossRef]
15. Khoo, D.Y.; Kok, W.-M.; Mukti, R.R.; Mintova, S.; Ng, E.-P. Ionothermal approach for synthesizing AlPO-5 with hexagonal thin-plate morphology influenced by various parameters at ambient pressure. *Solid State Sci.* **2013**, *25*, 63–69. [CrossRef]
16. Smith, J.G. *Organic Chemistry*, 6th ed.; McGraw-Hill Education: New York, NY, USA, 2017.
17. Majano, G.; Ng, E.-P.; Lakiss, L.; Mintova, S. Nanosized molecular sieves utilized as an environmentally friendly alternative to antioxidants for lubricant oils. *Green Chem.* **2011**, *13*, 2435–2440. [CrossRef]
18. Wong, J.-T.; Ng, E.-P.; Adam, F. Microscopic Investigation of Nanocrystalline Zeolite L Synthesized from Rice Husk Ash. *J. Am. Ceram. Soc.* **2011**, *95*, 805–808. [CrossRef]
19. Cundy, C.S.; Cox, P.A. The hydrothermal synthesis of zeolites: Precursors, intermediates and reaction mechanism. *Microporous Mesoporous Mater.* **2005**, *82*, 1–78. [CrossRef]
20. Auwal, I.; Khoerunnisa, F.; Dubray, F.; Mintova, S.; Ling, T.; Wong, K.-L.; Ng, E.-P. Effects of Synthesis Parameters on the Crystallization Profile and Morphological Properties of SAPO-5 Templated by 1-Benzyl-2,3-Dimethylimidazolium Hydroxide. *Crystals* **2021**, *11*, 279. [CrossRef]
21. Van Heyden, H.; Mintova, S.; Bein, T. Nanosized SAPO-34 Synthesized from Colloidal Solutions. *Chem. Mater.* **2008**, *20*, 2956–2963. [CrossRef]
22. Zhao, C.; Yang, Y.; Chen, W.; Wang, H.; Zhao, D.; Webley, P.A. Hydrothermal synthesis of novel AlPO4-5 brooms and nano-fibers and their templated carbon structures. *CrystEngComm* **2009**, *11*, 739–742. [CrossRef]
23. IZA-SC Database of Zeolite Structures. Available online: https://www.izastructure.org/databases/ (accessed on 5 July 2022).
24. Barthomeuf, D. Topological model for the compared acidity of SAPOs and SiAl zeolites. *Zeolites* **1994**, *14*, 394–401. [CrossRef]
25. Salmasi, M.; Fatemi, S.; Hashemi, S. MTO reaction over SAPO-34 catalysts synthesized by combination of TEAOH and morpholine templates and different silica sources. *Sci. Iran.* **2012**, *19*, 1632–1637. [CrossRef]
26. Baerlocher, C.; Meier, W.M.; Olson, D.H. *Atlas of Zeolite Structure Types*, 5th ed.; Elsevier: Amsterdam, The Netherlands, 1996.
27. Concepción, P.; Nieto, J.M.L.; Mifsud, A.; Perez-Pariente, J. Preparation and characterization of Mg-containing AFI and chabazite-type materials. *Zeolites* **1996**, *16*, 56–64. [CrossRef]
28. *The Molecular Dimension of 1-Propylpyridinium Cation Was Estimated Using Hyperchem™-Release 7.03 for Windows Molecular Modeling System*; Hypercube, Inc.: Lancaster, TX, USA, 2002.
29. Newalkar, B.L.; Kamath, B.V.; Jasra, R.V.; Bhat, S.G.T. The effect of gel pH on the crystallization of aluminophosphate molecular sieve AlPO4-5. *Zeolites* **1997**, *18*, 286–290. [CrossRef]
30. Ghrear, T.M.A.; Rigolet, S.; Daou, T.J.; Mintova, S.; Ling, T.C.; Tan, S.H.; Ng, E.-P. Synthesis of Cs-ABW nanozeolite in organotemplate-free system. *Microporous Mesoporous Mater.* **2019**, *277*, 78–83. [CrossRef]
31. Cheong, Y.-W.; Rigolet, S.; Daou, T.J.; Wong, K.-L.; Ling, T.C.; Ng, E.-P. Crystal growth study of nanosized K-MER zeolite from bamboo leaves ash and its catalytic behaviour in Knoevenagel condensation of benzaldehyde with ethyl cyanoacetate. *Mater. Chem. Phys.* **2020**, *251*, 123100. [CrossRef]
32. Jhung, S.H.; Chang, J.-S.; Hwang, J.S.; Park, S.-E. Selective formation of SAPO-5 and SAPO-34 molecular sieves with micro-wave irradiation and hydrothermal heating. *Microporous Mesoporous Mater.* **2003**, *64*, 33–39. [CrossRef]
33. Newsam, J.M. The Zeolite Cage Structure. *Science* **1986**, *231*, 1093–1099. [CrossRef]
34. Cundy, C.S.; Cox, P.A. The Hydrothermal Synthesis of Zeolites: History and Development from the Earliest Days to the Present Time. *Chem. Rev.* **2003**, *103*, 663–702. [CrossRef]
35. Francis, R.J.; O'Hare, D. The kinetics and mechanisms of the crystallisation of microporous materials. *J. Chem. Soc. Dalton Trans.* **1998**, 3133–3148. [CrossRef]
36. Wong, S.-F.; Deekomwong, K.; Wittayakun, J.; Ling, T.C.; Muraza, O.; Adam, F.; Ng, E.-P. Crystal growth study of K-F nanozeolite and its catalytic behavior in Aldol condensation of benzaldehyde and heptanal enhanced by microwave heating. *Mater. Chem. Phys.* **2017**, *196*, 295–301. [CrossRef]
37. Askari, S.; Halladj, R.; Sohrabi, M. An overview of the effects of crystallization time, template and silicon sources on hydro-thermal synthesis of sapo-34 molecular sieve with small crystals. *Rev. Adv. Mater. Sci.* **2012**, *32*, 83–93.
38. Kruk, M.; Jaroniec, M.; Sayari, A. Adsorption Study of Surface and Structural Properties of MCM-41 Materials of Different Pore Sizes. *J. Phys. Chem. B* **1997**, *101*, 583–589. [CrossRef]
39. Derewinski, M.; Briend, M.; Peltre, M.J.; Main, P.P.; Barthomeuf, D. Changes in the environment of si and al in SAPO-37 zeolite during acidity measurements. *J. Phys. Chem.* **1993**, *97*, 13730–13735. [CrossRef]

40. Silbey, R.J.; Alberty, R.A.; Papadantonakis, G.A.; Bawendi, M.G. *Physical Chemistry*, 5th ed.; John, Wiley & Sons, Inc.: Hoboken, NJ, USA, 2022.
41. Beh, J.J.; Lim, J.K.; Ng, E.P.; Ooi, B.S. Synthesis and size control of zeolitic imidazolate framework-8 (ZIF-8): From the perspective of reaction kinetics and thermodynamics of nucleation. *Mater. Chem. Phys.* **2018**, *216*, 393–401. [CrossRef]
42. Li, T.; Chung, S.-H.; Nastase, S.; Galilea, A.; Wang, Y.; Mukhambetov, I.; Zaarour, M.; de Miguel, J.C.N.; Cazemier, J.; Dokania, A.; et al. Influence of active-site proximity in zeolites on Brønsted acid-catalyzed reactions at the microscopic and mesoscopic levels. *J. Chem. Catal.* **2023**, *3*, 100540. [CrossRef]
43. Mandari, V.; Devarai, S.K. Biodiesel production using homogeneous, heterogeneous, and enzyme catalysts via transesterification and esterification reactions: A critical review. *BioEnergy Res.* **2022**, *15*, 935–961. [CrossRef]
44. Shokouhimehr, M. Magnetically separable and sustainable nanostructured catalysts for heterogeneous reduction of nitroaromatics. *Catalysts* **2015**, *5*, 534–560. [CrossRef]
45. Appaturi, J.N.; Andas, J.; Ma, Y.-K.; Phoon, B.L.; Batagarawa, S.M.; Khoerunnisa, F.; Hussin, M.H.; Ng, E.-P. Recent advances in heterogeneous catalysts for the synthesis of alkyl levulinate biofuel additives from renewable levulinic acid: A comprehensive review. *Fuel* **2022**, *323*, 124362. [CrossRef]

Disclaimer/Publisher's Note: The statements, opinions and data contained in all publications are solely those of the individual author(s) and contributor(s) and not of MDPI and/or the editor(s). MDPI and/or the editor(s) disclaim responsibility for any injury to people or property resulting from any ideas, methods, instructions or products referred to in the content.

Article

Evidence of Synergy Effects between Zinc and Copper Oxides with Acidic Sites on Natural Zeolite during Photocatalytic Oxidation of Ethylene Using Operando DRIFTS Studies

Norberto J. Abreu [1,2,3,*], Héctor Valdés [1,*], Claudio A. Zaror [4], Tatianne Ferreira de Oliveira [5], Federico Azzolina-Jury [6] and Frédéric Thibault-Starzyk [6]

[1] Laboratorio de Tecnologías Limpias, Facultad de Ingeniería, Universidad Católica de la Santísima Concepción, Alonso de Ribera 2850, Concepción 4030000, Chile
[2] Departamento de Ingeniería Química, Facultad de Ingeniería y Ciencias, Universidad de la Frontera, Francisco Salazar 01145, Temuco 4780000, Chile
[3] Center of Waste Management and Bioenergy, Scientific and Technological Bioresources Nucleus, Universidad de la Frontera, Francisco Salazar 01145, Temuco 4780000, Chile
[4] Departamento de Ingeniería Química, Facultad de Ingeniería, Universidad de Concepción, Edmundo Larenas s/n, Concepción 4030000, Chile
[5] School of Agronomy, Federal University of Goiás, Goiânia 74690-900, Brazil
[6] Laboratoire Catalyse et Spectrochimie, Normandie Univ, ENSICAEN, UNICAEN, CNRS, 14050 Caen, France
* Correspondence: norberto.abreu@ufrontera.cl (N.J.A.); hvaldes@ucsc.cl (H.V.)

Abstract: In this article, the role of surface sites of modified zeolites with semiconductor nanoparticles as alternative photocatalysts for protecting post-harvest foodstuff from the detrimental effects of ethylene is addressed. Two single and one double catalyst based on zinc and copper oxides supported over modified zeolite samples were prepared. Physical, chemical, and surface properties of prepared materials were studied by several characterization methods. UV-Vis absorption spectra show that the applied modification procedures increase the optical absorption of light in the UV and visible regions, suggesting that an increase in the photocatalytic activity could take place mainly in the obtained co-impregnated catalyst. An ethylene conversion around 50% was achieved when the parent natural zeolite support was modified with both transition metal oxides, obtaining higher removal efficiency in comparison to single oxide catalysts. Adsorption and photocatalytic oxidation experiments were also performed using single and double catalysts supported over fumed silica, attaining lower ethylene conversion and thus highlighting the role of zeolite surfaces as adsorption sites for ethylene during photocatalytic reactions. *Operando* diffuse reflectance infrared Fourier transform spectroscopy (DRIFTS) studies reveal that a synergistic mechanism occurs, involving ethylene adsorption at acidic sites of zeolite and its photocatalytic oxidation due to the generation of radicals by the light activation of nanoparticles of zinc and copper oxides.

Keywords: adsorption; natural zeolite support; single and double metal oxide catalysts; photocatalysis; synergistic mechanism

Citation: Abreu, N.J.; Valdés, H.; Zaror, C.A.; de Oliveira, T.F.; Azzolina-Jury, F.; Thibault-Starzyk, F. Evidence of Synergy Effects between Zinc and Copper Oxides with Acidic Sites on Natural Zeolite during Photocatalytic Oxidation of Ethylene Using Operando DRIFTS Studies. *Catalysts* **2023**, *13*, 610. https://doi.org/10.3390/catal13030610

Academic Editors: De Fang and Yun Zheng

Received: 11 February 2023
Revised: 12 March 2023
Accepted: 14 March 2023
Published: 17 March 2023

Copyright: © 2023 by the authors. Licensee MDPI, Basel, Switzerland. This article is an open access article distributed under the terms and conditions of the Creative Commons Attribution (CC BY) license (https://creativecommons.org/licenses/by/4.0/).

1. Introduction

Ethylene is regarded as a gaseous multifunctional phytohormone that regulates growth and senescence of plants and fruits [1]. Even at a low concentration and temperature, ethylene can accelerate the ageing and spoiling of harvested fruits [2]. This powerful plant hormone cause effects at part-per-million (ppm) to part-per-billion (ppb) concentrations, affecting both the aesthetics and function of climacteric and non-climacteric fruit and vegetables [1]. For example, in the case of climacteric fruit such as bananas, tomatoes, apricots, and apples, the exposition to ethylene concentrations from 0.01 to 0.5 µL L^{-1} could induce shrinkage and decay. Moreover, high score registration of loss in fruit sensitive to ethylene was reported in the United Kingdom, where at least 1.4 million bananas and 1.5 million tomatoes were wasted daily in 2010, mainly caused by ethylene mediated effects [3]. Moreover, studies conducted in Germany during the same year mentioned that around 50% of food waste belongs to fruit and vegetables directly related with ripening induced by ethylene [4]. Hence, the control and removal of ethylene from fruits' environment is an important challenge for improving their quality and increasing their shelf life [5,6]. For such purpose, several methods have been focused on the removal of ethylene, including ventilation, recuperative adsorption, and techniques concerning destructive oxidation [7]. Among these methods, the combination of adsorbents with photocatalytic materials for the oxidation of ethylene was suggested as a very promising and cost-effective technique [7].

As a result of their specific properties, different structures of zeolites have been used for ethylene adsorption such as mordenite [8], clinoptilolite [9,10], and some synthetic zeolites [11–13]. Zeolites can be used as adsorbent materials or as supports in photocatalytic applications. Mechanical, physical, and chemical characteristics such as high mechanical resistance, individual microporous and mesoporous structure, the presence of Brønsted and Lewis acid sites, and a great ion-exchange capacity allow zeolites to be used in different applications [14,15]. It has been indicated that the exchange of compensating cations originally presented at zeolite surface by specific cations, increases the selectivity of the adsorption towards several compounds, including ethylene [16–19].

In the last years, the combined effect of photocatalysis and adsorption in the removal of pollutants has been of significant interest for the scientific community [14,20]. It has been claimed that such combination allows the concentration of the contaminants near photocatalytic active sites, promoting photocatalytic reactions, increasing the adsorption of the intermediaries, and the re-use of the adsorbent [15]. Although studies concerning both combined processes have been conducted [15,20], there is still a lack of experimental evidence about the role of the active sites in the reaction mechanism. Results presented here address this issue.

Semiconductor materials with a wide band gap, such as titanium dioxide (e.g., ~3.2 eV for anatase phase), has been extensively used as an effective photocatalyst due to its capacity to totally oxidize organic pollutants, its chemical stability, among some other properties [21,22]. However, it has been reported that TiO_2 tends to lose its activity during the photocatalytic oxidation of contaminants because of a decrease in the number of active sites on the catalyst surface, leading to the search of alternative materials [23,24]. Besides TiO_2, there are several materials that present excellent photocatalytic activities under UV light with a wide band gap energy. For instance, graphitic carbon nitride (g-C_3N_4), a photocatalyst with high stability and adequate light adsorption [25]; bismuth oxychloride (BiOCl), a novel semiconductor material used recently in the degradation of caffeic and gallic acids [26]; zinc oxide, an environmentally friendly catalyst used in the degradation and mineralization of environmental pollutants [27]; and some other semiconductor materials.

Among them, zinc oxide has emerged as a non-toxic, very common in nature n-type oxide semiconductor with a wide band gap (3.37 eV) and absorption bands in the UV spectra that can be used in several applications such as for cytotoxicity, antibacterial activities, solar cells, electrical and optical devices as well as in photocatalytic applications [28]. It has a high thermal and electro-chemical stability. It is also relatively cheap due to its high availability in nature [29]. Zinc oxide stands out for its high photocatalytic efficiency in the mineralization of organic contaminants in comparison to other similar semiconductors, including even TiO_2, as well as for its high stability [21]. The photocatalytic activity of ZnO has been proved several times, showing excellent results in the degradation of organic compounds such as rose bengal [30], methyl orange [31], rhodamine B [32], and some other emergent contaminants [33].

Recently, it has been suggested that the photocatalytic activity of zinc oxide can be enhanced and extended to the visible spectrum by doping with metal ions or by creating structures on the surface with lower energy gap semiconductors [34]. Copper oxide is a *p*-type semiconductor oxide with a narrow band gap (1.2 eV) [35] and can be excited even with radiation of visible spectrum [20]. Excellent photocatalytic results have come due to the modification of zinc oxide with copper oxide nanoparticles [36,37]. The high availability, excellent physico-chemical properties, and the low cost of copper make it an attractive material for applied research and encourage its recent use as a catalyst in several applications. Photocatalytic materials containing CuO have also been applied in the oxidation of organic pollutants such as methylene blue [38,39] and methyl orange [40] among several organic contaminants [41].

Double catalysts, such as zinc and copper oxides supported on modified natural zeolite, could enhance the photocatalytic activity by increasing the electron–hole pair separation efficiency and also by diminishing the recombination of generated electron–hole pairs by the mutual transfer of electrons or holes between both semiconductors [37], whereas acidic surface sites of zeolite could contribute to the retention of the adsorbed molecules near photocatalytic sites, enhancing catalytic performance.

Most of the research efforts for ethylene elimination have considered adsorption and photocatalysis as separated processes [7–11,42,43]. Although some authors used zinc oxide and copper oxide as single or bimetallic oxide heterojunctions supported on zeolites for photocatalysis [44–46], these works were mainly focused on removal rate and reaction kinetics. In this work, *operando* diffuse reflectance infrared Fourier transform spectroscopy (DRIFTS) studies were applied to unveil the chemical interactions that take place between ethylene molecules and active surface sites of modified zeolite with supported single and double zinc and copper oxides during the photocatalytic oxidation of ethylene. This work was focused on studying the synergistic ethylene adsorption/photocatalytic oxidation mechanism by considering the evolution of the infrared bands of the active sites under *operando* DRIFTS assays. Moreover, to assess the effect of the support in the reaction mechanism, fumed silica was used in control experiments as a support of metal oxides. Additionally, TiO_2 was mixed with the raw fumed silica and used as a benchmark photocatalyst for comparison. As a result, a surface reaction mechanism is proposed.

2. Results and Discussion
2.1. Physico-Chemical Characteristics

Chemical analyses of all samples are presented in Table 1. *Nitrogen adsorption isotherms* at 77 K reveal a combination of characteristic type I and IV isotherms, related to microporous and mesoporous structure behavior of the zeolite samples, respectively (see Figure S1 in Supplementary Materials) [47]. Due to the removal of impurities in the parent zeolite, the surface area increases and the Si/Al ratio diminishes by around 9% after the double ion exchange conducted using the $(NH_4)_2SO_4$ solution followed by a thermal outgassing at 623 K (see Table 1). However, subsequent modifications with transition metals diminish the surface areas, mainly in the samples impregnated with zinc oxides.

Table 1. Physical–chemical surface properties of parent zeolite and fumed silica supports and single and double metal oxide catalysts supported on modified natural zeolite and on fumed silica.

Samples	S_{BET} [a] [$m^2\,g^{-1}$]	V_{micro} [a] [$cm^3\,g^{-1}$]	V_{meso} [a] [$cm^3\,g^{-1}$]	Si/Al [b]	Si [b,c] [%]	Al [b,c] [%]	Ca [b,c] [%]	Na [b,c] [%]	Fe [b,c] [%]	Ti [b,c] [%]	Mg [b,c] [%]	Zn [b,c] [%]	Cu [b,c] [%]
Z_Nat	280.88	0.07	0.14	5.32	65.39	12.29	11.13	2.12	5.27	0.89	0.53	0.02	ND
ZH	358.24	0.12	0.10	5.44	71.13	13.07	6.83	0.38	5.66	0.90	0.30	0.02	ND
ZH_Cu	281.58	0.07	0.12	4.94	65.99	13.36	3.65	0.30	2.25	0.44	0.39	0.02	12.16
ZH_Zn	203.17	0.06	0.12	5.02	68.67	13.67	3.07	0.32	2.12	0.43	0.32	10.95	ND
ZH_Zn/Cu	261.40	0.04	0.13	5.01	68.42	13.65	3.03	0.41	2.16	0.44	0.31	5.42	5.42
FS	261.02	ND	0.32	ND	100.0	ND	ND	ND	ND	ND	ND	ND	ND
FS_Cu	242.35	ND	0.31	ND	92.20	ND	ND	ND	ND	ND	ND	ND	7.80
FS_Zn	224.80	ND	0.31	ND	92.11	ND	ND	ND	ND	ND	ND	7.89	ND
FS_Zn/Cu	220.29	ND	0.26	ND	92.07	ND	ND	ND	ND	ND	ND	3.78	4.15

[a] Determined by nitrogen adsorption/desorption at 77 K. [b] Zeolite supported samples composition determined by X-ray fluorescence. [c] Fumed silica supported samples composition determined by energy dispersive X-ray spectroscopy.

Results obtained from **XRD diffractograms** shown in Figure 1 reveal that the mineralogical and structural composition of the parent zeolite is not significantly changed after all the applied chemical and thermal treatments. The higher intensity peaks correspond to clinoptilolite (JCPDS 39-183), mordenite (JCPDS 29-1257), and quartz (JCPDS 46-1045). Such peaks remain with similar intensity after the applied modification procedures. Moreover, the peaks related to copper oxide (JCPDS 48-1548) and zinc oxide (JCPDS 36-1451) can be identified in the diffractograms of the modified samples.

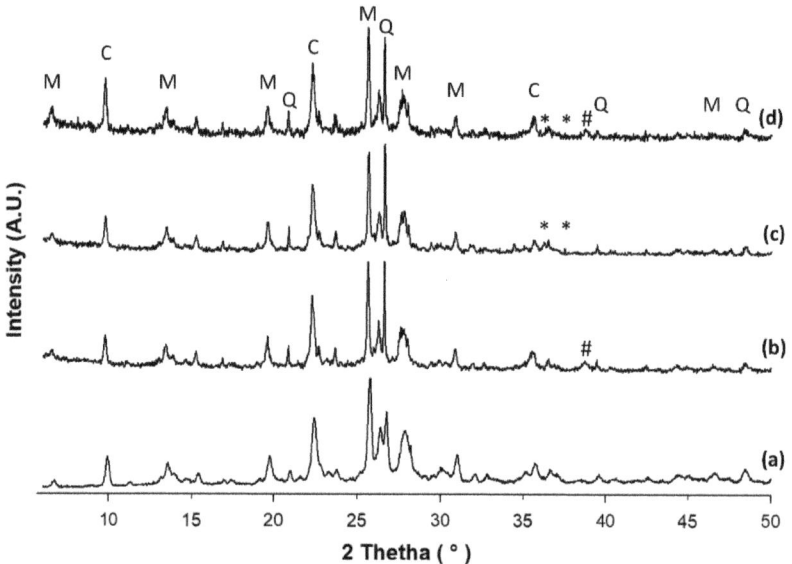

Figure 1. X-ray diffraction patterns of parent zeolite support and prepared single and double metal oxide catalysts supported on modified natural zeolite. (**a**) ZNAT; (**b**) ZH_Cu; (**c**) ZH_Zn; (**d**) ZH_Zn/Cu. M: Mordenite, C: Clinoptilolite, Q: Quartz, #: copper oxide, *: zinc oxide.

Morphological structures, as observed by *Scanning Electron Microscopy images*, are depicted for zeolite samples in Figure 2A,C–E. SEM images reveal porous lamellar morphologies with individual zeolite crystals around a 40–70 nm width. Similar results were reported in clinoptilolite pore network studies by other authors [48]. Meanwhile, for fumed silica samples (see Figure 2B,F–H), individual quasi-spherical shaped particles around 60 μm were observed with an evident porous structure. The porous characteristics were confirmed by nitrogen adsorption/desorption assays (see Table 1 and Figures S1 and S2 in Supplementary Materials); however, in those samples supported on fumed silica, only a mesoporosity network was found without a microporous structure. In the case of zeolite samples, the surface area reduction after conducting the modification with transition metals can be associated to the formation of metal oxide nanoparticles and their agglomeration, generating particles of a bigger size inside the zeolite porous structure that can block nitrogen diffusion. SEM and transmission electron microscopy (TEM) images corroborate such findings (Figures 2 and 3).

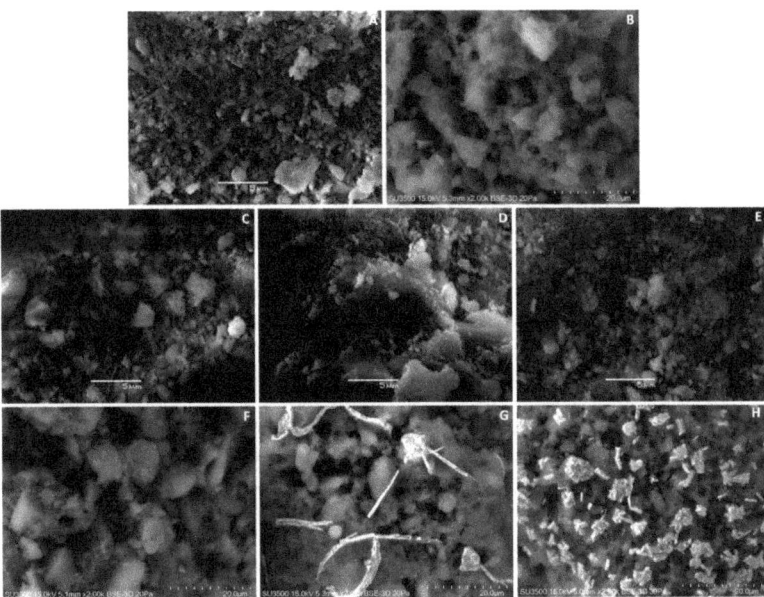

Figure 2. Scanning electron microscopy images of parent zeolite support and prepared single and double metal oxide catalysts supported on modified natural zeolite: (**A**) ZNAT; (**B**) FS; (**C**) ZH_Cu; (**D**) ZH_Zn; (**E**) ZH_Zn/Cu; (**F**) FS_Cu; (**G**) FS_Zn; (**H**) FS_Zn/Cu.

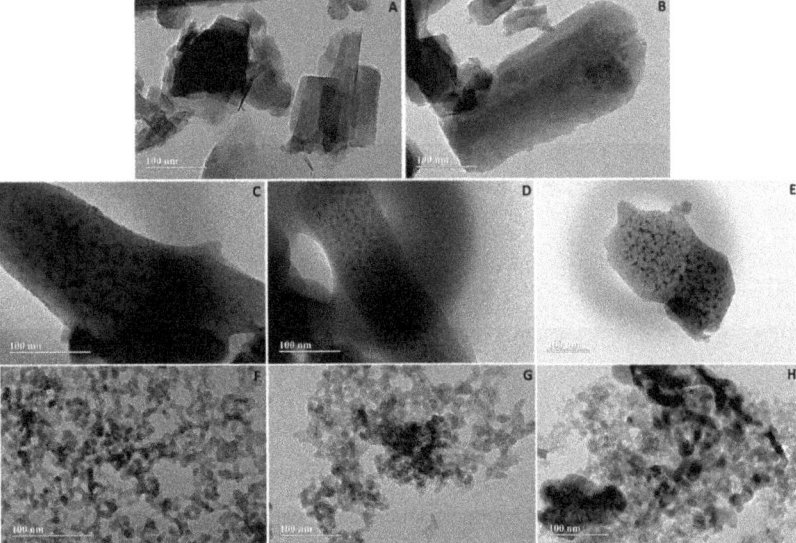

Figure 3. Transmission electron microscopy images of parent zeolite support and prepared single and double metal oxides supported on modified natural zeolite and fumed silica: (**A**) ZNAT; (**B**) ZH; (**C**) ZH_Cu; (**D**) ZH_Zn; (**E**) ZH_Zn/Cu; (**F**) FS_Cu; (**G**) FS_Zn; and (**H**) FS_Zn/Cu.

Additionally, *Transmission Electron Microscopy images* displayed in Figure 3 corroborate the lamellar morphology and depict the incorporation of metal oxide nanoparticles on the zeolite surface. Results show that the sequence of ion exchange and wet impregnation procedures using transition metal salts followed by the calcination treatment at 623 K do not only exchange compensating cations presented in the zeolite sample, but also generate nanoparticles of metal oxides (see Figure 3C–E). The formation of small oxide particles around 10.01 ± 2.5 nm of diameters are obtained when only copper salt is used (Figure 3C). However, when zinc nitrate is applied as a precursor for zinc oxide formation, metal oxide nanoparticles of slightly higher sizes can be observed with particle sizes of 12.2 ± 3.20 nm and 13.9 ± 3.65 nm (Figure 3D,E) for ZH_Zn and ZH_Zn/Cu, respectively. Such results suggest that during zinc oxide formation, particles flocculate, leading to the clustering of small particles and the formation of bigger metal oxide particles. During the impregnation step, water is evaporated and concentrated transition metal salts are deposited onto zeolite surface that later are decomposed during the calcination procedure, driving the generation of the observed metal oxide nanoparticles. The generation of aggregates of nanoparticles of a bigger size can explain the reduction in the surface area values listed in Table 1 for such samples. Moreover, in the oxide supported catalysts over fumed silica (see Figure 3F–H), a similar behavior was observed. However, in those cases, slightly higher particle sizes were observed in the TEM assay. Particles of 11.5 ± 4.1 nm, 13.2 ± 3.9 nm nm, and 14.7 ± 3.7 nm can be seen at FS_Cu, FS_Zn, and FS_Zn/Cu samples, respectively. Such findings could be related to the existence of larger pores in the fumed silica as compared to natural zeolite as it was observed during porosimetry measurements (see Table 1).

The *UV-visible* absorbance spectra of parent zeolite and fumed silica supports and prepared single and double metal oxide catalysts supported on modified natural zeolite and fumed silica are displayed in Figure 4A,B, respectively. Light absorption spectra acquired by ultraviolet-visible diffuse reflectance spectroscopy (UV-Vis DRS) were transformed to Tauc plots to determine the optical band gap energy of each assessed material (Figure 4C,D) [49]. The band gap energy was calculated by extrapolating the linear part of the vertical region of the curve [50,51].

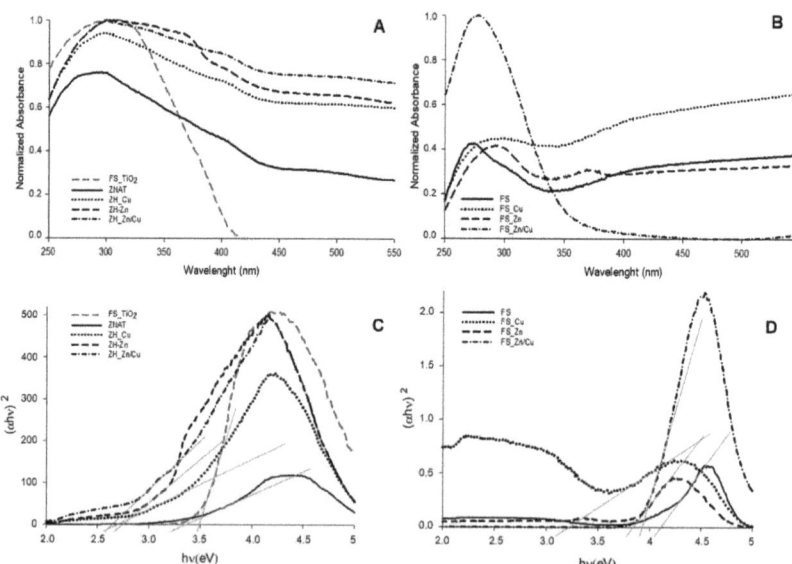

Figure 4. UV-visible absorbance spectra (upper part) and Tauc plots (bottom part) of parent supports and prepared single and double metal oxide catalysts supported on modified natural zeolite (A,C), and fumed silica (B,D). Additionally, TiO_2 (A,C) is included as a benchmarked photo-catalyst.

As it can be seen in Figure 4A, the incorporation of copper and zinc oxides onto the surface of natural zeolite increases the optical absorption in the UV and visible regions.

Thus, the values of the band gap energy in the zeolite samples decrease in the following order: ZH_Zn/Cu < ZH_Cu < ZH_Zn < ZNat (see Figure 4C and Table 2). Moreover, a similar behavior regarding optical absorption was observed in the catalysts supported on fumed silica; however, the band gap energies are higher than those obtained from the catalysts supported on modified zeolite. In such cases, the band gap energies present the following trend: FS_Cu < FS_Zn < FS_Zn/Cu < FS (see Figure 4D and Table 2).

Table 2. Band gap energies of prepared photo-catalysts, including their respective supports. Results were obtained by UV-Vis DRS spectroscopy.

Samples	Band Gap Values (eV)
Z_Nat	3.35
ZH_Cu	2.70
ZH_Zn	2.75
ZH_Zn/Cu	2.65
FS	4.20
FS_Cu	3.20
FS_Zn	3.85
FS_Zn/Cu	3.90
FS_TiO$_2$	3.50

It is important to notice that zeolite supported samples present higher light absorbance than fumed silica samples. Fumed silica is a transparent material which transmit most of incident light, whereas zeolite is an opaquer material, increasing the light absorption in the support. Hence, copper-based materials used here present different trends regarding band gap measurements when they are deposited over the different supports. Indeed, double catalyst samples also show differences. When zeolite is used as a support material (ZH_Zn/Cu), a better particle dispersion was obtained, suggesting the formation of an heterostrucure where exited electrons from the higher band gap material are trapped by the surface of the deposited dopant (CuO), resulting in a lesser band gap energy [51]. Meanwhile, when fumed silica is applied as a support material (FS_Zn/Cu) the band gap energy is similar to the higher band gap material (FS_Zn), suggesting that the formation of the observed cluster in this sample affected the transition of the electrons generated from ZnO to CuO particles.

2.2. In Situ FTIR Characterisation

Additionally, acidic characteristics of studied samples were determined by **in situ FTIR assays** of pyridine adsorption and desorption under vacuum. Table 3 lists the concentration of Brønsted acid sites (BAS) and Lewis acid sites (LAS) of natural and modified zeolites as well as prepared single and double metal oxides supported on fumed silica. In this sense, after conducting the modification of natural zeolite by ion exchange with ammonium sulfate and thermally outgassed at 623 K (ZH sample), as it was expected, BAS concentration increased; meanwhile, LAS concentration diminished as compared to natural zeolite results. However, after the introduction of the copper and zinc oxides on previously modified zeolites, BAS were not detected by pyridine adsorption, and the amount of LAS is lower than in the other zeolite samples. Such results can be attributed to diffusion limitations of pyridine to access such acidic sites present in the natural zeolite and those created by ion exchange with ammonium, because of the blockage caused by the formation of aggregates of metal oxide nanoparticles or to dissolution and washing of the extra-framework phase during the process of metal introduction. Aggregates of metal oxide particles of bigger size can be responsible for the blockage of micropores, decreasing the catalyst surface area considerably, as shown in Table 1 [52].

Table 3. Characterization of the nature and strength of acidic surface sites of parent supports and prepared single and double metal oxide catalysts by in situ FTIR analyses using pyridine.

Samples	Concentration of Brønsted Acidic Sites [a] [µmol Py g^{-1}]	Concentration of Lewis Acidic Sites [a] [µmol Py g^{-1}]	Strength Distribution of Lewis Acidic Sites (%) [b]		
			Weak	Mild	Strong
ZNAT	17.21	84.69	49.47	9.08	41.45
ZH	78.00	37.00	0	0	100
ZH_Cu	ND	56.20	61.93	26.33	11.74
ZH_Zn	ND	34.20	43.10	4.59	52.31
ZH_Zn/Cu	ND	45.64	48.30	23.47	28.23
FS	ND	ND	ND	ND	ND
FS_Cu	ND	3.00	100	0	0
FS_Zn	ND	106.00	61.30	29.20	9.50
FS_Zn/Cu	ND	194.00	55.15	15.46	29.39

[a] Determined by in situ FTIR using pyridine as a probe molecule. [b] Determined by in situ FTIR after thermal-programmed desorption of adsorbed pyridine.

Finally, after the pyridine adsorption experiment, a temperature-programmed desorption was applied in order to obtain a rough estimate of the strength of the Lewis acidic sites. Results listed in Table 3 suggest that zinc sites enhance the Lewis acidity strength; whereas, copper presents the opposite trend in the modified zeolite samples. Moreover, the characterization of raw and modified fumed silica samples show a similar trend with an increased strength of LAS in the samples modified with zinc oxide. However, in this case, a higher LAS strength is observed when the double metal oxide catalyst was supported over the fumed silica (FS_ZnO/CuO).

2.3. Photocatalytic Oxidation of Ethylene onto Parent Zeolite Support and Prepared Single and Double Metal Oxide Catalysts Supported on Modified Natural Zeolite

The effect of the physical–chemical properties of the parent zeolite support and prepared single and double metal oxide catalysts supported on modified natural zeolite on the photocatalytic oxidation of ethylene are shown in Figures 5 and 6. Moreover, results of control experiments using pure raw fumed silica, prepared single and double metal oxide catalysts supported on fumed silica are also presented. Fumed silica consists of finely dispersed amorphous silicon dioxide with a surface covered by silanol groups [53]. Its use is widespread in industrial applications as a catalyst support and as an adsorbent [54,55]. Such kinds of experiments allow determination by comparison the role of acidic sites of zeolite surface in the photocatalytic reaction. Additionally, a benchmarked assay using TiO$_2$ P25 mixed with raw fumed silica is also displayed to contrast the photocatalytic performance of the composite materials under similar reaction conditions.

Curves of the ethylene dimensionless concentration at the reactor outlet stream as the function of time are displayed in Figure 5. The first 600 min of every curve represents the variation of ethylene dimensionless concentration during the adsorption step until saturation is reached (without irradiation). Adsorbed amounts of ethylene during the dynamic adsorption step onto the studied samples (µmol g^{-1}) are reported in Figure 6. Afterwards, the photocatalytic oxidation step takes place under irradiation conditions. As it can be seen, the stationary state of the photocatalytic step is attained after a few minutes. The photocatalytic removal of ethylene for each sample is also presented in Figure 6 and was calculated considering the concentration of ethylene in the reactor inlet and outlet streams in the stationary step.

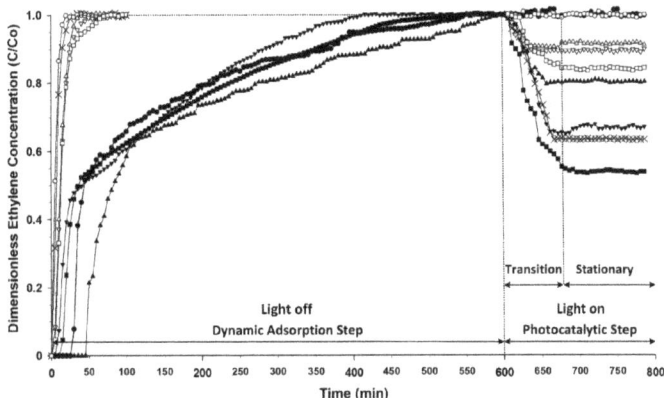

Figure 5. Ethylene adsorption and photocatalytic degradation using parent zeolite support. Prepared single and double metal oxide catalysts supported on modified natural zeolite, raw fumed silica, prepared single and double metal oxide catalysts supported on fumed silica, and TiO$_2$ mixed with raw fumed silica: ● ZNAT, ▲ ZH_Cu, ▼ ZH_Zn, ■ ZH_Zn/Cu, ○ FS, △ FS_Cu, ▽ FS_Zn, □ FS_Zn/Cu, X FS_TiO$_2$.

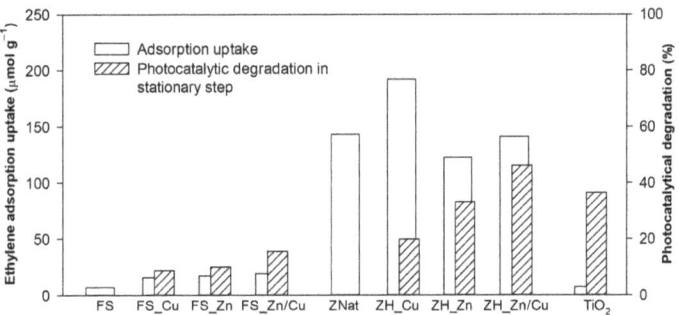

Figure 6. Ethylene adsorption and photocatalytic degradation results using parent zeolite support (ZNat), single metal oxide catalysts supported on modified natural zeolite (ZH_Cu; ZH_Zn), copper and zinc oxides supported on modified natural zeolite (ZH_Zn/Cu), raw fumed silica (FS), single metal oxide catalysts supported on fumed silica (FS_Cu; FS_Zn), and copper and zinc oxides supported on fumed silica (FS_Zn/Cu), and TiO$_2$ P25 mixed with fumed silica (FS_TiO$_2$).

Results displayed in Figures 5 and 6 provide evidence that the applied modification procedures slightly affect the adsorption capacity of natural zeolite and fumed silica toward ethylene molecules, mainly in those cases where samples are modified with copper salts. During the adsorption step, it can be observed that ethylene adsorption is almost negligible for raw and modified fumed silica samples compared to samples where zeolite is used as a support. Such differences could be related to the favorable physical–chemical surface properties of zeolite compared to fumed silica. The higher surface area, the presence of microporous and mesoporous structure with several Si-OH, and Si-OH-Al and cationic active sites contribute to the observed increase in the adsorption of ethylene in comparison to fumed silica. In the case of prepared single and double metal oxide catalysts supported on modified natural zeolite, the adsorption of ethylene during this step increases according to the following order: ZH_Zn_ < ZH_Zn/Cu < ZNat < ZH_Cu. Such results agree with those reported in a previous study [10]. Copper oxide supported on modified natural zeolite, leads to an enhancement in the adsorption capacity toward ethylene. The good dispersion of the copper oxides over the zeolite surface as reported in Figure 3C could favor its interaction with ethylene. Ethylene adsorption also takes place not only by the interaction with Si-OH-Al bridge groups of zeolite surface but also by the interactions between π-electrons of ethylene with metal cations deposited in the zeolite surface [5,10]. A lower adsorption of ethylene is obtained on ZH_Zn. Such a result could be related to (i) the observed formation of bigger aggregates of nanoparticles in this sample, as indicated by TEM images (see Figure 3D), (ii) the decrease in the surface area, as reported by N_2 adsorption assays (see Table 1), and (iii) the low accessibility to Brønsted and Lewis sites, as shown by pyridine adsorption (Table 3). The amount of adsorbed ethylene seems directly related to the surface area of the samples and to the surface concentration of Lewis sites.

Results displayed in Figure 6 confirm the good adsorption capacity of the modified natural zeolite used, as well as the single metal oxide catalysts supported on them. Some other ethylene adsorbents have been studied previously including activated carbon-based materials, carbon nanoballs, and synthetic and natural zeolites [7,9,12]. Such studies have reported ethylene adsorption capacities around 10 to 80 µmol g^{-1} [7] at similar ethylene concentrations. Regarding photocatalysis results, the highest degradation rate is achieved when the parent natural zeolite support is modified with both transition metal oxides (ZH_Zn/Cu). The single metal oxides supported on modified natural zeolite (ZH_Zn and ZH_Cu) are less active.

No photocatalytic oxidation is registered using the parent natural zeolite support (ZNat). A similar trend is attained during photocatalytic oxidation of ethylene using fumed silica as a support for metal oxides (FS_Zn/Cu > FS_Zn > FS_Cu > FS). No photodegradation is catalyzed by the raw fumed silica material (FS). Lower degradation values are obtained when the fumed silica is used as a support for metal oxides rather than the zeolite. In agreement with the stated working hypothesis, a higher photocatalytic degradation is accomplished with the double metal oxides supported on modified natural zeolite obtained by co-impregnation, being even higher than the benchmark TiO_2 sample. Such behaviors agree with those results obtained by UV-Vis spectroscopy characterization (see Figure 4). A higher degradation rate is reached in the sample that presents the highest UV light absorption as it is the case of ZH_Zn/Cu sample. Moreover, several studies have reported that ZnO possesses an excellent photocatalytic activity similar or even greater than that of TiO_2, in the degradation of organic compounds [30,32,33]. In addition, it was also reported that CuO presents a lower photocatalytic activity under UV irradiation than ZnO [56,57]. It is worth noting that the highest ethylene conversions are obtained when the catalysts based on both transition metal oxides are used, even when different materials are applied as supports.

In this heterostructure material zinc oxide could absorb UV photons, forming electron–hole (e–h) pairs, whereas copper oxide nanoparticles could reduce the recombination of e–h pairs, increasing the absorption of photons with lower energy [37]. It is well known that (e–h) pairs react with oxygen and water molecules, generating superoxide and hydroxyl radicals, respectively [22,42,58]. Such active species usually act as powerful oxidants, leading to the oxidation of close adsorbed ethylene molecules. Previous studies have reported the photocatalytic conversion of ethylene from 10 to 90% using TiO_2 based photocatalysts [7]. However, operational conditions were different from those applied in this study. It is worth noting that during photocatalytic oxidation processes, parameters such as wavelength irradiation, light intensity, catalyst load, among other factors could have an impact in the removal efficiency [59,60]. Thus, TiO_2 was used here as a benchmark photocatalyst as a way of comparison.

In addition, studies concerning other ethylene oxidation techniques such as $KMnO_4$ have shown almost full ethylene remotion at the first hours of use; however, it has a rapid loss of reaction efficiency during the time of storage [61]. Similarly, it has been mentioned that $KMnO_4$ can cause product contamination [62].

Moreover, the observed degradation rate in the single and double metal oxide catalysts supported on modified natural zeolite could have been favored by the presence of several adsorption sites where ethylene molecules are adsorbed and oxidized by the radicals generated in the nearby semiconductor oxide nanoparticles. More detailed information on such kinds of interactions is followed by *operando* DRIFTS studies and is presented in Figure 7.

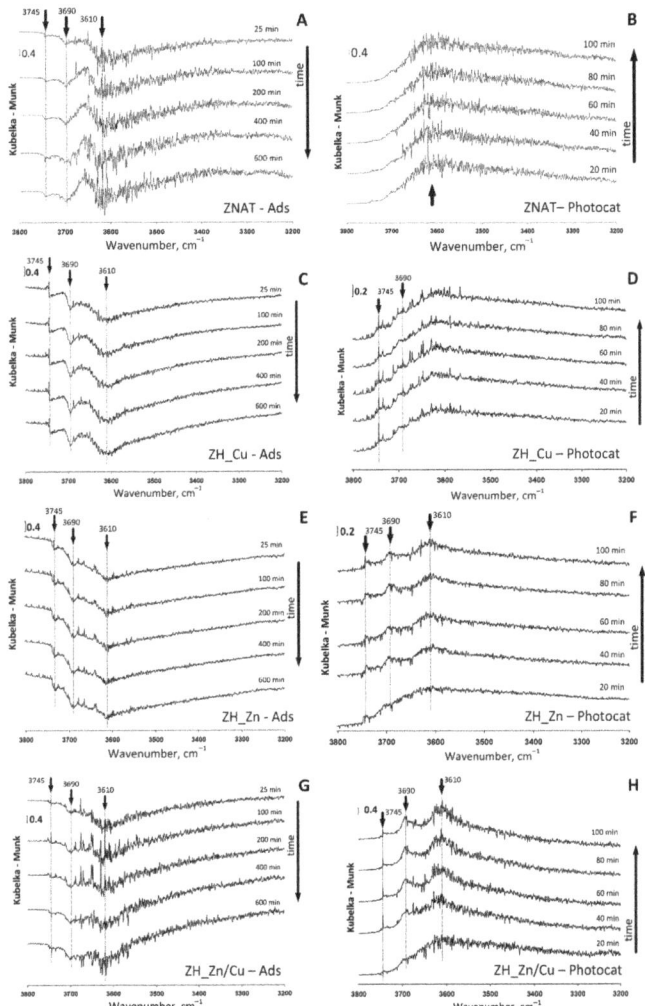

Figure 7. Evolution of subtracted DRIFTS spectra of the OH stretching vibration zone during the adsorption step (left) and photocatalytic degradation step of ethylene (right) onto parent zeolite support. Prepared single and double metal oxide catalysts supported on modified natural zeolite: ZNAT (**A**,**B**), ZH_Cu; (**C**,**D**); ZH_Zn (**E**,**F**); ZH_Zn/Cu; (**G**,**H**).

The variations of *operando* DRIFTS spectra during the adsorption step (without irradiation) of the experimental results presented in Figures 5 and 6 of ethylene onto single and double metal oxide catalysts supported on zeolite and fumed silica samples are shown in Figure 7A,C,E,G and Figure 8A,C,E,G, respectively. Additionally, *operando* DRIFTS spectra of TiO_2 interactions with ethylene through the adsorption step are also included in Figure 9A. Results displayed in Figures 7–9 represent the subtractions of the obtained spectrum for each sample as a function of ethylene exposition time and the respective spectrum of activated samples prior to ethylene contact. As was previously reported [10,63], in the ν(OH) region, between 4000 and 3300 cm^{-1}, the peak at 3745 cm^{-1} could be assigned to silanol groups (Si-OH) vibration. Moreover, the band at 3690 cm^{-1} comes from extra-framework phase, and the peak at 3610 cm^{-1} is the usual zeolitic Bronsted acidic OH group known as the bridge silanol (Si-OH-Al). The progressive loss on the intensities of the IR bands at 3590 cm^{-1}, 3690 cm^{-1}, and 3745 cm^{-1} related to OH groups are observed for the parent zeolite support, the prepared single and double metal oxide catalysts supported on modified natural zeolite. However, the higher difference is obtained at 3610 cm^{-1}, suggesting that most of ethylene adsorption occurs by the interaction with acidic Si-OH-Al groups. Although an increase in the adsorption of ethylene is observed for the zeolite samples modified with copper oxide (see Figure 6), no other IR vibrations are observed in the DRIFTS spectra beside those already mentioned during the adsorption step. This can be due to the low concentration of ethylene on the surface of the catalyst during the experiment and the fast decomposition and desorption in the photocatalytic step.

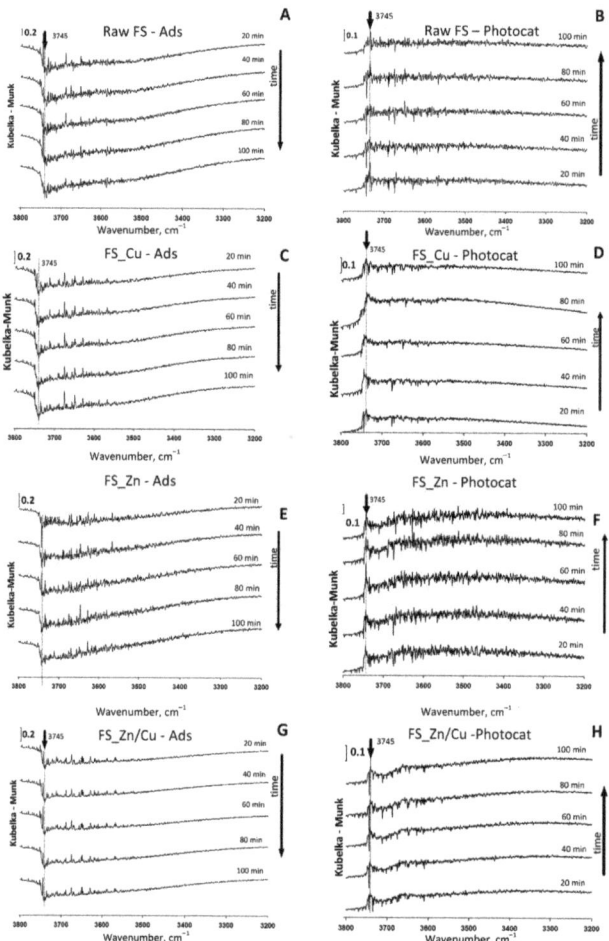

Figure 8. Evolution of subtracted DRIFTS spectra of the OH stretching vibration zone during the adsorption step (left) and photocatalytic degradation step of ethylene (right) onto raw fumed silica. Prepared single and double metal oxide catalysts supported on fumed silica: raw FS (**A**,**B**), FS_Cu (**C**,**D**), FS_Zn (**E**,**F**), and FS_Zn/Cu (**G**,**H**).

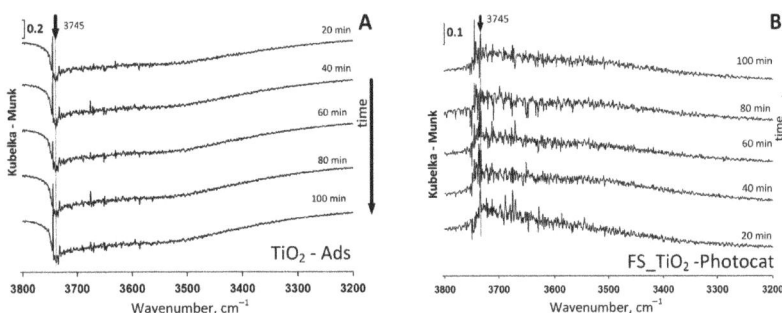

Figure 9. Evolution of subtracted DRIFTS spectra of the OH stretching vibration zone during the adsorption step (**A**) and photocatalytic degradation step of ethylene (**B**) onto TiO$_2$ mixed with raw fumed silica (FS_TiO$_2$).

In the case of the fumed silica sample (Figure 8A), recorded spectra show only intensity changes in the band located at 3745 cm^{-1} related to the presence of silanol groups in the raw material and its interaction with ethylene molecules during adsorption. Such sites are occupied quickly and remain invariable after the saturation is reached. A similar trend is observed for single and double metal oxide catalysts supported on fumed silica (Figure 8C,E,G) and for TiO$_2$ mixed with raw fumed silica at 8% of weight (Figure 9A). Such results agree with the almost negligible ethylene adsorption observed for these samples compared to zeolite supported catalysts with acidic Brønsted sites.

Spectra reported in Figures 7 and 8 are obtained at different time periods as subtraction of the registered spectra at specific times from the spectrum acquired when the saturation is reached at the final step of the adsorption stage of each sample. During the photocatalysis stage, as can be seen in Figure 7H, the ZH_Zn/Cu double catalyst show restored OH vibration bands (3745 cm^{-1}, 3690 cm^{-1}, and 3610 cm^{-1}) which had been consumed during the adsorption stage (see Figure 7G). Under irradiation, they are released again and became unoccupied due to the photocatalytic oxidation of adsorbed ethylene, leading to a new adsorption equilibrium. Analogous behavior can be observed on the single zinc oxide supported on modified natural zeolite (Figure 7F) and to a lesser extent, on the single copper oxide supported on modified natural zeolite (Figure 7D). Ethylene molecules adsorbed in such sites could be oxidized by radicals generated at nanoparticles of copper and zinc oxides. Hence, some recovered adsorption sites allow new ethylene molecules to be adsorbed and oxidized photocatalytically. However, in the case of the parent zeolite support, subtracted DRIFTS spectra (Figure 7B) do not show any evolution in the vibration of OH characteristic bands during the experiment, suggesting that in the absence of the semiconductor particles, ethylene cannot be eliminated. Such results agree with those obtained by monitoring the concentration of ethylene at the inlet and outlet stream of the photocatalytic reaction cell using GC analyses (see Figures 5 and 6). The double metal oxide catalyst based on copper and zinc oxides supported on modified natural zeolite shows the highest removal efficiency toward ethylene oxidation in the series, and the parent natural zeolite support (ZNAT) does not show any C$_2$H$_4$ removal.

Operando DRIFTS assays during the photocatalytic step with fumed silica samples confirm that there is no contribution of the raw fumed silica during the photocatalytic oxidation step and are shown in the supplementary materials. In the other fumed silica samples, a slight change can be perceived in the IR band at 3745 cm^{-1}, but only at the beginning of the irradiation, It cannot be attributed to photocatalytic oxidation, but rather, just to the effect of heating during this short transition period that leads to a fast elimination of the few ethylene adsorbed in these samples. Afterwards, in the stationary state, the surface of those samples remains unchanged.

Similar changes in the OH bands were observed when TiO$_2$ mixed with raw fumed silica was used (Figure 9). The spectra of such sample showed only the disappearance of the IR vibration band at 3745 cm^{-1} related to OH groups during the adsorption step. The same signal came back after the photocatalytic step, indicating less adsorption sites for ethylene compared to samples containing modified zeolite.

Results obtained here account for different factors that are involved during the photocatalytic oxidation of ethylene. When metal oxide semiconductors are deposited in a support with almost negligible adsorption capacity, as with fumed silica compared to zeolite, ethylene oxidation efficiencies are below 20% (see Figures 5 and 6). The observed outcomes suggest that zinc oxide is mainly responsible for photocatalytic degradation. When zinc oxide is combined with copper oxide, a material with higher photocatalytic activity is obtained. Other researchers similarly reported that in heterostructure catalysts, where zinc and copper oxides are combined, the electron–hole pair separation efficiency is enhanced, increasing the photocatalytic activity [37,58,64]. However, when metal oxide semiconductors are deposited over a material with a high adsorption capacity, such as modified natural zeolite, the retention of ethylene molecules at adsorption sites in the vicinity of semiconductor particles could be the main factor for the observed increase in the overall oxidation efficiency. The DRIFTS assay of the ZH_Zn/Cu sample provides evidence that adsorbed ethylene molecules at acidic surface sites are photocatalytically removed once the semiconductors are irradiated. In this way, the oxidation ability of superoxide and hydroxyl radicals generated due to the e–h pairs activation and their interactions with oxygen and water molecules could be more efficient when the semiconductor materials are supported over a material with high content of acidic surface sites that allow radicals to oxidize ethylene molecules adsorbed in their vicinity.

2.4. Mechanistic Approach

Results reported here in addition to previous studies [10] contribute to validate the hypothesis of ethylene elimination by a synergic mechanism that includes adsorption and photocatalytic oxidation. DRIFTS analyses enable to track acidic groups of zeolite surface and their behavior during the process. The use of modified natural zeolite as a support for zinc and copper oxides allows ethylene to be mainly adsorbed at bridged hydroxyl and at external and internal silanol sites as it is represented by Equation (1). Moreover, in hydrophilic zeolites such as the natural zeolite used here, water molecules also interact mainly with bridged hydroxyl groups affiliated mainly with aluminum in the zeolite lattice (Equation (2)).

$$C_2H_4 + * \leftrightarrow C_2H_4* \tag{1}$$

$$H_2O + *BAS \leftrightarrow H_2O*BAS \tag{2}$$

Under irradiation, zinc and copper oxides seem to absorb UV photons, forming electron–hole (e–h) pairs (Equation (3)).

$$ZnO/CuO + h\nu \rightarrow n(h^+ + e^-) \tag{3}$$

In the nanoparticles of semiconductors, zinc oxide increases the optical absorption; whereas, copper oxide nanoparticles decrease the optical band gap of the material, increasing the absorption of photons with lower energy and reducing the e–h recombination rate [65–68]. It is generally accepted that generated electrons and electrons' holes interact with oxygen and water molecules, producing superoxide and hydroxyl radicals [60] (Equations (4)–(8)) increasing the efficiency of the photocatalytic process [60,69,70].

$$h^+ + H_2O* \rightarrow OH^\circ * + H^+ \tag{4}$$

$$e^- + O_2 \rightarrow O_2^{-\circ} \tag{5}$$

$$O_2^{-\circ} + H^+ \rightarrow HO_2^\circ \tag{6}$$

$$H^+ + O_2^{-\circ} + HO_2^\circ \rightarrow H_2O_2 + O_2 \tag{7}$$

$$H_2O_2 + h\nu \rightarrow 2OH^° \tag{8}$$

Finally, such radical species seems to be the responsibility of photocatalytic degradation of adsorbed ethylene at the neighboring site, generating intermediary oxidation by-products [60] (Equations (9)–(10)) that finally lead to ethylene mineralization, as summarized by Equation (11).

$$OH^° * + C_2H_4* \rightarrow \text{Intermediates} \tag{9}$$

$$O_2^{-°} + C_2H_4* \rightarrow \text{Intermediates} \tag{10}$$

$$C_2H_4 + 3O_2 \xrightarrow[cat]{h\nu} 2CO_2 + 2H_2O \tag{11}$$

Comparative experiments conducted using fumed silica as a support with poor adsorption capacity towards ethylene proved the important role of acidic surface sites of zeolite, increasing the adsorption of ethylene and its photocatalytic oxidation in the vicinity of zinc and copper oxides. Further experiments are needed in order to unveil the details of the formation of intermediary oxidation by-products of higher molecular mass than water and CO_2. Dynamic studies of ethylene photocatalysis presented here were conducted in a continuous packed bed cell set in a FT/IR spectrometer using only one initial ethylene concentration in order to compare the catalysts' activity. For the purpose to gather kinetic information, further extensive experimentation must be conducted using ethylene at different initial concentrations.

3. Materials and Methods

3.1. Materials and Reagents

A natural zeolite (53% clinoptilolite, 40% mordenite, and 7% quartz) was used in this study as a parent support material and was provided by the Chilean mining company "Minera Formas". Ethylene (0.01% in Ar) was supplied by Indura (Santiago, Chile), whereas argon (>99.9% purity) and oxygen (>99.5% purity) were supplied by Praxair (Santiago, Chile). Inorganic chemicals (>99.5% purity) such as $(NH_4)_2SO_4$ and $Cu(NO_3)_2·3H_2O$ were purchased from Merck (Darmstadt, Germany). $Zn(NO_3)_2·6H_2O$ was supplied by Sigma-Aldrich Corporation (St. Louis, MO, USA). Solutions were prepared using deionized water (≥ 18.0 MΩ cm) produced in a Thermo Scientific Barnstead Easypure II RF system (Thermo Fisher Scientific, Waltham, MA, USA).

3.2. Modification of Parent Support

Natural zeolite modification was carried out in consecutive stages. First, raw material was ground and sieved to a range of 300–425 μm particle size. Then, it was rinsed with deionized water and dried at 398 K for 24 h, using a LabTech® oven model LDO-080F (Daihan LabTech Co., Ltd., Gyeonggi-do, Korea).

In the pre-treatment stage, raw zeolite was modified by ion exchange with ammonium sulphate (0.1 mol dm^{-3}) at 363 K for two hours in a temperature-controlled water bath, using a mass volume ratio of 0.1 g of zeolite per cm^3 of salt solution and then it was washed with deionized water during 4 h to eliminate the excess of salt as it has been reported in previous work [10]. This procedure was repeated once after rinsing. Samples were dried at 398 K for 24 h. Afterwards, samples were out-gassed at 623 K using a home-made tubular furnace (heating rate of 3 K min^{-1}) during two hours under argon flow (100 cm^3 min^{-1}).

During the metal loading stage, pre-treated samples were modified by wet impregnation using an IKA® RV 10 vacuum rotary evaporator (IKA®-Werke, Staufen, Germany) at 363 K. Thus, two single metal oxide and one double metal oxide catalysts were obtained by single impregnation with $Cu(NO_3)_2·3(H_2O)$ (0.13 mol dm^{-3}), single impregnation with $Zn(NO_3)_2·6(H_2O)$ (0.13 mol dm^{-3}), and co-impregnation using a combination of both salts at 0.065 mol dm^{-3}, respectively. Salt solution concentrations were calculated to obtain single metal oxide catalysts with 8% of the desired metal oxide mass impregnated in the zeolite surface and a double metal oxide catalyst with 4% of each metal oxide in the co-impregnated sample. Subsequently, samples were dried using the same conditions mentioned before. As the last modification step, samples were calcined under oxygen flow (100 cm^3 min^{-1}) at 623 K (heating rate of 1 K min^{-1}) for 4 h and stored in a desiccator until their further use. Additionally, fumed silica was used as a support of metal oxides in control experiments. Raw fumed silica was modified using the same sequential procedure described before for the metal loading stage, including the followed modification steps. The performance of the samples prepared using natural zeolite and fumed silica were compared to understand the key role of acidic sites of zeolite surface in the photocatalytic process.

According to the applied modification methodology, samples were named as follows:

- Natural zeolite thermally outgassed at 623 K (ZNAT);
- H form of natural zeolite (ZH), obtained from natural zeolite modified by ion exchanged with ammonium and thermally outgassed at 623 K;
- Copper oxide supported on modified natural zeolite, obtained from natural zeolite ion exchanged with ammonium followed by a single wet impregnation with copper salt and later calcined at 623 K (ZH_Cu);
- Zinc oxide supported on modified natural zeolite, obtained from natural zeolite ion exchanged with ammonium followed by a single wet impregnation with zinc salt and later calcined at 623 K (ZH_Zn);
- Copper and zinc oxides supported on modified natural zeolite, obtained from natural zeolite ion exchanged with ammonium followed by a wet co-impregnation with copper and zinc salts and later calcined at 623 K (ZH_Zn/Cu);
- Raw fumed silica thermally outgassed at 623 K (FS);
- Copper oxide supported on fumed silica, obtained from fumed silica modified by single wet impregnation with copper salt and later calcined at 623 K (FS_Cu);
- Zinc oxide supported on fumed silica, obtained from fumed silica modified by single wet impregnation with zinc salt and later calcined at 623 K (FS_Zn);
- Copper and zinc oxides supported on fumed silica, obtained from fumed silica modified by wet co-impregnation with copper and zinc salts and later calcined at 623 K (FS_Zn/Cu).

Furthermore, TiO$_2$ (99.9% purity) supplied by Evonik (Nuremberg, Germany) was mixed with the raw fumed silica at 8% weight and thermally outgassed at 623 K and used as a benchmark catalyst for comparison (FS_TiO$_2$).

3.3. Characterisation of the Parent Zeolite Support and of the Single and Double Metal Oxide Catalysts

The parent zeolite support and the single and double metal oxide catalysts supported on modified natural zeolite were characterized using different analytical techniques. X-ray powder diffraction (XRD) patterns were obtained using a Bruker Endeavor diffractometer model D4/MAX-B with a copper cathode lamp (λ = 1.541 Å), operated at 20 mA and 40 kV. X-ray diffraction patterns were collected with a 0.02° resolution from 4° to 80° and a time interval of 1 s. The specific surface areas were determined by nitrogen adsorption at 77 K, using Micromeritics ASAP 2000 equipment (Norcross, GA, USA). Samples were previously outgassed at 623 K during 2 h under inert flow (70% N$_2$, 30% He). The surface area values were determined using the single-point surface area method, as described elsewhere [71]. The chemical composition of the parent zeolite support, single and double metal oxide catalysts supported on modified natural zeolite were determined by X-ray fluorescence (XRF) using a Rigaku model ZSX Primus II spectrometer. Scanning electron microscopy (SEM) images were obtained in a Mira L.M.H. microscope operated at 30 kV and 177 µA. Samples were ground, deposited over a carbon film, and covered with a platinum coat. A high vacuum (<3 × 10^{-4} mbar) was established in the microscope before image acquisition. Transmission electron microscopy (TEM) studies were carried out in a JEOL, JEM 1200 EX-II device equipped with a Gatan 782 camera and using an accelerating voltage of 120 kV. Samples were ground and deposited over a carbon covered mesh for imagery.

In situ IR assays of pyridine adsorption under vacuum were conducted to determine the concentration and strength of acidic sites present in the parent supports and in the prepared single and double metal oxide catalysts. Pyridine interactions with Brønsted and Lewis acid sites were monitored by Fourier transform infrared spectroscopy (FTIR). Samples were ground to powder and compacted as self-supported pellets of 20 mg and 16 mm diameter. Pellets were heated inside a thermo-regulated FTIR glass cell to 623 K (heating rate of 3 K min^{-1}) and activated under high vacuum (<7 × 10^{-5} mbar) for 12 h. Pyridine adsorption experiments were carried out at 423 K in a Nicolet 5700 spectrometer (Thermo Fisher Scientific Inc., Waltham, MA, USA) equipped with an MCT/A detector. Spectra were averaged from 32 consecutive acquisitions obtained in the range from 4000–1100 cm^{-1} with a resolution of 4 cm^{-1}, using the OMNIC software V 9.2.86. After adsorption experiments, thermal-programmed desorption of pyridine was performed under vacuum. FTIR spectra were collected after heating the pyridine saturated samples from 463 K to 723 K to study strength of acidic sites. Thus, the percentage of pyridine desorbed at different temperature ranges from 463 K to 523 K, from 523 K to 623 K, and from 623 K to 723 K are related to weak, mild, and strong acidic sites, respectively [72].

The amount of Brønsted acid sites (BAS) and Lewis acid sites (LAS), expressed in terms of μmol of adsorbed pyridine per gram of catalyst, were determined from the integrated intensity of the absorption bands at 1545 cm^{-1} and at 1445–1450 cm^{-1}, respectively, using the commonly used modified Beer–Lambert–Bouguer law (Equation (12)):

$$A = \varepsilon \frac{n^{Py}}{S} \quad (12)$$

where n^{Py} is the number of Py species (in μmol) adsorbed at the acidic site (Lewis or Brønsted), A is the peak area (in cm^{-1}), and S is the cross-section area of the zeolite disc ($S = 2$ cm^2). ε is the integrated molar absorption coefficient (in cm μmol^{-1}). The values for ε were taken from the work of Zholobenko et al. [73], as 1.38 and 1.87 cm μmol^{-1} for Brønsted and Lewis acid sites, respectively, in mordenites.

UV-visible light absorption properties of the prepared photo-catalysts were assessed by UV-Vis DRS spectroscopy. Experiments were carried out in a VARIAN 4000 UV-Vis spectrophotometer equipped with an integration sphere. Solid samples were located in the sample holder and spectra were collected in the range from 200–800 nm in absorbance mode. The optical band gap energy was determined applying the Tauc Plot methodology by extrapolating the linear part of the vertical region of the curve [41,51].

3.4. Photocatalytic Degradation of Ethylene

Photocatalytic experiments were conducted in a commercial Praying Mantis™ DRIFTS cell (Harrick Scientific, New York, NY, USA) containing a quartz window and set in a JASCO FT/IR 4700 spectrometer equipped with an MCT/M detector (JASCO International Co., Ltd., Tokyo, Japan). Dynamic photocatalytic experiments were performed using 0.08 g of sample deposited in the DRIFTS chamber. Prior to any experiment, zeolite samples were activated for 2 h at 623 K (heating rate of 1 K min^{-1}) under vacuum. Irradiation during photo-catalytic experiments was supplied by a polychromatic light emitted by an LC8 spot light source (L10852, 200 W) from Hamamatsu Photonics (Shizuoka, Japan). A light guide from Hamamatsu was placed in the quartz window of the Praying Mantis DRIFTS cell. The irradiation intensity was measured using a light power meter from Hamamatsu Photonics (Shizuoka, Japan). The experimental setup used in this study (see Figure 10) was adapted from one used previously [10].

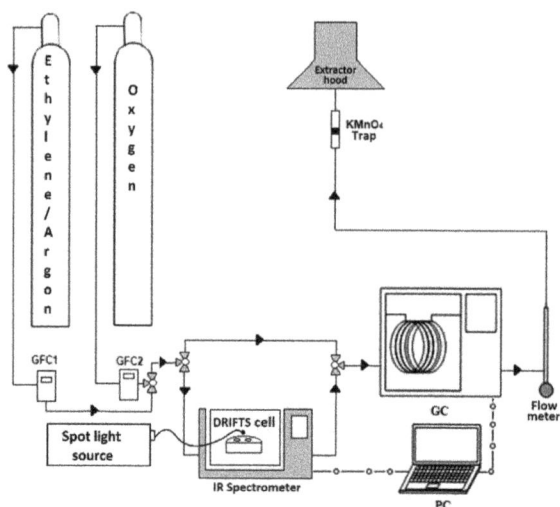

Figure 10. Schematic representation of the experimental set-up used for photocatalytic oxidation of ethylene.

In a typical experiment, a total flow of 25 cm^3 min^{-1} of a gas mixture composed of 80 ppmv of C_2H_4 and 20,000 ppmv of O_2 balanced in argon was continuously supplied over the sample bed (simulating the atmosphere around stored fruits). For each sample, an adsorption step was initially conducted with the light turned off until saturation was reached. Once the ethylene saturation was achieved, the light was turned on and the photocatalytic step was initiated. The temperature of the photocatalytic reactor cell was kept constant at 293 K, using a water jacketed heat exchanger. During the whole experimentation, spectra were collected as a function of time in the range of 4000–400 cm^{-1} with a resolution of 1 cm^{-1} using the spectra manager software V 2.14.02 (JASCO International Co., Ltd., Tokyo, Japan). At the same time, the concentrations of ethylene and CO_2 at the reactor outlet were monitored on-line by gas chromatography (Perkin Elmer Clarus 500 gas chromatographer, Waltham, MA, USA) using a flame ionization detector (FID) and a thermal conductivity detector (TCD), respectively. More detailed information about analytic techniques can be found elsewhere [10].

Adsorption capacities obtained during the first step (with the light turned off) toward ethylene using the parent zeolite support. The prepared single and double metal oxide catalysts supported on modified natural zeolite and fumed silica were determined by the integration of the breakthrough curves until reaching saturation conditions (Equation (13)), as follows:

$$q_{ethylene} = \frac{FC_{in}}{m} \int_0^{t_s} \left(1 - \frac{C_{out}}{C_{in}}\right) dt \qquad (13)$$

where $q_{ethylene}$ ($\mu mol_{ethylene}$ g^{-1}) represents the total amount of ethylene adsorbed per gram of sample, F (cm^3 min^{-1}) stands for the gas flow rate, m (g) is the mass of sample inside the DRIFTS cell, t_s (min) is the adsorption time to reach saturation, C_{in} and C_{out} (μmol dm^{-3}) are the inlet and outlet concentrations of ethylene as a function of time, respectively.

The percentage of photocatalytic removal of ethylene obtained during the second step (with the light turned on) was calculated from experiments at steady state conditions, taking into consideration the amount of ethylene in the reactor inlet and outlet streams.

4. Conclusions

Modified natural zeolites could be used as support for copper and zinc oxides leading to efficient photocatalytic oxidation of ethylene. DRIFTS results evidenced a synergistic effect of adsorption and photocatalytic oxidation over the catalyst surface. Among the functional groups of zeolite surface, acidic Al-OH-Si bridges are the most influential for ethylene adsorption. Such surface groups concentrate ethylene molecules near to the photocatalyst semiconductor nanoparticles of zinc and copper oxides that promote ethylene oxidation under light irradiation. A photocatalyst based on nanoparticles of zinc and copper oxides supported o modified natural zeolite provides the highest contribution to photocatalytic activity toward ethylene oxidation. The increase in light absorption combined with a reduction in the band gap energy contributes to an enhancement in the photocatalytic oxidation of adsorbed ethylene. The development of new composite materials comprising semiconductors supported over zeolite for photocatalytic oxidation of ethylene emissions from fruit warehouses ought to consider the chemical surface interactions among the combined materials and ethylene. Moreover, further experiments should be performed using a light source with a wavelength corresponding to visible spectra in order to prove that better photocatalytic efficiencies are obtained using zeolites modified with copper and zinc oxide nanoparticles that take the advantage of the visible spectrum of solar radiation.

Supplementary Materials: The following supporting information can be downloaded at: https://www.mdpi.com/article/10.3390/catal13030610/s1, Figure S1: Nitrogen adsorption/desorption isotherms at 77 K of parent zeolite support and prepared single and double metal oxide catalysts supported on modified natural zeolite, Figure S2: Nitrogen adsorption/desorption isotherms at 77 K of fumed silica support and prepared single and double metal oxide catalysts supported on fumed silica.

Author Contributions: Conceptualization, N.J.A., H.V., F.A.-J. and C.A.Z.; methodology, N.J.A., H.V., F.A.-J. and T.F.d.O.; formal analysis, N.J.A., H.V. and C.A.Z.; investigation, N.J.A. and H.V.; resources, N.J.A., F.A.-J. and H.V.; data curation, N.J.A.; writing—original draft preparation, N.J.A. and H.V.; writing—review and editing, N.J.A., H.V. and F.T.-S.; visualization, N.J.A., H.V., T.F.d.O. and F.A.-J.; supervision, H.V.; project administration, H.V.; funding acquisition, H.V. and N.J.A. All authors have read and agreed to the published version of the manuscript.

Funding: This research was funded by CONICYT, FONDECYT/Regular (Grant 1170694); ANID, FONDECYT/postdoctorate (Grant 3210158) and CONICYT, PCHA/Doctorado Nacional 2015 (Grant 21150082).

Informed Consent Statement: Not applicable.

Data Availability Statement: Data are contained within the article tables and supplementary material.

Acknowledgments: The authors are grateful to the SMARTC Center from Scientific and Technological Bioresources Nucleus for their collaboration with assays and imagery. The authors also want to thank Mohamad El-Roz from *CNRS Laboratoire Catalyse et Spectrochimie* for his helpful discussion and Víctor A. Solar from *Laboratorio de Tecnologías Limpias, Universidad Católica de la Santísima Concepción* for his valuable collaboration.

Conflicts of Interest: The authors declare no conflict of interest.

References

1. Iqbal, N.; Khan, N.A.; Ferrante, A.; Trivellini, A.; Francini, A.; Khan, M.I.R. Ethylene role in plant growth, development and senescence: Interaction with other phytohormones. *Front. Plant Sci.* **2017**, *8*, 475. [CrossRef]
2. Aprianti, S.; Bintoro, N. The effect of concentrations and exposure durations of ethylene gas on the respiration rate of tomato fruit (*Solanum lycopersicum*). *IOP Conf. Ser. Earth Environ. Sci.* **2021**, *653*, 012021. [CrossRef]
3. Blanke, M. Challenges of Reducing Fresh Produce Waste in Europe—From Farm to Fork. *Agriculture* **2015**, *5*, 389–399. [CrossRef]
4. Blanke, M.M.; Shekarriz, R. Ethylene levels along the food supply chain—A key to reducing food waste? *Acta Hortic.* **2015**, *1091*, 101–106. [CrossRef]
5. Saltveit, M.E. Effect of ethylene on quality of fresh fruits and vegetables. *Postharvest Biol. Technol.* **2021**, *15*, 279–292. [CrossRef]
6. Kader, A.A.; Cavalieri, R.; Ferguson, I. A Perspective on Postharvest Horticulture (1978–2003). *Hortscience* **2006**, *38*, 1004–1008. [CrossRef]
7. Keller, N.; Ducamp, M.N.; Robert, D.; Keller, V. Ethylene removal and fresh product storage: A challenge at the frontiers of chemistry. Toward an approach by photocatalytic oxidation. *Chem. Rev.* **2013**, *113*, 5029–5070. [CrossRef] [PubMed]

8. Kim, S.I.; Aida, T.; Niiyama, H. Binary adsorption of very low concentration ethylene and water vapor on mordenites and desorption by microwave heating. *Sep. Purif. Technol.* **2005**, *45*, 174–182. [CrossRef]
9. Erdoğan, B.; Sakizci, M.; Yörükoğullari, E. Characterization and ethylene adsorption of natural and modified clinoptilolites. *Appl. Surf. Sci.* **2008**, *254*, 2450–2457. [CrossRef]
10. Abreu, N.J.; Valdés, H.; Zaror, C.A.; Azzolina-Jury, F.; Meléndrez, M.F. Ethylene adsorption onto natural and transition metal modified Chilean zeolite: An operando DRIFTS approach. *Microporous Mesoporous Mater.* **2019**, *274*, 138–148. [CrossRef]
11. Golipour, H.; Mokhtarani, B.; Mafi, M.; Moradi, A.; Godini, H.R. Experimental Measurement for Adsorption of Ethylene and Ethane Gases on Copper-Exchanged Zeolites 13X and 5A. *J. Chem. Eng. Data* **2020**, *65*, 3920–3932. [CrossRef]
12. Cisneros, L.; Gao, F.; Corma, A. Silver nanocluster in zeolites. Adsorption of ethylene traces for fruit preservation. *Microporous Mesoporous Mater.* **2019**, *283*, 25–30. [CrossRef]
13. Van Zandvoort, I.; van Klink, G.P.M.; de Jong, E.; van der Waal, J.C. Selectivity and stability of zeolites [Ca]A and [Ag]A towards ethylene adsorption and desorption from complex gas mixtures. *Microporous Mesoporous Mater.* **2018**, *263*, 142–149. [CrossRef]
14. Guo, Y.; Zu, B. Zeolite-based Photocatalysts: A Promising Strategy for Efficient Photocatalysis. *J. Thermodyn. Catal.* **2013**, *4*, e120. [CrossRef]
15. Bahrami, M.; Nezamzadeh-Ejhieh, A. Effect of the supported ZnO on clinoptilolite nano-particles in the photodecolorization of semi-real sample bromothymol blue aqueous solution. *Mater. Sci. Semicond. Process.* **2015**, *30*, 275–284. [CrossRef]
16. Erdoğan, B. A comparative adsorption study of C_2H_4 and SO_2 on clinoptilolite-rich tuff: Effect of acid treatment. *J. Hazard. Mater.* **2013**, *262*, 627–633. [CrossRef]
17. Sue-aok, N.; Srithanratana, T.; Rangsriwatananon, K.; Hengrasmee, S. Study of ethylene adsorption on zeolite NaY modified with group I metal ions. *Appl. Surf. Sci.* **2010**, *256*, 3997–4002. [CrossRef]
18. Kuz'min, I.V.; Sokolova, N.A.; Subbotina, I.R.; Zhidomirov, G.M. Ethylene adsorption and transformation on zeolite Ga^+/ZSM5. *Russ. Chem. Bull.* **2015**, *64*, 278–283. [CrossRef]
19. Despres, J.; Koebel, M.; Kröcher, O.; Elsener, M.; Wokaun, A. Adsorption and desorption of NO and NO_2 on Cu-ZSM-5. *Microporous Mesoporous Mater.* **2003**, *58*, 175–183. [CrossRef]
20. Phanichphant, S.; Nakaruk, A.; Chansaenpak, K.; Channei, D. Evaluating the photocatalytic efficiency of the $BiVO_4$/rGO photocatalyst. *Sci. Rep.* **2019**, *9*, 16091. [CrossRef]
21. Li, R.; Li, T.; Zhou, Q. Impact of titanium dioxide (TiO_2) modification on its application to pollution treatment—A review. *Catalysts* **2020**, *10*, 804. [CrossRef]
22. El-Roz, M.; Bazin, P.; Daturi, M.; Thibault-Starzyk, F. On the mechanism of methanol photooxidation to methylformate and carbon dioxide on TiO_2: An operando-FTIR study. *Phys. Chem. Chem. Phys.* **2015**, *17*, 11277–11283. [CrossRef]
23. Kozlov, D. Titanium dioxide in gas-phase photocatalytic oxidation of aromatic and heteroatom organic substances: Deactivation and reactivation of photocatalyst. *Theor. Exp. Chem.* **2014**, *50*, 133–154. [CrossRef]
24. Abbas, N.; Hussain, M.; Russo, N.; Saracco, G. Studies on the activity and deactivation of novel optimized TiO_2 nanoparticles for the abatement of VOCs. *Chem. Eng. J.* **2011**, *175*, 330–340. [CrossRef]
25. Shao, M.; Chen, W.; Ding, S.; Lo, K.H.; Zhong, X.; Yao, L.; Ip, W.F.; Xu, B.; Wang, X.; Pan, H. WX_y/G-C_3N_4 (WX_y = W_2C, WS_2, or W_2N) Composites for Highly Efficient Photocatalytic Water Splitting. *ChemSusChem* **2019**, *12*, 3355–3362. [CrossRef] [PubMed]
26. Mera, A.C.; Rodriguez, C.A.; Pizarro-Castillo, L.; Melendrez, M.F.; Valdés, H. Effect of Temperature and Reaction Time during Solvothermal Synthesis of BiOCl on Microspheres Formation: Implications in the Photocatalytic Oxidation of Gallic Acid under Simulated Solar Radiation. *J. Sol-Gel Sci. Technol.* **2020**, *95*, 146–156. [CrossRef]
27. Li, D.; Haneda, H. Morphologies of Zinc Oxide Particles and Their Effects on Photocatalysis. *Chemosphere* **2003**, *51*, 129–137. [CrossRef]
28. Al-Ariki, S.; Yahya, N.A.A.; Al-A'nsi, S.A.; Jumali, M.H.H.; Jannah, A.N.; Abd-Shukor, R. Synthesis and comparative study on the structural and optical properties of ZnO doped with Ni and Ag nanopowders fabricated by sol gel technique. *Sci. Rep.* **2021**, *11*, 11948. [CrossRef]
29. Ayoub, I.; Kumar, V.; Abolhassani, R.; Sehgal, R.; Sharma, V.; Sehgal, R.; Swart, H.C.; Mishra, Y.K. Advances in ZnO: Manipulation of defects for enhancing their technological potentials. *Nanotechnol. Rev.* **2022**, *11*, 575–619. [CrossRef]
30. Hanif, M.A.; Kim, Y.S.; Ameen, S.; Kim, H.G.; Kwac, L.K. Boosting the Visible Light Photocatalytic Activity of ZnO through the Incorporation of N-Doped for Wastewater Treatment. *Coatings* **2022**, *12*, 579. [CrossRef]
31. Adeel, M.; Saeed, M.; Khan, I.; Muneer, M.; Akram, N. Synthesis and characterization of Co–ZnO and evaluation of its photocatalytic activity for photodegradation of methyl orange. *ACS Omega* **2021**, *6*, 1426–1435. [CrossRef]
32. Ma, S.; Huang, Y.; Hong, R.; Lu, X.; Li, J.; Zheng, Y. Enhancing Photocatalytic Activity of ZnO Nanoparticles in a Circulating Fluidized Bed with Plasma Jets. *Catalysts* **2021**, *11*, 77. [CrossRef]
33. Mao, T.; Liu, M.; Lin, L.; Cheng, Y.; Fang, C. A Study on Doping and Compound of Zinc Oxide Photocatalysts. *Polymers* **2022**, *14*, 4484. [CrossRef]
34. Modwi, A.; Ghanem, M.A.; Al-Mayouf, A.M.; Houas, A. Lowering energy band gap and enhancing photocatalytic properties of Cu/ZnO composite decorated by transition metals. *J. Mol. Struct.* **2018**, *1173*, 1–6. [CrossRef]
35. Dhineshbabu, N.R.; Rajendran, V.; Nithyavathy, N.; Vetumperumal, R. Study of structural and optical properties of cupric oxide nanoparticles. *Appl. Nanosci.* **2016**, *6*, 933–939. [CrossRef]

36. Bekru, A.G.; Tufa, L.T.; Zelekew, O.A.; Goddati, M.; Lee, J.; Sabir, F.K. Green Synthesis of a CuO-ZnO Nanocomposite for Efficient Photodegradation of Methylene Blue and Reduction of 4-Nitrophenol. *ACS Omega* **2022**, *7*, 30908–30919. [CrossRef]
37. Wei, A.; Xiong, L.; Sun, L.; Liu, Y.J.; Li, W.W. CuO nanoparticle modified ZnO nanorods with improved photocatalytic activity. *Chin. Phys. Lett.* **2013**, *30*, 046202. [CrossRef]
38. Maraj, M.; Raza, A.; Wang, X.; Chen, J.; Riaz, K.N.; Sun, W. Mo-Doped CuO Nanomaterial for Photocatalytic Degradation of Water Pollutants under Visible Light. *Catalysts* **2021**, *11*, 1198. [CrossRef]
39. Dasineh Khiavi, N.; Katal, R.; Kholghi Eshkalak, S.; Masudy-Panah, S.; Ramakrishna, S.; Jiangyong, H. Visible Light Driven Heterojunction Photocatalyst of CuO–Cu$_2$O Thin Films for Photocatalytic Degradation of Organic Pollutants. *Nanomaterials* **2019**, *9*, 1011. [CrossRef] [PubMed]
40. Durán, B.; Saldías, C.; Villarroel, R.; Hevia, S.A. In Situ Synthesis of CuO/Cu$_2$O Nanoparticle-Coating Nanoporous Alumina Membranes with Photocatalytic Activity under Visible Light Radiation. *Coatings* **2023**, *13*, 179. [CrossRef]
41. Mai, X.T.; Bui, D.N.; Pham, V.K.; Pham, T.H.T.; Nguyen, T.T.L.; Chau, H.D.; Tran, T.K.N. Effect of CuO Loading on the Photocatalytic Activity of SrTiO$_3$/MWCNTs Nanocomposites for Dye Degradation under Visible Light. *Inorganics* **2022**, *10*, 211. [CrossRef]
42. Hauchecorne, B.; Tytgat, T.; Verbruggen, S.W.; Hauchecorne, D.; Terrens, D.; Smits, M.; Vinken, K.; Lenaerts, S. Photocatalytic degradation of ethylene: An FTIR in situ study under atmospheric conditions. *Appl. Catal. B Environ.* **2011**, *105*, 111–116. [CrossRef]
43. Chen, L.; Xie, X.; Song, X.; Luo, S.; Ye, S.; Situ, W. Photocatalytic degradation of ethylene in cold storage using the nanocomposite photocatalyst MIL101(Fe)-TiO$_2$-rGO. *Chem. Eng. J.* **2021**, *424*, 130407. [CrossRef]
44. Montero, J.; Welearegay, T.; Thyr, J.; Stopfel, H.; Dedova, T.; Acik, I.O.; Österlund, L. Copper–zinc oxide heterojunction catalysts exhibiting enhanced photocatalytic activity prepared by a hybrid deposition method. *RSC Adv.* **2021**, *11*, 10224–10234. [CrossRef] [PubMed]
45. Luévano-Hipólito, E.; Torres-Martínez, L.M.; Fernández-Trujillo, A. Ternary ZnO/CuO/Zeolite composite obtained from volcanic ash for photocatalytic CO$_2$ reduction and H$_2$O decomposition. *J. Phys. Chem. Sol.* **2021**, *151*, 109917. [CrossRef]
46. Iazdani, F.; Nezamzadeh-Ejhieh, A. The photocatalytic rate of ZnO supported onto natural zeolite nanoparticles in the photodegradation of an aromatic amine. *Environ. Sci. Pollut. Res.* **2021**, *28*, 53314–53327. [CrossRef]
47. Condon, J.B.; James, B. *Surface Area and Porosity Determinations by Physisorption Measurements and Theory*, 1st ed.; Elsevier: Amsterdam, The Netherlands, 2006; pp. 55–90.
48. Sprynskyy, M.; Golembiewski, R.; Trykowski, G.; Buszewski, B. Heterogeneity and hierarchy of clinoptilolite porosity. *J. Phys. Chem. Sol.* **2010**, *71*, 1269–1277. [CrossRef]
49. Likhachev, D.; Malkova, N.; Poslavsky, L. Modified Tauc-Lorentz dispersion model leading to a more accurate representation of absorption features below the bandgap. *Thin Solid Films* **2015**, *589*, 844–851. [CrossRef]
50. Tauc, J. Optical properties and electronic structure of amorphous Ge and Si. *Mat. Res. Bull.* **1968**, *3*, 37–46. [CrossRef]
51. Pughazhenthiran, N.; Murugesan, S.; Valdés, H.; Selvaraj, M.; Sathishkumar, P.; Smirniotis, P.G.; Anandan, S.; Mangalaraja, R.V. Photocatalytic oxidation of ceftiofur sodium under UV-Visible irradiation using plasmonic porous Ag-TiO$_2$ nanospheres. *J. Ind. Eng. Chem.* **2022**, *105*, 384–392. [CrossRef]
52. Marakatti, V.S.; Halgeri, A.B.; Shanbhag, G. Metal ion-exchanged zeolites as solid acid catalysts for the green synthesis of nopol from Prins reaction. *Catal. Sci. Technol.* **2014**, *4*, 4065–4074. [CrossRef]
53. Barthel, H.; Rosch, L.; Weis, J. Fumed Silica-Production, Properties, and Applications. In *Organosilicon Chemistry II: From Molecules to Materials*; Auner, N., Weis, J., Eds.; Wiley: Hoboken, NJ, USA, 1996; Volume 1, pp. 761–778. [CrossRef]
54. Aboelfetoh, E.F.; Pietschnig, R. Preparation, characterization and catalytic activity of MgO/SiO$_2$ supported vanadium oxide based catalysts. *Catal. Lett.* **2014**, *144*, 97–103. [CrossRef]
55. Parida, S.K.; Dash, S.; Patel, S.; Mishra, B.K. Adsorption of organic molecules on silica surface. *Adv. Colloid Interface Sci.* **2006**, *121*, 77–110. [CrossRef]
56. Sabzehei, K.; Hadavi, S.H.; Bajestani, M.G.; Sheibani, S. Comparative evaluation of copper oxide nano-photocatalyst characteristics by formation of composite with TiO$_2$ and ZnO. *Solid State Sci.* **2020**, *107*, 106362. [CrossRef]
57. Sorbiun, M.; Shayegan Mehr, E.; Ramazani, A.; Fardood, S.T. Green Synthesis of Zinc Oxide and Copper Oxide Nanoparticles Using Aqueous Extract of Oak Fruit Hull (Jaft) and Comparing Their Photocatalytic Degradation of Basic Violet 3. *Int. J. Environ. Res.* **2018**, *12*, 29–37. [CrossRef]
58. Esmaili-Hafshejani, J.; Nezamzadeh-Ejhieh, A. Increased photocatalytic activity of Zn(II)/Cu(II) oxides and sulfides by coupling and supporting them onto clinoptilolite nanoparticles in the degradation of benzophenone aqueous solution. *J. Hazard. Mater.* **2016**, *316*, 194–203. [CrossRef]
59. Reza, K.M.; Kurny, A.; Gulshan, F. Parameters affecting the photocatalytic degradation of dyes using TiO$_2$: A review. *Appl. Water Sci.* **2017**, *7*, 1569–1578. [CrossRef]
60. De Lasa, H.; Serrano, B.; Salaices, M. Establishing Photocatalytic Kinetic Rate Equations: Basic Principles and Parameters. In *Photocatalytic Reaction Engineering*; Springer: Boston, MA, USA, 2005. [CrossRef]
61. Álvarez-Hernández, M.H.; Martínez-Hernández, G.B.; Avalos Belmontes, F.; Castillo-Campohermoso, M.A.; Contreras-Esquivel, J.C.; Artés-Hernández, F. Potassium Permanganate-Based Ethylene Scavengers for Fresh Horticultural Produce as an Active Packaging. *Food Eng. Rev.* **2019**, *11*, 159–183. [CrossRef]

62. Patdhanagul, N.; Srithanratana, T.; Rangsriwatananon, K.; Hengrasmee, S. Ethylene adsorption on cationic surfactant modified zeolite NaY. *Microporous Mesoporous Mater.* **2010**, *131*, 97–102. [CrossRef]
63. Bordiga, S.; Lamberti, C.; Bonino, F.; Travert, A.; Thibault-Starzyk, F. Probing zeolites by vibrational spectroscopies. *Chem. Soc. Rev.* **2015**, *44*, 7262–7341. [CrossRef]
64. Karimi Shamsabadi, M.; Behpour, M. Fabricated CuO–ZnO/nanozeolite X heterostructure with enhanced photocatalytic performance: Mechanism investigation and degradation pathway. *Mater. Sci. Eng. B Solid State Mater. Adv. Technol.* **2021**, *269*, 115170. [CrossRef]
65. Yu, Z.; Moussa, H.; Liu, M.; Schneider, R.; Wang, W.; Moliere, M.; Liao, H. Development of photocatalytically active heterostructured MnO/ZnO and CuO/ZnO films via solution precursor plasma spray process. *Surf. Coat. Technol.* **2019**, *371*, 107–116. [CrossRef]
66. Zhu, L.; Li, H.; Liu, Z.; Xia, P.; Xie, Y.; Xiong, D. Synthesis of the 0D/3D CuO/ZnO Heterojunction with Enhanced Photocatalytic Activity. *J. Phys. Chem. C* **2018**, *122*, 9531–9539. [CrossRef]
67. Bharathi, P.; Harish, S.; Archana, J.; Navaneethan, M.; Ponnusamy, S.; Muthamizhchelvan, C.; Shimomura, M.; Hayakawa, Y. Enhanced charge transfer and separation of hierarchical CuO/ZnO composites: The synergistic effect of photocatalysis for the mineralization of organic pollutant in water. *Appl. Surf. Sci.* **2019**, *484*, 884–891. [CrossRef]
68. Chabri, S.; Dhara, A.; Show, B.; Adak, D.; Sinha, A.; Mukherjee, N. Mesoporous CuO-ZnO p-n heterojunction based nanocomposites with high specific surface area for enhanced photocatalysis and electrochemical sensing. *Catal. Sci. Technol.* **2016**, *6*, 3238–3252. [CrossRef]
69. Ameta, R.; Solanki, M.S.; Benjamin, S.; Ameta, S.C. Photocatalysis. In *Advanced Oxidation Processes for Wastewater Treatment: Emerging Green Chemical Technology*; Ameta, S.C., Ameta, R., Eds.; Elsevier: Amsterdam, The Netherlands, 2018; Volume 1, pp. 135–175. [CrossRef]
70. Parrino, F.; Livraghi, S.; Giamello, E.; Ceccato, R.; Palmisano, L. Role of Hydroxyl, Superoxide, and Nitrate Radicals on the Fate of Bromide Ions in Photocatalytic TiO_2 Suspensions. *ACS Catal.* **2020**, *10*, 7922–7931. [CrossRef]
71. Azzolina Jury, F.; Polaert, I.; Estel, L.; Pierella, L.B. Synthesis and characterization of MEL and FAU zeolites doped with transition metals for their application to the fine chemistry under microwave irradiation. *Appl. Catal. A Gen.* **2013**, *453*, 92–101. [CrossRef]
72. Deng, C.; Zhang, J.; Dong, L.; Huang, M.; Li, B.; Jin, G.; Gao, J.; Zhang, F.; Fan, M.; Zhang, L.; et al. The effect of positioning cations on acidity and stability of the framework structure of Y zeolite. *Sci. Rep.* **2016**, *6*, 23382. [CrossRef]
73. Zholobenko, V.; Freitas, C.; Jendrlin, M.; Bazin, P.; Travert, A.; Thibault-Starzyk, F. Probing the acid sites of zeolites with pyridine: Quantitative AGIR measurements of the molar absorption coefficients. *J. Catal.* **2020**, *385*, 52–60. [CrossRef]

Disclaimer/Publisher's Note: The statements, opinions and data contained in all publications are solely those of the individual author(s) and contributor(s) and not of MDPI and/or the editor(s). MDPI and/or the editor(s) disclaim responsibility for any injury to people or property resulting from any ideas, methods, instructions or products referred to in the content.

Article

Catalytic Characterization of Synthetic K⁺ and Na⁺ Sodalite Phases by Low Temperature Alkali Fusion of Kaolinite during the Transesterification of Spent Cooking Oil: Kinetic and Thermodynamic Properties

Mohamed Adel Sayed [1,2], Jamaan S. Ajarem [3], Ahmed A. Allam [4], Mostafa R. Abukhadra [1,5,*], Jianmin Luo [6], Chuanyi Wang [7] and Stefano Bellucci [8]

1. Materials Technologies and Their Applications Lab, Geology Department, Faculty of Science, Beni-Suef University, Beni-Suef 62514, Egypt
2. Department of Chemistry, Faculty of Science, Beni-Suef University, Beni Suef 62514, Egypt
3. Zoology Department, College of Science, King Saud University, Riyadh 11362, Saudi Arabia
4. Zoology Department, Faculty of Science, Beni-Suef University, Beni-Suef 62514, Egypt
5. Geology Department, Faculty of Science, Beni-Suef University, Beni-Suef 65211, Egypt
6. School of Chemistry and Civil Engineering, Shaoguan University, Shaoguan 512005, China
7. School of Environmental Science and Engineering, Shaanxi University of Science and Technology, Xi'an 710021, China
8. INFN-Laboratori Nazionali di Frascati, Via E. Fermi 54, 00044 Frascati, Italy
* Correspondence: abukhadra89@science.bu.edu.eg

Abstract: The mineral raw Egyptian kaolinite was used as a precursor in the synthesis of two sodalite phases (sodium sodalite (Na.SD) and potassium sodalite (K.SD)) according to the low alkali fusion technique. The synthesized Na.SD phase demonstrates enhanced total basicity (6.3 mmol OH/g), surface area (232.4 m²/g), and ion exchange capacity (126.4 meq/100 g) compared to the K.SD phase (217.6 m²/g, 96.8 meq/100 g (ion exchange capacity), 5.4 mmol OH/g (total basicity). The catalytic performance of the two sodalite phases validates the higher activity of the sodium phase (Na.SD) than the potassium phase (K.SD). The application of Na.SD resulted in biodiesel yields of 97.3% and 96.4% after 90 min and 60 min, respectively, while the maximum yield using K.SD (95.7%) was detected after 75 min. Robust base-catalyzed reactions using Na.SD and K.SD as catalysts were suggested as part of an operated transesterification mechanism. Moreover, these reactions exhibit pseudo-first order kinetics, and the rate constant values were estimated with consideration of the change in temperature. The estimated activation energies of Na.SD (27.9 kJ·mol⁻¹) and K.SD (28.27 kJ·mol⁻¹) reflected the suitability of these catalysts to be applied effectively under mild conditions. The essential thermodynamic functions, such as Gibb's free energy (65.16 kJ·mol⁻¹ (Na.SD) and 65.26 kJ·mol⁻¹ (K.SD)), enthalpy (25.23 kJ·mol⁻¹ (Na.SD) and 25.55 kJ·mol⁻¹ (K.SD)), and entropy (−197.7 J·K⁻¹·mol⁻¹ (Na.SD) and −197.8 J·K⁻¹·mol⁻¹ (K.SD)), display the endothermic and spontaneous nature of the two transesterification systems.

Keywords: alkali ions; sodalite; transesterification; kinetics; thermodynamic; mechanism

Citation: Sayed, M.A.; Ajarem, J.S.; Allam, A.A.; Abukhadra, M.R.; Luo, J.; Wang, C.; Bellucci, S. Catalytic Characterization of Synthetic K⁺ and Na⁺ Sodalite Phases by Low Temperature Alkali Fusion of Kaolinite during the Transesterification of Spent Cooking Oil: Kinetic and Thermodynamic Properties. *Catalysts* **2023**, *13*, 462. https://doi.org/10.3390/catal13030462

Academic Editors: De Fang and Yun Zheng

Received: 21 January 2023
Revised: 17 February 2023
Accepted: 20 February 2023
Published: 22 February 2023

Copyright: © 2023 by the authors. Licensee MDPI, Basel, Switzerland. This article is an open access article distributed under the terms and conditions of the Creative Commons Attribution (CC BY) license (https://creativecommons.org/licenses/by/4.0/).

1. Introduction

The greenhouse effect, environmental side effects of utilizing fossil fuels, depletion of fossil fuel supplies, and rising crude oil prices exhibit notable negative consequences on the world economy and the safety of our ecosystem. Therefore, developing and introducing additional and sustainable energy sources with eco-friendly and renewable properties are the main concern and interest of researchers, governments, and environmental authorities [1,2]. Generally, biofuels such as biodiesel were assessed in several studies as clean, low-cost, sustainable, effective, and non-toxic alternative fuels [3]. As a fuel, the commonly developed biodiesel products exhibit notable high lubricant properties, viscosity, and an acceptable flash point (>130 °C) [4]. Moreover, as biodiesel products are sulfur and aromatic-free fuels, they burn cleaner than commercial diesel, giving them significant environmental value [5,6].

Recently, numerous studies have been introduced to utilize fatty acid methyl esters (FAMEs) to produce manufactured products such as lubricants [7], stabilizing agents for polyvinyl chloride [8], plasticizer [9], surfactants, corrosion inhibitors, and water repellents [10,11]; additionally, it can be used for the production of gas, fatty alcohol, and hydrocarbon, which form the standard raw materials for chemical industries [11]. Chemically, biodiesel is known as a series of fatty acid methyl esters (FAME) that can be obtained by the facile transesterification processes of edible or non-edible vegetable oils as well as animal fats as sources of triglyceride with short-chain alcohols (methyl alcohol or ethyl alcohol) in the presence of appropriate heterogeneous or homogenous catalysts [12,13]. However, several triglyceride feedstocks were evaluated during the production of biodiesel (spent oil [14], palm oil [15], virgin cottonseed oil [16], and rubber seed oil [17]. Their availability, accessibility, cost, and suitability to various climatic conditions are essential factors to consider during the selection of the feedstock [18]. Therefore, several studies were introduced in later periods to assess the possible extraction of biodiesel from spent or waste cooking oil, which can act as low-cost, commercial, and recyclable precursors [19].

Most of the previous studies demonstrated a controlling effect of the incorporated catalyst on the efficiencies and rate of the transesterification reactions, in addition to the physical properties of the obtained biodiesel [20]. H_2SO_4, NaOH, and KOH are common and effective catalysts for homogeneously catalyzed transesterification reactions, and have resulted in significant biodiesel yields after acceptable periods and at moderate temperature values [21,22]. However, homogeneous transesterification is an effective process; it is associated with huge quantities of toxic effluent as a byproduct, complex separation processes, and a low recyclability value, which pose several environmental restrictions on its commercial applications [23]. The reported technical and environmental advantages of heterogeneous catalysts, which are known as multi-phase catalysts, over homogeneous catalysts strongly make them an essential part in the generation of biodiesel [24]. These advantages involve their high reusability, the facile separation process involved, their low energy consumption, and the fact that it creates no toxic effluents as by-products [25,26]. Therefore, great efforts have been made to introduce new species of effective heterogeneous catalysts or to enhance the catalytic performances of the commonly used catalysts [3].

Natural and synthetic structures of alkali-enriched (K^+, Na^+, and Ca^{2+}) aluminosilicate, especially the zeolite phases, are in a highly recommended class of multi-phase catalysts in the potential transesterification of various types of vegetable oils as well as their spent products [13,27]. This is due to their affordable costs, simple production processes, the availability of their precursors, and their significant physicochemical properties [12,28,29]. The synthetic phases of zeolite display a notable and significant micro-porous structure, surface area, ion exchangeability, thermal stability, surface reactivity, chemical and crystalline flexibility, dispersion properties, and mechanical stability [30–32]. Generally, zeolite is a crystalline, microporous, and hydrated aluminosilicate material that is enriched in alkaline earth and/or alkaline ions. Structurally, it consists of SiO_4 and Al_2O_4 tetrahedral units and their connection by three-dimensional corner oxygen sharing, forming a series of linked cages with highly ordered structural nanochannels or pores [33,34]. Several types of synthetic zeolite (sodalite, zeolite-A, cancrinite, zeolite-P, zeolite-Y, and zeolite-X) were studied as catalysts; they are still part of an active and attractive research area considering the insufficient studies that were introduced to describe the kinetics and mass transfer properties of their transesterification catalytic systems [35]. Moreover, several studies demonstrated notable changes in their morphology as well as their main physicochemical properties, including their porosity, crystalline degree, surface area, ion exchange capacity, basicity, and adsorption affinity in terms of the type of the incorporated raw materials and synthesis conditions (temperature, time interval, degree of alkalinity, and the type of alkaline solution) [27,30,36]. Recently, the production of synthetic zeolite phases by alkali fusion methods followed by a hydrothermal treatment was recommended strongly for the synthesis phases of zeolite with enhanced stability, crystallinity, basicity, and ion exchange capacity [37,38].

Based on the previous consideration, the presented study here focuses on the kinetic and thermodynamic properties of synthetic zeolite-based catalysts in the transesterification of waste cooking oil. The assessed zeolite phases are two sodalite forms (sodium sodalite (Na.SD) and potassium sodalite (K.SD)), which were produced from natural Egyptian kaolinite by the alkali fusion method. The catalytic performances of the two sodalite phases were determined experimentally and theoretically based on detailed kinetic and thermodynamic studies considering the types of dominant alkaline ions.

2. Results and Discussion

2.1. Characterization of the Catalysts

The following formation of the zeolite phases, the XRD patterns of the raw kaolinite, as well as the synthetic samples, were presented in Figure 1. The starting precursor exhibited the typical pattern of highly crystalline kaolinite with its notable peaks at about 12.3° (001), 24.9° (002), and 26.6° (111) (XRD. No. 04-012-5104). The zeolite samples, prepared either by using NaOH or by KOH, showed the typical patterns of sodalite (Ref. Code. 04-009-5259) (Figure 1A). The identification peaks of sodalite, which was prepared using NaOH (Na.SD), were marked at 14.33° (110), 24.69° (221), 31.84° (310), 35° (222), 38.2° (321), and 43.09° (330) (Figure 1A). Those of the synthetic sodalite prepared using KOH (K.SD) were marked at 14.48° (110), 24.76° (221), 31.98° (310), 35.09° (222), 38.44° (321), and 43.22° (330) (Figure 1A). The slight deviation in the observed positions of the diffraction peaks of K.SD, compared to those of Na.SD, reflects the effect of the alkaline ions on the structure of sodalite considering the ionic radius of these ions (1.94 Å (Na^+) and 1.34 Å (K^+)) and the substitution capacity within the zeolite structure. This effect was also made evident by the average crystallite's size calculated according to the Scherrer equation ($D = 0.9\lambda/W \cos\theta$), where W is the full width at half maximum in radians, θ is the Bragg's angle, and λ is the X-ray wavelength ($CuK\alpha$ = 0.15405 nm). The synthetic K.SD phase exhibited a smaller crystallite size (34.2 nm) than the synthetic Na.SD phase did (40.3 nm).

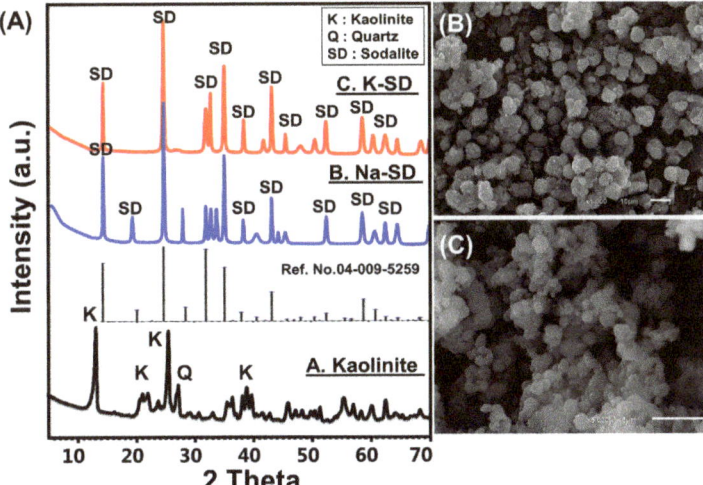

Figure 1. XRD patterns of kaolinite and the synthetic sodalite phases (**A**), SEM image of Na-sodalite (Na.SD) (**B**), and SEM image of K-sodalite (K.SD) (**C**).

The formation of the sodalite was confirmed also by the morphological transformation of the raw kaolinite material from characteristic pseudo-hexagonal flakes into different forms (Figure 1B,C). The two obtained sodalite phases exhibited notable changes in morphology. While the Na.SD particles showed well-developed spherical grains commonly separated from each other, the K.SD particles appeared as aggregates of agglomerated grains with spherical to cubic shapes (Figure 1C).

Moreover, the determined FT-IR spectra reflect the significant change from the kaolinite structure into the zeolitic tectosilicate structure of sodalite (Figure 2). The spectrum of kaolinite demonstrates the existence of structural Si-OH (3689 cm^{-1}), structural Al-OH (3622 and 912 cm^{-1}), adsorbed O-H (1641 cm^{-1}), Si-O (787 and 456 cm^{-1}), Si-O-Al (526 and 680 cm^{-1}), and Si-O-Si (1020 cm^{-1}) as the characteristic chemical groups of natural kaolinite minerals (Figure 2A) [38]. The estimated spectra of Na.SD and K.SD reveal that there was considerable shifting in the corresponding bands of the essential aluminosilicate chemical groups. This strongly signifies the impact of the alkaline alteration processes, which is associated with significant leaching effects on the structural Si and Al ions in addition to the remarkable exposure of the active siloxane groups [39]. Moreover, the detected bands within the range from 630 to 635 cm^{-1} denote the known symmetric stretching of the Si-O-Si group of the structural units of zeolite (Figure 2B,C) [40]. The formation of zeolite was also confirmed by the identified band around 1475 cm^{-1}, which signifies the presence of trapped zeolitic water within the internal channels of the zeolite structure [41].

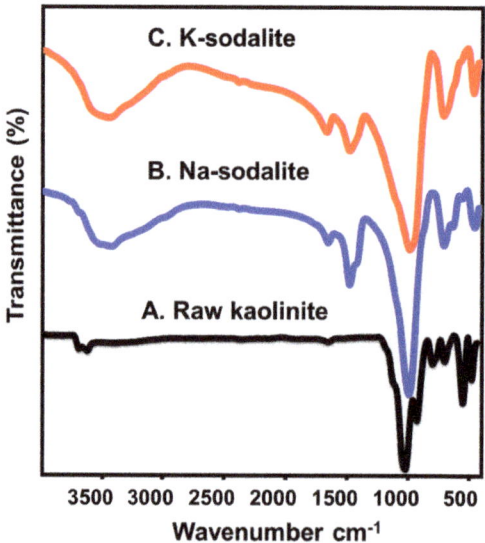

Figure 2. The FT–IR spectra of kaolinite (**A**), synthetic Na.SD (**B**), and synthetic K.SD (**C**).

The reported changes in the morphology as well as in the chemical and crystalline structure significantly affected the textural and physicochemical properties of the synthetic sodalite phases (Table 1). The synthetic Na.SD phase exhibited a higher surface area (232.4 m^2/g), ion exchange capacity (126.4 meq/100 g), and total basicity (6.3 mmol OH/g) than the determined values of K.SD (217.6 m^2/g (surface area), 96.8 meq/100 g (ion exchange capacity), 5.4 mmol OH/g (total basicity)) (Table 1). While the high surface area and total basicity of the Na.SD compared to those of the K.SD induced the former's catalytic activity, its high ion exchange capacity might have been caused by adverse and saponification effects during the transesterification reactions.

Table 1. The textural properties of kaolinite, K.SD, and Na.SD.

Sample	Surface Area	Total Pore Volume	Average Pore Size	Cation Exchange Capacity	Total Basicity
Kaolinite	10 m^2/g	0.072 cm^3/g	43.2 nm	—	—
K.SD	217.6 m^2/g	0.214 cm^3/g	9.7 nm	96.8 meq/100 g	6.3 mmol OH/g
Na.SD	232.4 m^2/g	0.247 cm^3/g	7.4 nm	126.4 meq/100 g	5.4 mmol OH/g

2.2. Transesterification Results

2.2.1. Effect of the Experimental Variables

Effect of Transesterification Intervals at Different Temperatures

The experimental impact of the transesterification duration on the formation rate and percentages of methyl ester was studied regularly from 15 min up to 120 min. This was evaluated with consideration of the different values for the transesterification temperature (40 °C up to 80 °C) and the fixed values for the other affecting parameters (methanol-to-oil molar ratio: 12:1 and the catalyst load: 2.5 wt.%) (Figure 3A,B). A satisfactory transesterification duration is essential to ensure the effective interaction and miscibility between the different reactants, and doing so turn achieved the best conversion rate and efficiency [42]. Therefore, the short intervals observed in the K.SD and Na.SD transesterification systems were associated with poor homogeneity, low miscibility, and high mass resistance in the conversion systems [13,43]. This resulted in the notably low conversion rates of TG into FAME during the initial intervals of the reactions at all the investigated temperature values, either in the presence of Na.SD or K.SD (Figure 3A,B). Consequently, expanding the reaction duration at a significant rate prompts the miscibility properties in the system, which enhances the formation percentages of FAME. This can be observed up to certain intervals at which the used Na.SD (90 min) and K.SD (75 min) achieved their best catalytic activities in terms of the determined yields (Figure 3A,B). Beyond the previously mentioned intervals, the Na.SD- and K.SD-based transesterification systems showed a considerable decrease in the quantities of the produced FAME regardless of the temperature. This adverse effect was attributed to the reversible nature of the methanolysis processes in addition to the expected acceleration of some of the side reactions, which include the re-dissolving of the existing glycerol into alcohol [12,42].

The experimental influence of the transesterification temperature on the activity of the K.S D and Na.SD catalysts was assessed experimentally from 40 °C to 80 °C based on the measured yields of FAME (Figure 3A,B). The measured FAME and the total yields accelerated strongly with the increase in the transesterification temperature up to 70 °C, using both Na.SD (94.8% yield) and K.SD (88.6% yield), which is close to the reported boiling point of the methyl alcohol (Figure 3A,B). This demonstrates that the transesterification processes were endothermic reactions and that their threshold kinetic energies enhanced the mass transfer properties within the system between the different reactants [44–46]. The significant loss in the methanol content as a result of its evaporation from the system during the conduction of the tests up to 70 °C had significant adverse impacts on the efficiency of the production of FAME [47].

The determined yields of FAME as well as the recognized catalytic properties of Na.SD and K.SD as functions of the assessed factors (temperature and reaction duration) validate the significant differences between the two sodalite phases in their catalytic activities. The sodium-rich phase (Na.SD) exhibited higher catalytic properties than the potassium-rich phase (K.SD) did at lower intervals. This was credited to the determined high basicity and surface area of the Na.SD compared to the K.SD, which reflect the significant effect of the species of the used alkaline solutions on the synthesis of the sodalite catalyst.

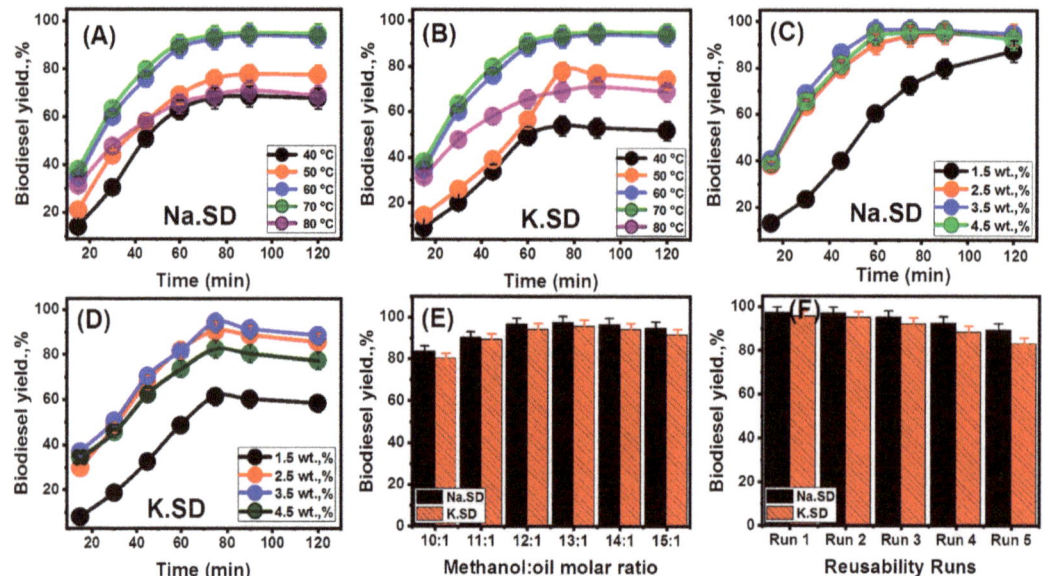

Figure 3. The effect of (**A**) the transesterification duration in different temperature conditions on the catalytic activity of Na.SD, (**B**) the transesterification duration on the catalytic activity of K.SD, (**C**) the effect of the catalyst loading of Na.SD on the obtained biodiesel yield, (**D**) the effect of the catalyst loading of K.SD on the obtained biodiesel yield, (**E**) the methanol-to-oil molar ratio, and (**F**) the recyclability properties of Na.SD and K.SD as heterogeneous catalysts.

Effects of the Catalysts Dosages on FAME Yield

The impact of the Na.SD and K.SD dosages on the efficiency of the production of FAME was studied from 1.5 wt.% up to 4.5 wt.%. This was evaluated with consideration of the different values of the transesterification duration (15 min up to 120 min) and of the fixed values of the other affecting parameters (methanol-to-oil molar ratio: 12:1 and temperature: 70 °C). Considering the best identified transesterification durations in the presence of Na.SD (90 min) and K.SD (75 min), the incorporated dosages of them exhibited a notable enhancement of the produced FAME as well as of the rate of the production of FAME (Figure 3C,D). This was detected within the assessed range from 1.5 wt.% (79.7% yield (Na.SD)) and 60.2% yield (K.SD)) to 3.5 wt.% (96.8% yield (Na.SD) and 91.4% yield (K.SD)) (Figure 3C,D). The incorporation of Na.SD and K.SD at satisfactorily high dosages were associated with a significant enhancement of the availability of the effective catalytic sites as well as of the interactive interfaces between the catalyst grains and the liquid phases [44]. This resulted in the reported enhancement of the generation quantities of FAME as well as of the actual transformation rates of triglycerides. However, the conducted tests in the presence of a dosage of 4.5 wt.% of both Na.SD and K.SD demonstrated observable adverse effects on the quantities of formed FAME (Figure 3C,D). This was reported in the literature as a result of the negative impacts of the presence of dosages beyond the threshold in the system on the viscosity, homogeneity, and mass transfer resistance between the spent oil, methanol, the sodalite particles as reactants [20,48,49].

Effect of Methanol-to Oil Molar Ratio

The molar ratio of methyl alcohol to triglyceride content is a crucial parameter that controls the efficiency of the generation of biodiesel during the conversion of the incorporated spent cooking oil [1,50]. According to stoichiometry, the best transesterification efficiency can be achieved by the successful reaction between 3 moles of methyl alcohol and 1 mole of triglyceride, which normally results in 3 moles of FAME in addition to 1 mole of glycerol [47]. According to Lee Chatelier's concept, the alcohol content in the systems should be preserved at levels higher than the reported stoichiometric value to keep the reaction progressing in the forward direction and to avoid the reversible properties of such transesterification processes [1,44]. The actual impact of the molar ratio of the alcohol to the oil on the Na.SD and K.SD transesterification systems were studied from 10:1 to 15:1 (Figure 3E). This was evaluated with consideration of the best values of the other affecting parameters obtained from the previously mentioned tests (sodalite dosage: 3.5 wt.%; temperature: 70 °C; and duration time: 90 min (Na.SD) and 75 min (K.SD)).

The efficiency of the production of FAME as well as the rate of its production were enhanced notably in terms of the increase in the adjusted molar ratio of the used alcohol up to 13:1 either in the presence of Na.SD (97.3%) or K.SD (95.7%) (Figure 3E). An excess in methanol within the recommended stoichiometric levels significantly enhances the miscibility properties of the system by reducing the viscosity as well as the mass transfer resistance [50]. Furthermore, high methanol ratios exhibit remarkable acceleration effects on the collision or the interaction processes between the triglycerides as dissolved molecules and the surface of the sodalite solid particles [44]. However, this was detected up to a certain ratio (14:1) in the experiment, and then the excess in the used alcohol caused observable adverse properties that resulted in low FAME yields, especially at the highest level of the adjusted methanol/oil ratio (15:1) (94.8% (Na.SD) and 91.6% (K.SD)) (Figure 3E). This reversible effect was documented in previous studies and was illustrated as being a result of the predicted re-dissolving of the glycerol molecules within the excess methanol beyond the stoichiometric levels. This negatively affected the balance between the biodiesel and glycerol in the transesterification reaction and effective phase separation, causing reversible reactions [13,51]. Moreover, the free alcohol molecules might create deactivation effects on the essential active catalytic chemical groups of the sodalite particles. Also, this might be associated with the significant conversion of them into non-preferred emulsifier centers after the expected inversion of the polar groups [52,53].

Recyclability of Na.SD and K.SD Catalysts

The stability and recyclability potentials of the prepared Na.SD and K.SD phases as solid heterogeneous catalysts were studied as critical parameters during the assessment of the products for realistic and commercial applications. The extracted Na.SD and K.SD particles after the transesterification processes were washed firstly with methanol as a common solvent of adsorbed organic molecules (fatty acid and glycerol) on their surfaces. This step was repeated for five runs, each run taking about 10 min, and then the particles were washed with distilled water for 15 min. After the washing step, the obtained Na.SD and K.SD were dried at 80 °C for about 12 h and then re-used again in a new transesterification test. All the recyclability tests were adjusted to the experimentally detected best values (sodalite dosage: 3.5 wt.%; temperature: 70 °C; methanol/oil ratio: 13:1; and duration time: 90 min (Na.SD) and 75 min (K.SD)) (Figure 3F).

Based on the measured results, Na.SD showed considerable recyclability properties and stability as an incorporated basic catalyst in the heterogeneous catalytic transesterification systems compared to K.SD considering the number of runs (5 runs) in which it was reused. In terms of the reusability of Na.SD, it reflected its ability to obtain FAME yields beyond 97% after two runs, beyond 95% after three runs, and above 89% after five runs (Figure 3F); additionally, the reusability properties of K.SD allowed it to obtain FAME yields higher than 95% and 92% after two and three runs, respectively, yet it is accompanied by a noticeable decrease at the fifth run, with a FAME yield of 83% (Figure 3F). The observable dwindling in the FAME yield and the activities of both the Na.SD and K.SD with the regular repetition of the reusing runs might be credited to the expected leaching of some of the exchangeable Na+ and K+ ions within the structure of sodalite during the washing and transesterification processes [32,54]. Furthermore, the over-accumulation of glycerol on the surfaces of their particles might have negatively affected the exposure of the active sites and might have deactivated them as active centers [12].

2.2.2. Physical and Chemical Properties of The Obtained Biodiesel

Both the ASTM D-6751 and EN 14214 biodiesel international standards were used to assess the physical qualifications of the biodiesel products obtained by the Na.SD and K.SD transesterification systems to be used as safe and suitable biofuel (Table 2). According to the presented criteria, the products by obtained both the Na.SD and K.SD exhibited acceptable viscosity and density for their use directly as suitable fuel. Additionally, the cetane index of the two biodiesel products is more than 45, signifying that their combustion quality and ignition delay time are acceptable for their use as effective and safe fuels in engines [13]. The determined flashpoint values as well as the calorific values of two biodiesel products validate the significant safety of their properties, as commercial products, for their handling and transportation, which is in agreement with the low values of the measured pour and cloud points (Table 2).

Table 2. The physical properties of the biodiesel products using Na.SD and K.SD catalysts.

Contents	ASTM D-6751	EN 14214	Na.SD	K.SD
Viscosity (mm^2/s)	1.9–6	3.5–5	3.72	3.24
Moisture content (wt.(%))	<0.05	<0.05	0.041	0.032
Flash point (°C)	>93	>120	134.2	129.8
Calorific value (MJ/kg)	—	>32.9	37.4	36.5
Cloud point (°C)	−3 to 15	—	5.7	5.33
Pour point (pp)	−5 to 10	—	6.2	5.7
Cetane number	≥47	≥51	54.3	52.5
Density (g/cm^3)	0.82–0.9	0.86–0.9	0.87	0.84
Acid value (Mg/KOH/g)	≤0.5	≤0.5	0.42	0.37

The determined species of the formed esters, according to the GC-MS analysis, demonstrated the existence of oleic acid, palmitoleic acid, and linoleic acid methyl esters as the dominant phases of the generated fatty acid methyl esters (FAME) (Table S1). Moreover, other phases were detected but at small percentages, such as myristic acid, palmitic acid, eicosanoic acid, stearic acid, caprylic acid, and behenic acid methyl esters (Table S1).

2.2.3. The Suggested Mechanism

By considering the differential spectra of the fresh Na.SD and K.SD as well as their used products after the performed experiments, their effective transesterification mechanisms can be illustrated. The notable observation of the identified FT-IR bands of the CH_2 aliphatic group as well as of the ester carbonyl group in the spectra of the used products validates the considerable adsorption of triglyceride molecules by the active sites of both Na.SD and K.SD. The non-detection of any bands related to the formation of sodium or potassium methoxides suggested no effects of the interaction between the active sites of the two sodalite phases and the methanol molecules on their mechanistic behaviors during the transesterification processes [12]. Therefore, it is suggested here that the transesterification on the surfaces of the Na.SD and K.SD initiated significant interactions between the basic alkali bonds of the sodalite structures (Na-O and K-O) and the triglycerides of the transformed oil (Figure 4) [44]. This suggestion is supported by the previously identified CH_2 aliphatic and ester carbonyl groups [55,56]. The previous step was followed with surface interaction processes between the triglyceride molecules and the alcohol molecules at the active functional sites on the surfaces of the Na.SD and K.SD (Figure 5). During this interaction, the position of the O in Na-O and K-O was substituted with the methoxide molecules (CH_3O-), forming fatty acid methyl ester. After that, the effective capturing of the hydrogen ions from the system by the starting glycerol backbone resulted in the formation of stable glycerol molecules that desorbed immediately from the surfaces of the Na.SD and K.SD

(Figure 4) [1]. This caused the evacuation of the effective active catalytic centers and their transformation into free sites for new runs of triglyceride interaction. Considering the previously reported mechanistic steps, the high surface area and total basicity of the Na.SD accelerated the transesterification rate at a higher efficiency compared to K.SD, reflecting the significant impacts of the used types of alkali on the procedures of the synthesis of sodalite from kaolinite as well as on the applied synthesis methodology.

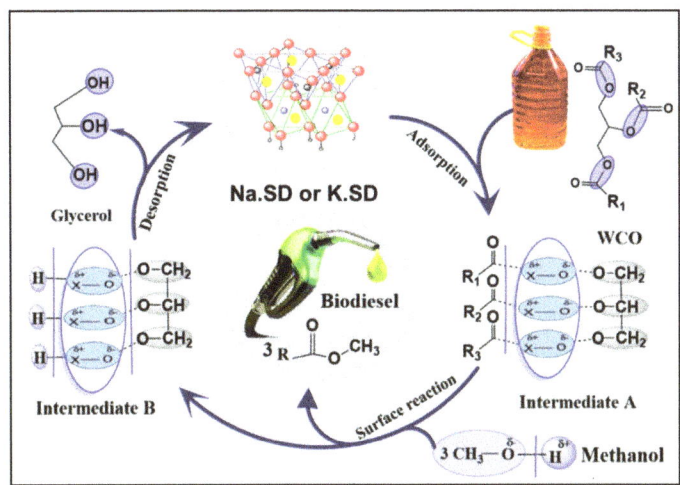

Figure 4. Schematic diagram of the transesterification of waste cooking oil into biodiesel using the synthetic Na.SD and K.SD catalysts.

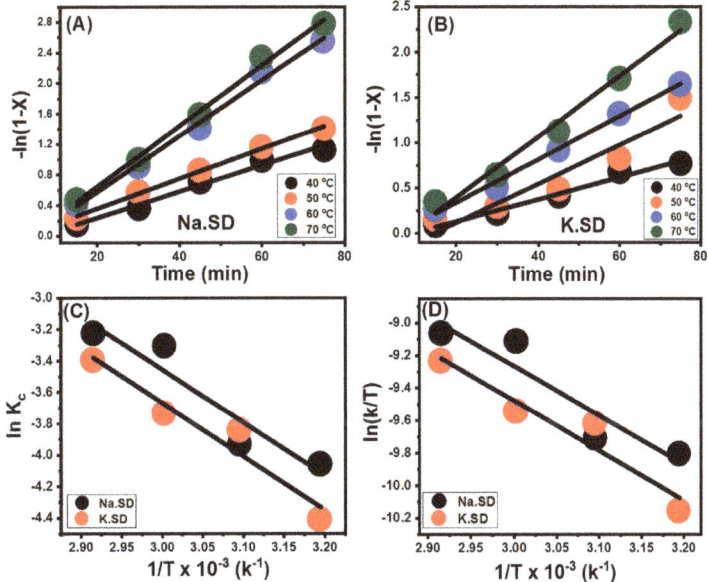

Figure 5. Results of the fitting of the biodiesel yields of Na.SD and K.SD with the pseudo-first-order kinetic model (**A**,**B**), with the Arrhenius equation (**C**), and with the Eyring–Polanyi equation (**D**).

2.2.4. Kinetics and Thermodynamics

The kinetic and thermodynamic functions of the Na.SD- and K.SD-based transesterification systems as well as their significance were evaluated based on the reported procedures in Section 2. The essential kinetic functions that were investigated in this study were the activation energy and reaction rate constant, while the obtained thermodynamic functions were the activation entropy, activation enthalpy, and Gibb's free energy.

Kinetic Studies

The Reactions Rate Constants

The rate constant (Kc) values of the occurred transesterification reactions based on the Na.SD and K.SD systems were obtained as mathematical parameters for the linear regression relations between -ln (1-X) vs. the reaction duration in min (Figure 5A,B). This was assessed with consideration of the experimental range of temperatures from 40 °C to 70 °C and the duration range from 15 min to 75 min after the fixation of the other factors (methanol-to-oil molar ratio: 12:1 and the catalyst load: 2.5 wt.). The fitting processes demonstrated the pseudo-first-order kinetic properties of the reactions that occurred during the conversion of waste cooking oil based on the Na.SD and K.SD transesterification systems with considerably high determination coefficients ($R^2 > 0.87$) (Figure 5A,B). This kinetic behavior validates the occurrence of the reaction according to the three significant steps of adsorption, surface interaction, and desorption progressive processes that are in notable agreement with the obtained findings of the FT-IR analysis of the used Na.SD and K.SD particles [29,42]. The observable increases in the theoretically estimated values of the Kc according to the temperature of the tests exhibited notable agreement with the observed experimental behaviors, including the positive influence of the temperature on the catalytic systems of the Na.SD and K.SD (Table 3). Moreover, the Kc values of the reactions that occurred in the presence of Na.SD were higher than those of the accomplished reactions in the presence of K.SD, validating the higher catalytic activity of the sodium sodalite phase.

Table 3. The estimated values of the determination coefficient and rate constant of the pseudo-first-order kinetic model.

Catalyst	Temperature (°C)	Kc (min^{-1})	(R^2)
Na.SD	40	0.01723	0.97813
	50	0.01958	0.99603
	60	0.03666	0.99031
	70	0.0397	0.99457
K.SD	40	0.01214	0.97264
	50	0.02143	0.871
	60	0.02386	0.9912
	70	0.03353	0.97781

The Reaction Activation Energy and Pre-Exponential Values

The regression relations between the rate constants and the values of 1/T previously estimated according to the Arrhenius equation were used to estimate the values of the activation energy (Ea) of the transformation reactions of the spent oil in the Na.SD and K.SD transesterification systems (Figure 5C; Table 4). The calculated activation energies during the transesterification processes in presence of the Na.SD and K.SD catalysts were 27.9 kJ.mol^{-1} and K.SD 28.27 kJ.mol^{-1}, respectively, while the obtained pre-exponential factors for the occurred reactions by Na.SD and K.SD were 758.94 min^{-1} and 688.33 min^{-1}, respectively (Table 4). It was reported that the transesterification systems that exhibit low activation energy were characterized by low energy barriers and could achieve signified yields and efficiencies at low to moderate operating temperatures [57]. Regarding the pre-exponential factor, the system which showed high values exhibited enhanced collision properties and enhanced chances of interaction between the different reactants and the exposed active catalytic centers of sodalite, and this in turn induced the methylation of the triglyceride molecules. This, in addition to the previously mentioned physicochemical properties, illustrates the higher catalytic activity of Na.SD compared to K.SD.

Table 4. The activation energy and pre-exponential values of the performed transesterification reactions using Na.SD and K.SD.

Parameters	Na.SD	K.SD
Slope	−3.36	−3.40
Intercept	6.63	6.534
Determination coefficient (R^2)	0.848	0.907
Activation energy (ΔE^*) (kJ.mol^{-1})	27.9	28.27
Pre-exponential value (A) (min^{-1})	758.94	688.33

The Thermodynamic Functions (Enthalpy, Entropy, and Gibb's Free Energy)

The main thermodynamic functions of the Na.SD- and K.SD-based transesterification reactions were obtained as mathematical theoretical parameters of the fitting results with the Eyring–Polanyi equation with consideration of the linear regression relations between ln (Kc/T) and 1/T (Figure 5D, Table 5). Based on the notable positively signed values of enthalpy (ΔH^*) (25.233 kJ.mol^{-1} (Na.SD) and 25.55 kJ.mol^{-1} (K.SD)) and Gibb's free energy (ΔG^*) (61.9–67.836 kJ.mol^{-1} (Na.SD) and 61.93–67.87 kJ.mol^{-1} (K.SD)) validate the spontaneous and endothermic behaviors of the two transesterification systems (Table 5). Moreover, the negatively signed entropy (ΔS^*) values of both Na.SD (−197.7 J.K^{-1}.mol^{-1}) K.SD (−197.8 J.K^{-1}.mol^{-1}) reflects a considerable reduction in the randomness of the reactions that occurred with the elevation in the transesterification temperature.

Table 5. The estimated thermodynamic functions of Na.SD and K.SD transesterification systems.

Thermodynamic Parameters		Na.SD	K.SD
Slope		−3.03508	−3.0734
Intercept		−0.16003	−0.25768
R^2		0.81843	0.88793
ΔH^* (kJ. mol^{-1})		25.233	25.55
ΔS^* (J.K^{-1}.mol^{-1})		−197.7	−197.8
(ΔG^*) (kJ. mol^{-1})	40 °C	61.905	61.936
	50 °C	63.882	63.914
	60 °C	65.859	65.892
	70 °C	67.836	67.87

2.2.5. Comparison Study

The catalytic activities of both Na.SD and K.SD were compared with other assessed basic heterogeneous catalysts in terms of the determined yields in certain experimental conditions. The assessed synthetic sodium and potassium sodalite phases obtained by the low-temperature alkaline fusion of natural kaolinite exhibited higher activity than several investigated products did, these products including synthetic apatite, nickel oxide-based catalysts, Cs modified silica, CaO, CaO/SiO$_2$, and kettle limescale (Table 6). Moreover, these results suggest that the alkaline fusion synthesis of sodalite using NaOH, rather than KOH, should be recommended to obtain a more effective catalyst that can achieve promising yields within reasonable time intervals in the presence of low solid dosages and low alcohol content.

Table 6. Comparison between the obtained biodiesel yield using Na.SD and K.SD catalysts and other catalysts in the literature.

Catalyst	Time	Temperature (°C)	Methanol/Oil Ratio	Dosage (wt.%)	Yield (%)	References
CaO/SiO$_2$	3 h	65	21:1	11	90.2	[58]
Kettle lime scale	15 min	61.7	3:1.7	8.9	93.4	[59]
Zeolite Na-X	8 h	65	6:1	3	83.5	[60]
CaO	3 h	65	20:1	5	95	[61]
Cesium modified silica	3 h	65	20:1	3	90	[62]
Ni/Fe carbonate-fuorapatite	2 h	70	8:1	10	97.5	[63]
Coconut coir husk	3 h	130	12:1	10	89.8	[64]
Diatomite/CaO/MgO	2 h	90	15:1	6	96.4	[50]
Ni/NiO@Diatomite	117 min	63.7	11.6:1	4	93.2	[20]
Na.SD	90 min	70	13:1	2.5	97.3	This study
K.SD	75 min	70	13:1	2.5	95.7	This study

3. Experimental Work

3.1. Materials

The kaolinite powder that was used as a precursor was delivered from the Central Metallurgical Research & Development Institute in Egypt after gentle beneficiation steps. NaOH scales (97%; Alfa Aesar, Egypt) as well as KOH pellets (90%; Sigma-Aldrich, Egypt) were used as the sources of the main alkaline ions during the alkali fusion of kaolinite. The spent cooking oil sample tested during the operated transesterification experiments represented a mixture of different commercial samples, which were obtained from different local restaurants. The composition of the incorporated spent cooking oil sample is detailed in Table S2.

3.2. Synthesis of Sodalite Catalysts

The two sodalite phases of sodium sodalite and potassium sodalite were obtained by the alkali fusion of the kaolinite, followed by a gentle hydrothermal alteration step. The kaolinite precursor was mixed in separate experiments with NaOH and KOH with consideration of the weight ratio at 1(kaolinite): 2(alkali hydroxide), and then the resulting mixtures were fused gently at 200 °C for 4 h. The fused products were ground carefully, and about 6 g of each fused product (NaOH/kaolinite and KOH/kaolinite) was homogenized within 100 mL of distilled water at an adjusted temperature of 70 °C while stirring for 2 h. This step was followed by the hydrothermal alteration of the mixtures at 90 °C for 4 h after transferring them into two Teflon-lined stainless steel reactors. By the end of the alteration interval, the reactors were cooled down and the synthesized sodalities fractions were separated from the residual alkaline solutions. Finally, the products were neutralized and washed from the excess alkali ions, dried (85 °C overnight), and labeled as Na.SD (sodium sodalite) and K.SD (potassium sodalite) (Figure 6).

3.3. Characterization Techniques

A transformation of the kaolinite into sodalite zeolite phases occurred, as revealed by the X-ray diffraction patterns determined using an X-ray diffractometer (PANalytical (Empyrean)) within the 5° to 70° determination range. The predicted changes in the chemical groups followed, as revealed by their FT-IR spectra, which were measured by a Fourier Transform Infrared spectrometer (Shimadzu FTIR−8400S) with a frequency range from 400 up to 4000 cm^{-1}. A scanning electron microscope (Gemini, Zeiss-Ultra 55) was used to determine and describe the morphological features of the raw and synthesized products based on the SEN images obtained at an accelerating voltage of 30 kV. The basicity properties were illustrated based on the determined values of the K.SD and Na.SD according to the reported methods by [33]. The ion exchange capacities of both the Na.SD and K.SD, as essential parameters in the transesterification processes, were measured by the BaCl$_2$ technique according to the reported procedures by [34]. The textural studies were assessed according to the BET surface area and porosity of the two sodalite phases, which were revealed by the adsorption/desorption isotherm curves obtained using the Beckman Coulter surface area analyzer (SA3100 type).

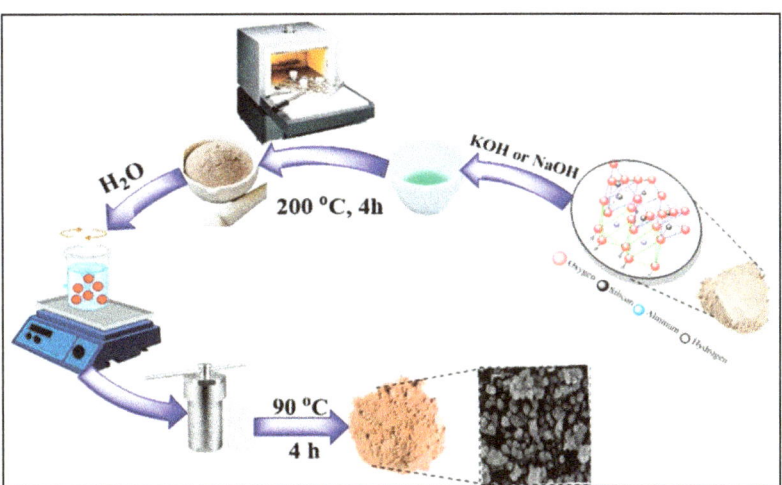

Figure 6. Schematic diagram of the synthesis procedures of Na.sodalite and K.sodalite.

3.4. Transesterification System

Stoichiometrically, each mol of triglyceride can be transesterified with three mol of methanol, producing one mol of pure glycerol and three mol of fatty acid methyl ester (FAME) (biodiesel). According to the previous stoichiometric base, the transesterification of the investigated spent oil sample was accomplished within a considerable experimental range for the incorporated reactants and operating conditions to attend to the appropriate ratios. All the performed reactions between the incorporated reactants were carried out in a specific reactor, the Teflon autoclave (150 mL), which was attached with a digital magnetic stirrer and hot plate to control the homogenization degree and the temperature according to the selected values. The main variables that were investigated during the study were the methanol molar ratio, sodalite loading, transesterification duration, and temperature, and the best obtained value of each test was considered during the operation of the next test.

The experimental procedures that were followed during the tests involved, firstly, the careful and effective filtration of the collected spent oil samples to get rid of the tough suspensions. Then, a certain volume of the filtrated oil sample was heated directly at 100 °C for 20 min to get rid of the present water molecules and to avoid the side effects of humidity. After cooling the heated oil sample, the sodalite catalysts (Na.SD and K.SD) were mixed separately with the oil at a certain dosage for 50 min followed by a gradual increase in the operating temperature up to a certain value according to the experimental design. This step is associated with the controlled incorporation of the methanol at an adjustable volume according to the pre-calculated molar ratio to the oil sample for a certain transesterification interval. By completing the reaction duration, the Na.SD and K.SD particles were separated by centrifugation from the liquid phases (biodiesel + glycerol) and were then separated from each other by a glass separating funnel. Then, the obtained sample was left for an additional 24 h to confirm the complete separation of the glycerol content, and this was followed by heating the sample for 3 h at about 70 °C to certify the effective evaporation of the rest of the methanol molecules. After that, the quantities as well as the types of the formed FAMEs were determined by gas chromatography coupled with a mass spectroscopy unit (GC-Mass) instrument (Agilent 7890A). The determined values were applied in the direct calculation of the experimentally obtained biodiesel yield (Equation (1)).

$$\text{Biodiesel yield (\%)} = \frac{(\text{wieght of biodiesel}) \times \% \text{ FAME}}{\text{weight of triglycerides}} \times 100 \qquad (1)$$

3.5. Analysis of the FAME Samples

The types, as well as the quantities of the formed fatty acid methyl esters during the transesterification reactions using both N.SD and K.SD, were measured by gas chromatography coupled with a mass spectroscopy unit (GC-Mass) instrument, Agilent-7890A. All the measurements were conducted with the existence of n-hexane as a non-polar solvent of the FAME molecules and methyl heptadecanoate (1 µL) as an analytical internal standard. The Agilent-7890A series gas chromatography system used was coupled with a flame ionization detector, split/splitless injector, and capillary column (DB WAX (30 m × 0.25 m × 0.25 µm)) saturated with a hydrogen carrier and with a 100:1split proportion. During its operation, the working temperature of the detector as well as the injector was maintained at about 280 °C, while the temperature of the oven was adjusted firstly to 120 °C and then raised up to 260 °C at an accelerating rate of 10 °C/min.

3.6. Kinetics Studies

The kinetic properties of the Na.SD and K.SD transesterification-based systems were observed within various intervals from 15 min up to 120 min and within an experimental temperature range from 40 °C until 70 °C with consideration of the other experimental conditions at certain values (catalyst dosage = 2.5 wt.% and methyl alcohol/spent oil molar ratio = 12:1). The transesterification process involves three essential consecutive reversible reactions and all of them include reactions between triacylglycerol- (TG) and methyl-producing diacylglycerols (DG*) (Equation (2)), monoacylglycerols (MG*) (Equation (3)), and fatty acid methyl esters (FAME) in addition to glycerol (GL) (Equation (4)) [65]. These reactions reflect the neglected impacts of the transesterification intermediates on the efficiency of the process as the FAMEs molecules can be produced directly during the reaction between the triglycerides and methyl alcohol. Therefore, the overall transesterification of triglycerides can be represented by one step according to Equation (5) [66]. Stoichiometrically, each mol of TG, DG*, and MG* react with one mol of alcohol (R-OH) to produce only one mol of GL and 3 mol of FAME [67].

$$TG + R - OH \rightleftarrows DG^* + FAME \qquad (2)$$

$$TG + R - OH \rightleftarrows MG^* + FAME \qquad (3)$$

$$TG + R - OH \rightleftarrows GL + FAME \qquad (4)$$

$$TG + 3R - OH \rightleftarrows GL + 3FAME \qquad (5)$$

This assumption was suggested based on four essential parameters as follows: (A) the incorporated main reactants of triglycerides, methanol, and solid catalyst particles is distributed homogenously within the system, (B) the possible occurrence of reverse reactions as well as the changes in the catalyst dosage can be neglected by adjusting the catalyst dosage at an appropriate amount, (C) the incorporated concentration of the methyl alcohol is assumed to be almost constant throughout the transesterification reaction, and (D) the generated intermediates during transesterification reactions are ignored. According to the reported mechanistic steps in the literature, the formation of the FAME by the heterogeneous solid catalyst involves (1) the adsorption of triacylglycerol by the active sites of sodalite as the initial step, (2) a surface reaction between the adsorbed molecules and active catalytic sites, and finally, (3) the desorption of glycerol [43].

3.6.1. Adsorption of Triglyceride by Active Sites of Sodalite

The transesterification efficiency of the adsorbed triacylglycerol molecules on the surfaces of both the Na.SD and K.SD as incorporated heterogeneous catalysts and the resulting fatty acid methyl esters are affected by the methanol content rather than by the concentrations of the triglycerides. This process depends essentially on the adsorption of significant quantities of TG molecules on the surfaces of both Na.SD and K.SD according to Equation (6), using the values of the free active catalytic sites (S) and the adsorbed quantities of TG (TG.S). Furthermore, the transesterification rate law can be expressed by Equation (7).

$$TG + S \rightleftarrows TG.S \tag{6}$$

$$r_{ad} = K_a[TG][S] - (K_{-a}[TG.S]) \tag{7}$$

3.6.2. Surface Reaction

This step involved significant reactions between the adsorbed TG and the alcohol molecules, where the TG molecules were affected by the nucleophilic attack of a methoxy group (CH3O-) forming a series of different species of fatty acid methyl esters in addition to the glycerol byproducts. The high affinity of the remaining free sites to the produced GL molecules causes the facile and significant diffusion of the AME molecules according to Equation (8), which uses the quantities of the GL molecules (GL.S) adsorbed by sodalite, and the rate of this surface reaction can be represented by Equation (9) [44].

$$R - OH + TG \rightleftarrows GL.S + FAME \tag{8}$$

$$r_s = K_s[R - OH][TG.S] - (K_{-s}[GL.S][FAME]) \tag{9}$$

3.6.3. Desorption of Glycerol

The release behavior of the adsorbed GL molecule (GL.S) can be determined according to Equation (10), and the rate law of this reaction can be represented by Equation (11). Based on this equation (Equation (10)) and the other representative equations of the mechanistic steps of transesterification (adsorption and surface reaction), the three steps can be represented by a general equation (Equation (12)) using the total quantities of the free active sites (S_{total}). Based on Equation (12), the rate of the surface reaction was modified, as in Equation (13), according to the values of the S_{total} instead of those of the TG.S and GL.S.

$$GL.S + GL \rightleftarrows S \tag{10}$$

$$r_d = K_d[GL.S][S] - (K_{-d}[GL][S]) \tag{11}$$

$$[TG.S].[S_{total}] = [S] + [TG.S] + [GL.S] \tag{12}$$

$$r_s = K_a.K_s[S_{total}]([TG][R-OH] - [GL][R-COOCH_3]/K_c)/(1 + K_A[TG] + [GL]/K_d) \tag{13}$$

The final representative equation of the general transesterification reaction rate is represented by Equation (14), which uses the rate constant (K_c) of the reaction, which was estimated according to Equation (15). The general reaction rate equation (Equation (14)) was presented with consideration of four essential factors, these being (1) the fact that the desorption of GL occurred at a higher rate than its adsorption did, (2) that both the K_a [TG] and [GL]/K_d exhibited values equal to zero, (3) that the concentration of the incorporated alcohol [R-OH] was higher than the present TG, and (4) that S_{total} values were constant for the studied heterogeneous catalyst. Using the variation in the TG content within the studied system from the start of the reaction ($[TG]_0$) up to a certain time interval ($[TG]_t$), the equation can be modified to the presented form in Equation (16). The methyl ester conversion ($X_{(FAME)}$) can be derived from the mass balance by Equation (17) and subsequently, Equation (18). The integration of the representative equation of the reaction rate as a function of the methyl ester conversion ($X_{(FAME)}$) resulted in the final equation of the system (Equation (19)). Therefore, the conversion efficiency as well as the transesterification rate were affected only by the availability or concentrations of TG molecules, suggesting only pseudo- first-order kinetic behavior [1].

$$r_s = d[TG]/dt = K_c[GL] \tag{14}$$

$$K_c = K_a \times K_s \times [R-OH] \times [S_{total}] \tag{15}$$

$$-ln([TG]_t/[TG]_0) = K_c \times t \tag{16}$$

$$X_{(R-COOCH_3)} = 1 - ([TG]_t/[TG]_0) \tag{17}$$

$$[TG]_t = [TG]_0 \times \left(1 - X_{(FAME)}\right) \tag{18}$$

$$-\ln(X_{(R-COOCH_3)}) = K_c \times t \tag{19}$$

3.7. The Activation Energy (Ea) and Thermodynamic Functions

The Arrhenius equation (Equation (20)) was used to estimate the activation energies (Ea) of the preformed transesterification reactions in the presence of both Na.SD and K.SD as heterogeneous basic catalysts. The values of Ea were obtained as mathematical parameters for the performance of the linear regression fitting processes of ln (K_c) versus the reciprocal values of the experimental activation temperature (1/T), where the slope indicates the Ea values and the intercept reflects the pre-exponential value (ln (A)).

$$K_c = A \exp(-Ea/RT) \tag{20}$$

3.8. The Thermodynamic Functions

The thermodynamic functions of the transesterification systems of Na.SD and K.SD that occurred were determined based on the Erying–Polanyi equation (Equation (21)). The included symbols, R, K, h, and K_b, identify the universal gas constant, rate constant, Planck's constant (6.626176 × 10^{-34} Js), and Boltzmann constant (1.3806 × 10^{-23} K^{-1}). The enthalpy (ΔH*) and entropy (ΔS*) values were obtained as parameters of the preformed linear regression plotting of ln (K/T) vs. (1/T) the slope (-H*/ΔRT) and intercept. However, the Gibbs free energy (ΔG*) was calculated using Equation (22).

$$\ln(K/T) = -(\Delta H^*/RT) + \ln(K_b/h) + (\Delta S^*/R) \tag{21}$$

$$\Delta G^* = \Delta H^* - T\Delta S^* \tag{22}$$

4. Conclusions

Two sodalite forms, the sodium-rich form (Na.SD) and potassium-rich form (K.SD), were prepared successfully by the low-temperature alkali fusion of kaolinite as potential basic catalysts in the fabrication of biodiesel by transesterification reactions. The Na.SD form confirms the considerable enhancement of catalytic activity and of the essential physicochemical parameters (total basicity: 6.3 mmol OH/g and surface area: 232.4 m^2/g) compared to K.SD (total basicity: 5.4 mmol OH/g and surface area: (217.6 m^2/g). The best yield obtained by using Na.SD, this being 97.3%, was maintained at 70 °C, with a 13:1 methanol: oil molar ratio and at a dose of 2.5 wt.% for 90 min, while the best yield that was obtained by K.SD was 95.7%, which was maintained at 70 °C with a 13:1 methanol: oil molar ratio and a dose of 2.5 wt.% for 75 min. Furthermore, the two catalyst showed enhanced recyclability during the four run;, the fifth run of Na.SD showed a slight decrease in the FAME yield to 89%, yet the fifth run of K.SD showed a remarkable decrease to 83%. The kinetic properties (pseudo-first order kinetics) reflected the operation of the transesterification processes according to the progressive steps of adsorption, surface reaction, and desorption. The activation energies of the two transesterification systems as well as the thermodynamic functions (Gibb's free energy, enthalpy, and entropy) signify the endothermic and spontaneous catalytic reactions that occur when using Na.SD and K.SD in addition to their significant performances under the low-temperature conditions.

Supplementary Materials: The following supporting information can be downloaded at: https://www.mdpi.com/article/10.3390/catal13030462/s1, Table S1. The determined fatty acid methyl esters in the obtained biodiesel sample over N.SD. Table S2. the Fatty acid content and physical properties of the inspected spent oil

Author Contributions: Conceptualization, M.R.A., S.B., J.S.A. and M.A.S.; methodology, M.A.S., J.L. and C.W.; Software, M.A.S., A.A.A. and J.L.; validation, M.R.A., S.B., J.S.A., A.A.A. and C.W.; formal analysis, M.A.S., A.A.A., J.L. and C.W.; investigation, M.R.A., S.B., J.S.A. and A.A.A.; resources, A.A.A., M.A.S., J.L. and C.W.; data curation, M.R.A., M.A.S., S.B., J.S.A., J.L. and C.W.; writing—original draft preparation, M.R.A., M.A.S., C.W., A.A.A., S.B., J.S.A., J.L. and C.W.; writing—review and editing, M.R.A., M.A.S., C.W., A.A.A., S.B., J.L. and C.W.; visualization, M.R.A., M.A.S. and S.B.; supervision, M.R.A., S.B., A.A.A. and C.W.; project administration, M.R.A., J.S.A. and A.A.A.; funding acquisition, J.S.A. All authors have read and agreed to the published version of the manuscript.

Funding: This research was funded by [King Saud University, Riyadh, Saudi Arabia] grant number [RSP2023R149], [Guangdong Basic and Applied Basic Research Foundation] grant number [2021A1515010060], and [Guangdong Province Scientific Research Platform] grant number [2022ZDZX4046].

Data Availability Statement: Data are available upon reasonable, by the Corresponding Authors.

Acknowledgments: The authors acknowledge the Researchers Supporting Project (RSP2023R149), King Saud University, Riyadh, Saudi Arabia. The financial support provided by Guangdong Basic and Applied Basic Research Foundation under Grant 2021A1515010060, and in part by Guangdong Province Scientific Research Platform Project under Grant 2022ZDZX4046, is highly appreciated.

Conflicts of Interest: The authors declare no conflict of interest.

References

1. Sayed, M.A.; Ahmed, S.A.; Othman, S.I.; Allam, A.A.; Al Zoubi, W.; Ajarem, J.S.; Abukhadra, M.R.; Bellucci, S. Kinetic, Thermodynamic, and Mechanistic Studies on the Effect of the Preparation Method on the Catalytic Activity of Synthetic Zeolite-A during the Transesterification of Waste Cooking Oil. *Catalysts* **2023**, *13*, 30. [CrossRef]
2. Rezania, S.; Oryani, B.; Park, J.; Hashemi, B.; Yadav, K.K.; Kwon, E.E.; Hur, J.; Cho, J. Review on transesterification of non-edible sources for biodiesel production with a focus on economic aspects, fuel properties and by-product applications. *Energy Convers. Manag.* **2019**, *201*, 112155. [CrossRef]
3. Orege, J.I.; Oderinde, O.; Kifle, G.A.; Ibikunle, A.A.; Raheem, S.A.; Ejeromedoghene, O.; Okeke, E.S.; Olukowi, O.M.; Orege, O.B.; Fagbohun, E.O.; et al. Recent advances in heterogeneous catalysis for green biodiesel production by transesterification. *Energy Convers. Manag.* **2022**, *258*, 115406. [CrossRef]
4. Ali, R.M.; Elkatory, M.R.; Hamad, H.A. Highly active and stable magnetically recyclable $CuFe_2O_4$ as a heterogenous catalyst for efficient conversion of waste frying oil to biodiesel. *Fuel* **2020**, *268*, 117297. [CrossRef]
5. Papargyriou, D.; Broumidis, E.; de Vere-Tucker, M.; Gavrielides, S.; Hilditch, P.; Irvine, J.; Bonaccorso, A.D. Investigation of solid base catalysts for biodiesel production from fish oil. *Renew. Energy* **2019**, *139*, 661–669. [CrossRef]
6. Kiehbadroudinezhad, M.; Merabet, A.; Hosseinzadeh-Bandbafha, H. A life cycle assessment perspective on biodiesel production from fish wastes for green microgrids in a circular bioeconomy. *Bioresour. Technol. Rep.* **2023**, *21*, 101303. [CrossRef]
7. Karmakar, G.; Ghosh, P.; Sharma, B.K. Chemically modifying vegetable oils to prepare green lubricants. *Lubricants* **2017**, *5*, 44. [CrossRef]
8. Danov, S.M.; Kazantsev, O.A.; Esipovich, A.L.; Belousov, A.S.; Rogozhin, A.E.; Kanakov, E.A. Recent advances in the field of selective epoxidation of vegetable oils and their derivatives: A review and perspective. *Catal. Sci. Technol.* **2017**, *7*, 3659–3675. [CrossRef]
9. Pleissner, D.; Lau, K.Y.; Zhang, C.; Lin, C.S.K. Plasticizer and surfactant formation from food-waste- and algal biomass-derived lipids. *ChemSusChem* **2015**, *8*, 1686–1691. [CrossRef]
10. Karis, D.; Cain, R.; Young, K.; Shand, A.; Holm, T.; Springer, E. Non-fuel uses for fatty acid methyl esters. *Biofuels Bioprod. Biorefining* **2022**, *16*, 1893–1908. [CrossRef]
11. Belousov, A.S.; Esipovich, A.L.; Kanakov, E.A.; Otopkova, K.V. Recent advances in sustainable production and catalytic transformations of fatty acid methyl esters. *Energy Fuels* **2021**, *5*, 4512–4545. [CrossRef]
12. Sayed, M.R.; Abukhadra, M.R.; Ahmed, S.A.; Shaban, M.; Javed, U.; Betiha, M.A.; Shim, J.-J.; Rabie, A.M. Synthesis of advanced MgAl-LDH based geopolymer as a potential catalyst in the conversion of waste sunflower oil into biodiesel: Response surface studies. *Fuel* **2020**, *282*, 118865. [CrossRef]

13. Basyouny, M.G.; Abukhadra, M.R.; Alkhaledi, K.; El-Sherbeeny, A.M.; El-Meligy, M.A.; Soliman, A.T.A.; Luqman, M. Insight into the catalytic transformation of the waste products of some edible oils (corn oil and palm oil) into biodiesel using MgO/clinoptilolite green nanocomposite. *Mol. Catal.* **2021**, *500*, 111340. [CrossRef]
14. Pirouzmand, M.; Anakhatoon, M.M.; Ghasemi, Z. One-step biodiesel production from waste cooking oils over metal incorporated MCM-41; positive effect of template. *Fuel* **2018**, *216*, 296–300. [CrossRef]
15. Murta, A.L.S.; De Freitas, M.A.V.; Ferreira, C.G.; Peixoto, M.M.D.C.L. The use of palm oil biodiesel blends in locomotives: An economic, social and environmental analysis. *Renew. Energy* **2021**, *164*, 521–530. [CrossRef]
16. Malhotra, R.; Ali, A. 5-Na/ZnO doped mesoporous silica as reusable solid catalyst for biodiesel production via transesterification of virgin cottonseed oil. *Renew. Energy* **2019**, *133*, 606–619. [CrossRef]
17. Sai, B.A.V.S.L.; Subramaniapillai, N.; Mohamed, M.S.B.K.; Narayanan, A. Optimization of continuous biodiesel production from rubber seed oil (RSO) using calcined eggshells as heterogeneous catalyst. *J. Environ. Chem. Eng.* **2020**, *8*, 103603. [CrossRef]
18. Hoseini, S.; Najafi, G.; Sadeghi, A. Chemical characterization of oil and biodiesel from Common Purslane (Portulaca) seed as novel weed plant feedstock. *Ind. Crops Prod.* **2019**, *140*, 111582. [CrossRef]
19. Ibrahim, M.L.; Nik Abdul Khalil, N.N.A.; Islam, A.; Rashid, U.; Ibrahim, S.F.; Sinar Mashuri, S.I.; Taufiq-Yap, Y.H. Preparation of Na2O supported CNTs nanocatalyst for efficient biodiesel production from waste-oil. *Energy Convers. Manag.* **2020**, *205*, 112445. [CrossRef]
20. Bin Jumah, M.N.; Ibrahim, S.M.; Al-Huqail, A.A.; Bin-Murdhi, N.S.; Allam, A.A.; Abu-Taweel, G.M.; Altoom, N.; Al-Anazi, K.M.; Abukhadra, M.R. Enhancing the catalytic performance of NiO during the transesterification of waste cooking oil using a diatomite carrier and an integrated Ni0Metal: Response surface studies. *ACS Omega* **2021**, *6*, 12318–12330. [CrossRef]
21. Banković-Ilić, I.B.; Stamenković, O.S.; Veljković, V.B. Biodiesel production from non-edible plant oils. *Renew. Sustain. Energy Rev.* **2012**, *16*, 3621–3647. [CrossRef]
22. Takase, M.; Zhang, M.; Feng, W.; Chen, Y.; Zhao, T.; Cobbina, S.J.; Yang, L.; Wu, X. Application of zirconia modified with KOH as heterogeneous solid base catalyst to new non-edible oil for biodiesel. *Energy Convers. Manag.* **2014**, *80*, 117–125. [CrossRef]
23. Sharma, Y.C.; Singh, B.; Korstad, J. Latest developments on application of heterogenous basic catalysts for an efficient and eco friendly synthesis of biodiesel: A review. *Fuel* **2011**, *90*, 1309–1324. [CrossRef]
24. Mukhtar, A.; Saqib, S.; Lin, H.; Shah, M.U.H.; Ullah, S.; Younas, M.; Rezakazemi, M.; Ibrahim, M.; Mahmood, A.; Asif, S.; et al. Current status and challenges in the heterogeneous catalysis for biodiesel production. *Renew. Sustain. Energy Rev.* **2022**, *157*, 112012. [CrossRef]
25. Lee, J.-S.; Saka, S. Biodiesel production by heterogeneous catalysts and supercritical technologies. *Bioresour. Technol.* **2010**, *101*, 7191–7200. [CrossRef]
26. Brahma, S.; Basumatary, B.; Basumatary, S.F.; Das, B.; Brahma, S.; Rokhum, S.L.; Basumatary, S. Biodiesel production from quinary oil mixture using highly efficient Musa chinensis based heterogeneous catalyst. *Fuel* **2023**, *336*, 127150. [CrossRef]
27. Fattahi, N.; Triantafyllidis, K.; Luque, R.; Ramazani, A. Zeolite-based catalysts: A valuable approach toward ester bond formation. *Catalysts.* **2019**, *9*, 758. [CrossRef]
28. Otieno, S.O.; Kowenje, C.O.; Okoyo, A.; Onyango, D.M.; Amisi, K.O.; Nzioka, K.M. Optimizing production of biodiesel catalysed by chemically tuned natural zeolites. *Mater. Today Proc.* **2018**, *5*, 10561–10569. [CrossRef]
29. Li, Z.; Ding, S.; Chen, C.; Qu, S.; Du, L.; Lu, J.; Ding, J. Recyclable Li/NaY zeolite as a heterogeneous alkaline catalyst for biodiesel production: Process optimization and kinetics study. *Energy Convers. Manag.* **2019**, *192*, 335–345. [CrossRef]
30. Nasief, F.; Shaban, M.; Alamry, K.A.; Abu Khadra, M.R.; Khan, A.A.P.; Asiri, A.M.; El-Salam, H.A. Hydrothermal synthesis and mechanically activated zeolite material for utilizing the removal of Ca/Mg from aqueous and raw groundwater. *J. Environ. Chem. Eng.* **2021**, *9*, 105834. [CrossRef]
31. Simanjuntak, W.; Pandiangan, K.D.; Sembiring, Z.; Simanjuntak, A.; Hadi, S. The effect of crystallization time on structure, microstructure, and catalytic activity of zeolite-A synthesized from rice husk silica and food-grade aluminum foil. *Biomass Bioenergy* **2021**, *148*, 106050. [CrossRef]
32. Abukhadra, M.R.; Othman, S.I.; Allam, A.A.; Elfayoumi, H. Insight into the catalytic properties zeolitized kaolinite/diatomite geopolymer as an environmental catalyst for the sustainable conversion of spent cooking oil into biodiesel; optimization and kinetics. *Sustain. Chem. Pharm.* **2021**, *22*, 100473. [CrossRef]
33. Rios, C.; Williams, C.; Fullen, M. Nucleation and growth history of zeolite LTA synthesized from kaolinite by two different methods. *Appl. Clay Sci.* **2009**, *42*, 446–454. [CrossRef]
34. Mokrzycki, J.; Fedyna, M.; Marzec, M.; Panek, R.; Szerement, J.; Marcińska-Mazur, L.; Jarosz, R.; Bajda, T.; Franus, W.; Mierzwa-Hersztek, M. The influence of zeolite X ion-exchangeable forms and impregnation with copper nitrate on the adsorption of phosphate ions from aqueous solutions. *J. Water Process Eng.* **2022**, *50*, 103299. [CrossRef]
35. Alismaeel, Z.T.; Al-Jadir, T.M.; Albayati, T.M.; Abbas, A.S.; Doyle, A.M. Modification of FAU zeolite as an active heterogeneous catalyst for biodiesel production and theoretical considerations for kinetic modeling. *Adv. Powder Technol.* **2022**, *33*, 103646. [CrossRef]
36. Salam, M.A.; Abukhadra, M.R.; Mostafa, M. Effective decontamination of As(V), Hg(II), and U(VI) toxic ions from water using novel muscovite/zeolite aluminosilicate composite: Adsorption behavior and mechanism. *Environ. Sci. Pollut. Res.* **2020**, *27*, 13247–13260. [CrossRef]

37. Ayele, L.; Pérez-Pariente, J.; Chebude, Y.; Díaz, I. Conventional versus alkali fusion synthesis of zeolite A from low grade kaolin. *Appl. Clay Sci.* **2016**, *132–133*, 485–490. [CrossRef]
38. Altoom, N.; Adlii, A.; Othman, S.I.; Allam, A.A.; Alqhtani, H.A.; Al-Otaibi, F.S.; Abukhadra, M.R. Synthesis and characterization of β-cyclodextrin functionalized zeolite-A as biocompatible carrier for Levofloxacin drug; loading, release, cytotoxicity, and anti-inflammatory studies. *J. Solid State Chem.* **2022**, *312*, 123280. [CrossRef]
39. Shaban, M.; Sayed, M.I.; Shahien, M.G.; Abukhadra, M.R.; Ahmed, Z.M. Adsorption behavior of inorganic- and organic-modified kaolinite for Congo red dye from water, kinetic modeling, and equilibrium studies. *J. Sol-Gel Sci. Technol.* **2018**, *87*, 427–441. [CrossRef]
40. Sakızcı, M.; Özer, M. The characterization and methane adsorption of Ag-, Cu-, Fe-, and H-exchanged chabazite-rich tuff from Turkey. *Environ. Sci. Pollut. Res.* **2019**, *26*, 16616–16627. [CrossRef]
41. Mostafa, M.; El-Meligy, M.A.; Sharaf, M.; Soliman, A.T.; AbuKhadra, M.R. Insight into chitosan/zeolite-A nanocomposite as an advanced carrier for levofloxacin and its anti-inflammatory properties; loading, release, and anti-inflammatory studies. *Int. J. Biol. Macromol.* **2021**, *179*, 206–216. [CrossRef]
42. Abukhadra, M.R.; Sayed, M.A. K^+ trapped kaolinite ($Kaol/K^+$) as low cost and eco-friendly basic heterogeneous catalyst in the transesterification of commercial waste cooking oil into biodiesel. *Energy Convers. Manag.* **2018**, *177*, 468–476. [CrossRef]
43. Arana, J.T.; Torres, J.J.; Acevedo, D.F.; Illanes, C.O.; Ochoa, N.A.; Pagliero, C.L. One-step synthesis of CaO-ZnO efficient catalyst for biodiesel production. *Int. J. Chem. Eng.* **2019**, *2019*, 1806017. [CrossRef]
44. Roy, T.; Sahani, S.; Sharma, Y.C. Study on kinetics-thermodynamics and environmental parameter of biodiesel production from waste cooking oil and castor oil using potassium modified ceria oxide catalyst. *J. Clean. Prod.* **2020**, *247*, 119166. [CrossRef]
45. Ayoub, M.; Bhat, A.H.; Ullah, S.; Ahmad, M.; Uemura, Y. Optimization of Biodiesel Production over Alkaline Modified Clay Catalyst. *J. Japan Inst. Energy.* **2017**, *96*, 456–462. [CrossRef]
46. Seela, C.R.; Alagumalai, A.; Pugazhendhi, A. Evaluating the feasibility of diethyl ether and isobutanol added Jatropha Curcas biodiesel as environmentally friendly fuel blends. *Sustain. Chem. Pharm.* **2020**, *18*, 100340. [CrossRef]
47. Ibrahim, S.M.; El-Sherbeeny, A.M.; Shim, J.-J.; AlHammadi, A.A.; Abukhadra, M.R. -SO_3H-functionalization of sub-bituminous coal as a highly active acidic catalyst during the transesterification of spent sunflower oil; characterization, application, and mechanism. *Energy Rep.* **2021**, *7*, 8699–8710. [CrossRef]
48. Salim, S.M.; Izriq, R.; Almaky, M.M.; Al-Abbassi, A.A. Synthesis and characterization of ZnO nanoparticles for the production of biodiesel by transesterification: Kinetic and thermodynamic studies. *Fuel* **2022**, *321*, 124135. [CrossRef]
49. Gardy, J.; Rehan, M.; Hassanpour, A.; Lai, X.; Nizami, A.-S. Advances in nano-catalysts based biodiesel production from non-food feedstocks. *J. Environ. Manag.* **2019**, *249*, 109316. [CrossRef]
50. Rabie, A.M.; Shaban, M.; Abukhadra, M.R.; Hosny, R.; Ahmed, S.A.; Negm, N.A. Diatomite supported by CaO/MgO nanocomposite as heterogeneous catalyst for biodiesel production from waste cooking oil. *J. Mol. Liq.* **2019**, *279*, 224–231. [CrossRef]
51. Bhatia, S.K.; Gurav, R.; Choi, T.-R.; Kim, H.J.; Yang, S.-Y.; Song, H.-S.; Park, J.Y.; Park, Y.-L.; Han, Y.-H.; Choi, Y.-K.; et al. Conversion of waste cooking oil into biodiesel using heterogenous catalyst derived from cork biochar. *Bioresour. Technol.* **2020**, *302*, 122872. [CrossRef]
52. Singh, V.; Bux, F.; Sharma, Y.C. A low cost one pot synthesis of biodiesel from waste frying oil (WFO) using a novel material, β-potassium dizirconate (β-$K_2Zr_2O_5$). *Appl. Energy* **2016**, *172*, 23–33. [CrossRef]
53. Abukhadra, M.R.; Soliman, S.R.; Bin Jumah, M.N.; Othman, S.I.; AlHammadi, A.A.; Alruhaimi, R.S.; Albohairy, F.M.; Allam, A.A. Insight into the sulfonation conditions on the activity of sub-bituminous coal as acidic catalyst during the transesterification of spent corn oil; effect of sonication waves. *Sustain. Chem. Pharm.* **2022**, *27*, 100691. [CrossRef]
54. Bellucci, S.; Eid, M.H.; Fekete, I.; Péter, S.; Kovács, A.; Othman, S.I.; Ajarem, J.S.; Allam, A.A.; Abukhadra, M.R. Synthesis of K^+ and Na^+ Synthetic Sodalite Phases by Low-Temperature Alkali Fusion of Kaolinite for Effective Remediation of Phosphate Ions: The Impact of the Alkali Ions and Realistic Studies. *Inorganics* **2023**, *11*, 14. [CrossRef]
55. Kamel, D.A.; Farag, H.A.; Amin, N.K.; Zatout, A.A.; Ali, R.M. Smart utilization of jatropha (*Jatropha curcas* Linnaeus) seeds for biodiesel production: Optimization and mechanism. *Ind. Crops Prod.* **2018**, *111*, 407–413. [CrossRef]
56. Kouzu, M.; Kasuno, T.; Tajika, M.; Sugimoto, Y.; Yamanaka, S.; Hidaka, J. Calcium oxide as a solid base catalyst for transesterification of soybean oil and its application to biodiesel production. *Fuel* **2008**, *87*, 2798–2806. [CrossRef]
57. Yang, X.-X.; Wang, Y.-T.; Yang, Y.-T.; Feng, E.-Z.; Luo, J.; Zhang, F.; Yang, W.-J.; Bao, G.-R. Catalytic transesterification to biodiesel at room temperature over several solid bases. *Energy Convers. Manag.* **2018**, *164*, 112–121. [CrossRef]
58. Chen, G.; Shan, R.; Li, S.; Shi, J. A biomimetic silicification approach to synthesize $CaO-SiO_2$ catalyst for the transesterification of palm oil into biodiesel. *Fuel* **2015**, *153*, 48–55. [CrossRef]
59. Aghel, B.; Mohadesi, M.; Ansari, A.; Maleki, M. Pilot-scale production of biodiesel from waste cooking oil using kettle limescale as a heterogeneous catalyst. *Renew. Energy* **2019**, *142*, 207–214. [CrossRef]
60. Babajide, O.; Musyoka, N.; Petrik, L.; Ameer, F. Novel zeolite Na-X synthesized from fly ash as a heterogeneous catalyst in biodiesel production. *Catal. Today* **2012**, *190*, 54–60. [CrossRef]
61. Maneerung, T.; Kawi, S.; Wang, C.-H. Biomass gasification bottom ash as a source of CaO catalyst for biodiesel production via transesterification of palm oil. *Energy Convers. Manag.* **2015**, *92*, 234–243. [CrossRef]
62. Amani, H.; Asif, M.; Hameed, B. Transesterification of waste cooking palm oil and palm oil to fatty acid methyl ester using cesium-modified silica catalyst. *J. Taiwan Inst. Chem. Eng.* **2016**, *58*, 226–234. [CrossRef]

63. Abukhadra, M.R.; Dardir, F.M.; Shaban, M.; Ahmed, E.A.; Soliman, M.F. Spongy Ni/Fe carbonate-fluorapatite catalyst for efficient conversion of cooking oil waste into biodiesel. *Environ. Chem. Lett.* **2018**, *16*, 665–670. [CrossRef]
64. Thushari, I.; Babel, S.; Samart, C. Biodiesel production in an autoclave reactor using waste palm oil and coconut coir husk derived catalyst. *Renew. Energy* **2019**, *134*, 125–134. [CrossRef]
65. Lima, A.C.; Hachemane, K.; Ribeiro, A.E.; Queiroz, A.; Gomes, M.C.S.; Brito, P. Evaluation and kinetic study of alkaline ionic liquid for biodiesel production through transesterification of sunflower oil. *Fuel* **2022**, *324*, 124586. [CrossRef]
66. Jain, S.; Sharma, M.; Rajvanshi, S. Acid base catalyzed transesterification kinetics of waste cooking oil. *Fuel Process. Technol.* **2011**, *92*, 32–38. [CrossRef]
67. Hassan, W.A.; Ahmed, E.A.; Moneim, M.A.; Shaban, M.S.; El-Sherbeeny, A.M.; Siddiqui, N.; Shim, J.-J.; Abukhadra, M.R. Sulfonation of Natural Carbonaceous Bentonite as a Low-Cost Acidic Catalyst for Effective Transesterification of Used Sunflower Oil into Diesel; Statistical Modeling and Kinetic Properties. *ACS Omega* **2021**, *6*, 31260–31271. [CrossRef]

Disclaimer/Publisher's Note: The statements, opinions and data contained in all publications are solely those of the individual author(s) and contributor(s) and not of MDPI and/or the editor(s). MDPI and/or the editor(s) disclaim responsibility for any injury to people or property resulting from any ideas, methods, instructions or products referred to in the content.

Article

Facial One-Pot Synthesis, Characterization, and Photocatalytic Performance of Porous Ceria

Amal A. Atran [1,2], Fatma A. Ibrahim [1,3], Nasser S. Awwad [4], Mohd Shkir [5] and Mohamed S. Hamdy [1,*]

[1] Catalysis Research Group (CRG), Department of Chemistry, College of Science, King Khalid University, P.O. Box 9004, Abha 61413, Saudi Arabia
[2] Department of Chemistry, College of Science, Najran University, P.O. Box 1988, Najran 11001, Saudi Arabia
[3] Department of Chemistry, Faculty of Women for Art, Science and Education, Ain Shams University, Cairo 11757, Egypt
[4] Department of Chemistry, College of Science, King Khalid University, P.O. Box 9004, Abha 61413, Saudi Arabia
[5] Department of Physics, College of Science, King Khalid University, P.O. Box 9004, Abha 61413, Saudi Arabia
* Correspondence: m.s.hamdy@gmail.com or mhsaad@kku.edu.sa; Tel.: +966-17-214-8892

Abstract: A facial one-step synthesis procedure was applied to prepare porous sponge-like ceria (CeO_2). The synthesis was performed by mixing cerium nitrate with citric acid, followed by thermal treatment. The produced solid material was characterized by several techniques, such as XRD, SEM, N_2 sorption measurement, DR-UV-vis, and Raman spectroscopy. The characterization data showed that the nanoparticles of the porous ceria were formed with a three-dimensional pore system. Moreover, the measured surface area of the porous sample was eight times higher than the commercially available ceria. The photocatalytic performance of the porous ceria was investigated in two different applications under visible light illumination. The first was the decolorization of a methyl green aqueous solution, while the second was the photocatalytic elimination of a gaseous mixture consisting of five short-chain hydrocarbons (C1–C3). The obtained results showed that the photocatalytic activity of porous ceria was higher than that of the commercial sample. Finally, the recycling of porous ceria showed low deactivation (less than 9%) after four consecutive runs.

Keywords: cerium oxide; photocatalysis; flash combustion; water treatment; air purification; visible light

1. Introduction

The main concern of increasing industrial activity around the world is the issue of environmental pollution, and in particular, the pollution of water sources and air contamination [1]. The influence of water pollution is not limited to human life only but extends to other living organisms and, indeed, ecological life [2]. Moreover, air contamination can deteriorate human health and plant health as well. In 1972, researchers developed a cost-effective technique that provides the possibility to remove most of the pollutants either from water or air, which is a photocatalytic technique [3]. Photocatalysis has drawn a lot of interest recently, especially if a sustainable light source is utilized to activate the photocatalyst [4]. Unlike conventional heterogeneous catalysis, in which intensive heating processes may be required, photocatalysis occurs at room temperature and in a simple way using just light as the energy source [5]. The ideal photocatalyst should be photoactive, cost-effective, stable, nontoxic, and able to use visible and/or near UV light [6].

During the past two decades, the demand for active materials that can be used as photocatalysts in the visible light range has increased. Metal oxides (semiconductors), such as TiO_2 and ZnO, were among the materials that attracted the most attention and were used as photocatalysts to remove organic pollutants with high efficiency [7]. However, most semiconductors can only utilize light from the UV region, which limits the use of these materials in real industrial applications [8]. Amongst semiconductors, cerium oxide (ceria) is a promising photocatalyst with moderate activity under visible light illumination, which is a big advantage when compared to other metal oxides [9]. Ceria has attracted a lot of interest, and several publications were reported to present its utilization in different photocatalytic applications. The number of publications that deal with ceria has increased dramatically from 466 papers in 2001 to 2367 papers in 2021. Ceria is a cubic fluorite-type oxide that belongs to the rare earth oxides. It exists in both trivalent and tetravalent oxidation states; however, the electronic structure $Ce^{4+}([Xe]4f^0)$ is more stable than Ce^{3+} ($[Xe]4f^1$) [10–12]. Moreover, ceria is considered an n-type semiconductor material with a broad bandgap of roughly 3.1–3.44 eV [13]. According to the literature, the overall reaction rate of ceria under sunlight illumination is small due to its reducing potential and also due to the fast recombination of the photogenerated electron-hole pairs [14]. In order to improve the photocatalytic efficiency of ceria, several tactics can be applied, such as doping with other elements, composite formation with other semiconductors, nanoparticle preparation, and the formation of oxygen-defected sites. Increasing the photocatalytic efficiency of ceria will positively reflect on its utilization in industrial processes, such as in water treatment, air purification, and solar cells.

Several attempts were reported to improve the physical and photochemical properties of ceria. Fagen et al. prepared ceria nanoparticles via a precipitation method and produced ceria with a surface area of 24 m^2/g [15]. In another study, Ramasamy et al. applied a sol–gel procedure to prepare ceria nanoparticles with an energy gap of ≈3.79 eV [16]. Moreover, Govindhasamy et al. applied a one-step chemical precipitation method to the facial synthesis of ceria nanoparticles with a bandgap energy of 3.2 eV [17]. Mostafa et al. applied the hydrothermal method to prepare ceria nanoparticles with a surface area of 57.5 m^2/g [18]. In addition, Yoki et al. synthesized ceria nanoparticles with a bandgap value of 3.48 eV via the sol–gel method [19]. In these studies, the photocatalytic efficiency of ceria under visible light illumination was not as high as expected, which indicates that more improvements are required for ceria nanoparticles to enhance their photocatalytic activity.

In the current study, porous ceria nanoparticles were prepared in a simple one-step thermal technique to improve photocatalytic performance. The produced materials were characterized by several techniques, such as X-ray diffraction (XRD), N_2 sorption measurements, ultraviolet–visible spectroscopy (UV–vis), scanning electron microscopy (SEM), and Raman spectrometer. The photocatalytic activity of the prepared porous ceria was evaluated in two different applications under visible light illumination. The first is the decolorization reaction of methyl green (MG) dye in an aqueous solution, while the second is the photocatalytic elimination of a gaseous mixture consisting of five short-chain hydrocarbons (C1–C3). Commercial ceria nanoparticles were involved in the study as a control sample. The kinetics and rate constant of each reaction were calculated, and the obtained results are reported and discussed herewith.

2. Results
2.1. The Characterization Data

The phase composition of the two samples: the porous and the commercial ceria, was studied through XRD analysis, and the obtained patterns are presented in Figure 1. All the diffraction patterns are well indexed to the face-centered cubic-fluorite structure of ceria [20,21], with characteristic indexing planes of (111), (200), (220), (311), (222), and (400). Moreover, the absence of any additional phases in the prepared samples indicates that cerium nitrate was completely transformed into ceria by the applied thermal treatment. It is noteworthy that the commercial ceria sample showed sharp peaks, indicating high crystalline bulky particles. A slight decrease in the intensity of the porous ceria peaks causes the full width to be increased by half the maximum, and thus the particle size is reduced. Further analysis of the lattice and structural parameters was performed. The lattice parameters and unit cell volume were calculated to be a = b = c = 5.41744 Å and

V = 158.99458 Å3 (for porous ceria) and 5.40656 Å and V = 158.03847 Å3 (for commercial ceria) using the relations [22,23]: $1/d^2 = (h^2 + k^2 + l^2)/a^2$, here ($a_{111} = \sqrt{3} \times d$), and volume $V = a^3$, which are well matched with the standard card JCPDS#01-0800 of cubic crystal system of ceria. The microstructural parameters, such as crystallite size, dislocation density, and strain values, were determined using the following equations: $L_{hkl} = 0.9\lambda/(\beta \cos\theta)$, $\delta_{hkl} = 1/L^2$, and $\varepsilon = \beta \cos\theta/4$ and were found to be 4.8 nm, 4.38×10^{-2} nm^{-1}, and 2.93×10^{-2}, respectively, for porous ceria and 19.4 nm, 2.93×10^{-3} nm^{-1}, and 7.25×10^{-3}, respectively, for the commercial ceria sample. These values show a great variation in crystallite size and hence in other values.

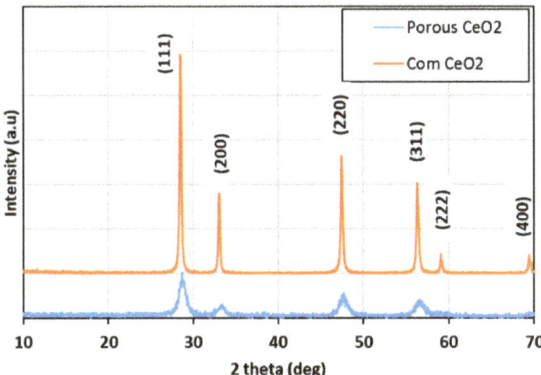

Figure 1. The XRD patterns of the porous ceria compared to that of commercial ceria.

Figure 2 represents the nitrogen sorption isotherm of the commercial ceria compared with that of the porous ceria. BET surface area was remarkably enhanced from 5.96 m^2/g in the commercial ceria to 47.18 m^2/g in the porous ceria, respectively. The N$_2$ isotherms of the investigated ceria samples are presented in Figure 2. A typical IV adsorption–desorption curve indicates the mesoporous properties of these prepared porous samples, while a typical II for the commercial ceria indicates nonporous properties. Moreover, the textural properties are summarized in Table 1. The pore size and pore volume of the prepared porous sample were 2.8 nm and 0.138 cm^3/g. The pore characteristics clearly illustrate the formation of mesopores in the ceria due to the use of citric acid during its synthesis.

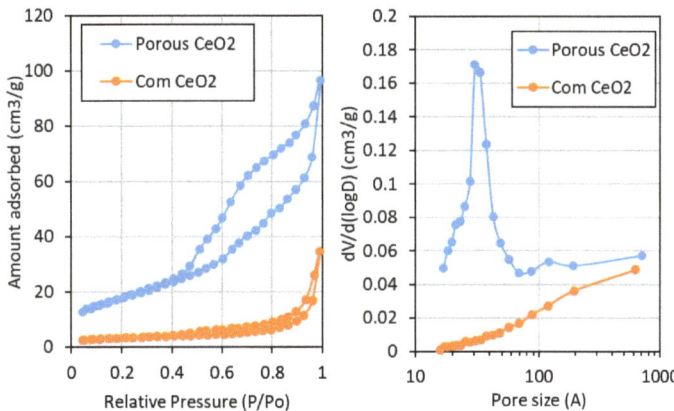

Figure 2. (**Left**) The N$_2$ sorption isotherms of the prepared porous ceria compared with a commercial ceria sample. (**Right**) the corresponding pore size distribution.

Table 1. The textural properties of the porous ceria compared with that of commercial ceria, as obtained from N$_2$ sorption measurements.

Sample	Surface Area (m^2/g)	Pore Volume (cm^3/g)	Pore Size (nm)
Com CeO$_2$	5.96	0.51	-
Porous CeO$_2$	47.18	0.138	2.8

The prepared porous ceria was investigated by using Raman spectroscopy (Figure 3). Theoretically, the fluorite-type cubic crystal structure of ceria exhibits only one Raman active fundamental mode at 464 cm^{-1}, which is the triply degenerate F$_{2g}$ mode that corresponds to a symmetric Ce-O stretching vibration in the Oh point group [24]. It can be regarded as a symmetric breathing mode of the six oxygen atoms around the central cerium ion [25]. In Figure 3, the porous ceria sample exhibited a broad peak compared to the commercial sample due to the stretching vibrations of the ceria nanoparticles. Therefore, Raman analysis added further evidence for the formation of ceria nanoparticles without contamination and/or phase change.

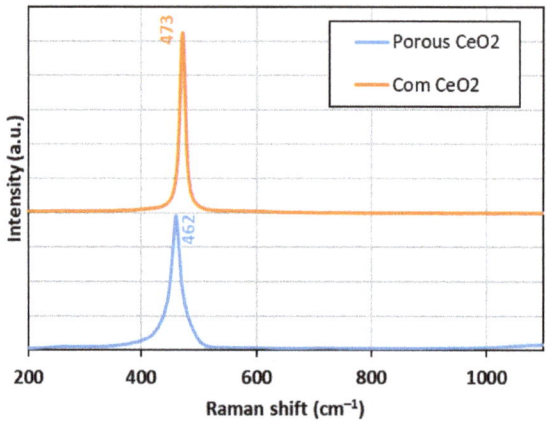

Figure 3. Raman spectrum for porous ceria compared to that of commercial ceria.

The morphological structure of the prepared porous ceria was explored by using SEM, and the obtained micrographs are presented in Figure 4. The SEM micrographs show a great morphological difference between the porous and commercial samples. The commercial ceria consists of agglomerated sheets or layers, while the prepared sample clearly shows a rough, porous nature with a sponge-like structure.

Diffuse reflectance UV-vis spectra are plotted and expressed in the Kubelka-Munk function as a function of the measured wavelength for the porous and commercial ceria samples (Figure 5). Generally speaking, the two samples have strong absorption bands in the UV region; this could be due to charge transfer during the O$_{2p}$-to-Ce$_{4f}$ transition with Ce^{4+} and O^{2-} states, which is higher than the 4f^1–5d^1 transition in ceria with mixed valences [26]. Moreover, the bandgap of the investigated samples was calculated from the equation E = h × c/λ, where h is Plank's constant (6.626 × 10^{-34} J s^{-1}), c is the speed of light (3.0 × 10^8 m s^{-1}), and λ is the cutoff wavelength (nanometers) [27]. However, the bandgap energy (Eg) for the porous ceria shows a slightly smaller bandgap energy (3.08 eV) than the commercial ceria (3.14 eV) despite their nanostructure and the small size of the crystallites, as characterized by XRD studies.

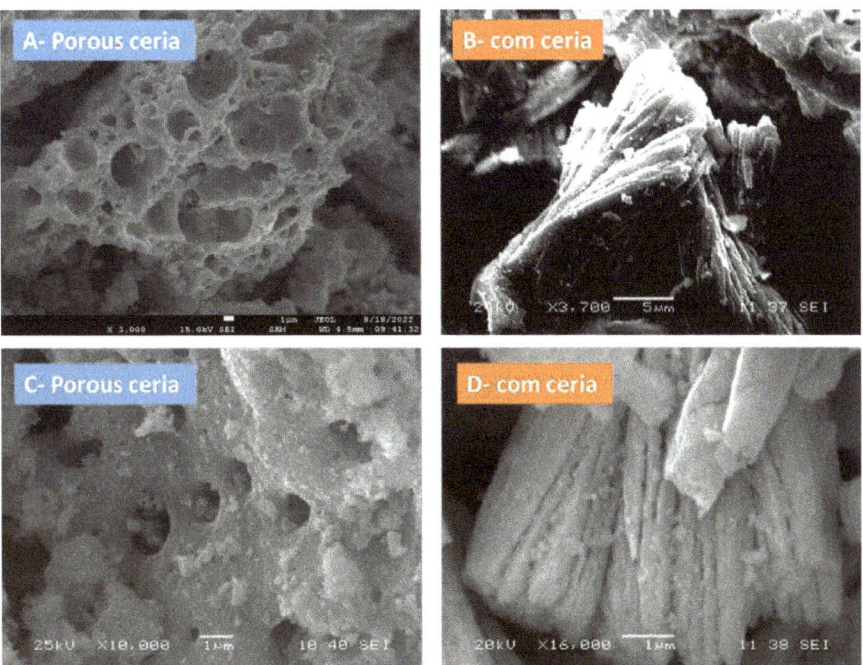

Figure 4. The SEM micrographs of the porous ceria sample (**A**,**C**) and the commercial ceria sample (**B**,**D**).

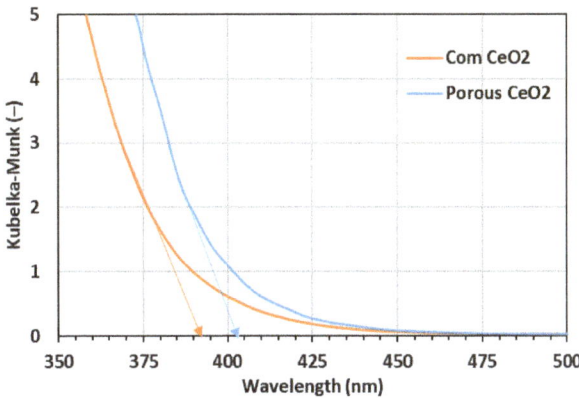

Figure 5. The adsorption behavior of the compared ceria samples expressed as Kubelka-Munk as a function of the measured wavelength (nm). Bandgap calculation was 3.12 and 3.01 eV for the commercial and porous ceria, respectively.

2.2. Photocatalytic Activity

In the photocatalytic study, the activity of the prepared porous ceria (porous CeO_2) was compared with two samples: the commercial ceria (com CeO_2) and the thermally treated commercial ceria (TT CeO_2). The first application, which was applied to investigate the photocatalytic activity of the porous ceria, was the decolorization reaction of methyl green dye under visible light illumination. Several blank experiments were performed at the beginning of the study, such as the photolysis of MG dye, the catalytic decolorization of the dye, and the adsorption of the dye over cerium oxide. The obtained results show that the dye is very stable, and no photolysis could be observed under the light. Moreover, the ceria could not activate the decolorization reaction without light. Finally, the adsorption affinity of ceria towards the investigated dye was insignificant.

The photocatalytic decolorization profiles of the applied dye over the investigated samples are plotted in Figure 6A. The obtained results show the difference in activity between the samples. The porous ceria exhibited the highest activity, with 40% dye decolorization achieved after 120 min, whereas the commercial ceria and the thermally treated sample exhibited a decolorization activity of almost 20% over the same time, i.e., 120 min. Moreover, the decolorization profiles fit perfectly with a first-order reaction rate model, as seen in Figure 6B, with R-squared values of 0.99 as an indication of the highly fitting property. Furthermore, the first-order rate constant (k) of the samples in the decolorization reaction of methyl green is shown in Figure 6C. The calculated rate constants of commercial (ZnO, CeO_2, and TiO_2) and porous ceria are 9×10^{-4}, 1×10^{-3}, 2.3×10^{-3}, and 3.35×10^{-3} min^{-1}, respectively. In other words, the obtained rate constant of the porous ceria was higher than the other three photocatalysts. The obtained high activity is strong evidence for the positive effect of morphological and textural properties on photocatalytic activity. The reusability study was carried out by using porous ceria for four consecutive runs without sample treatment; the obtained results are plotted in Figure 6D. The calculated (k) of the four reactions shows that the recycling of porous ceria showed small deactivation (less than 9%) after four consecutive runs.

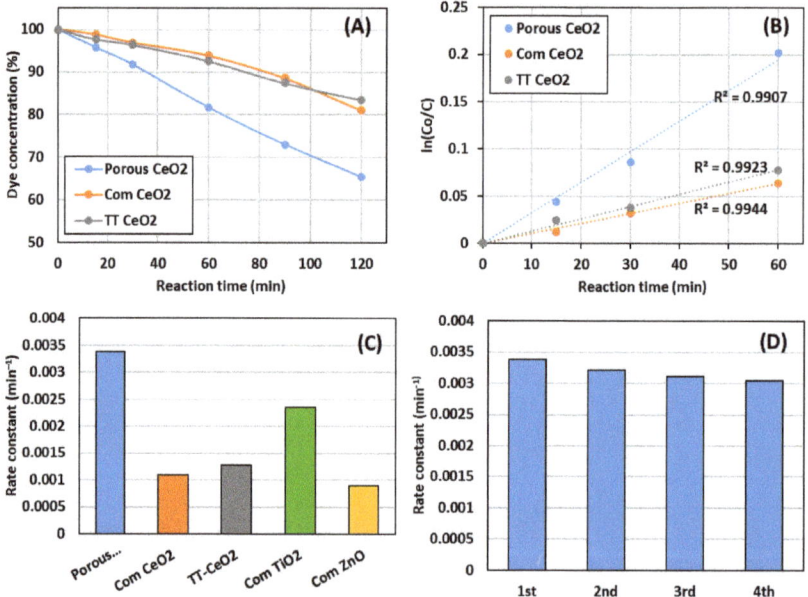

Figure 6. The obtained photocatalytic data for the decolorization reaction of methyl green dye under the illumination of visible light: (**A**) the decolorization profiles, (**B**) the first-order kinetic profiles, (**C**) the first-order rate constants for samples, and (**D**) the reusability of porous ceria sample in consecutive four runs.

In the second photocatalytic application, the elimination of a gas mixture over the porous and commercial ceria is shown in Figure 7. The gas mixture contained five different gases: methane, ethane, ethene, propane, and propene. Methane was very stable and did not show any elimination over the investigated photocatalysts. Ethane exhibited some resistance over the commercial ceria [28], but over the porous ceria, almost 36% of the

25 ppm ethane was eliminated. For ethene, the first unsaturated hydrocarbon in the gas mixture that was eliminated over the two samples, it took almost 60 min over commercial ceria and 45 min over porous ceria. Almost the same result was obtained for propane, the largest saturated hydrocarbon in the gas mixture. The last hydrocarbon in the mixture was propene, which is the second-most unsaturated hydrocarbon; over porous ceria, total elimination was observed after 30 min, while over commercial ceria, total elimination was obtained after 45 min.

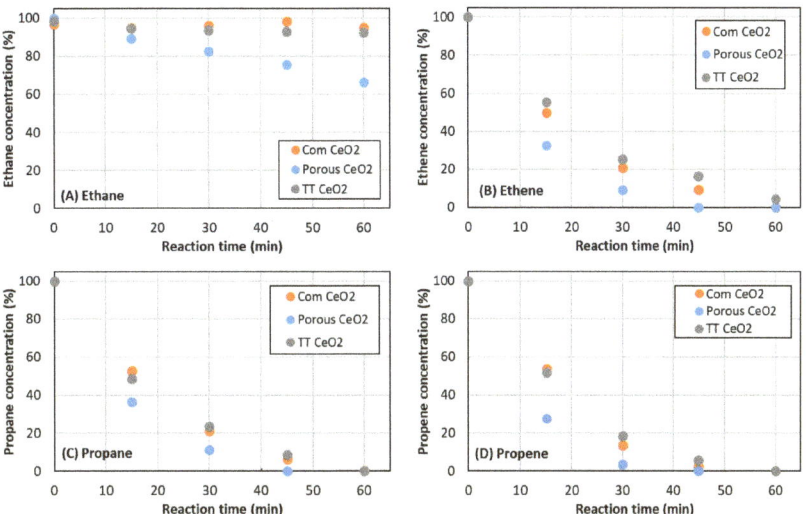

Figure 7. The elimination profile of (**A**) ethane, (**B**) ethene, (**C**) propane, and (**D**) propene over porous, commercial, and thermally treated ceria.

The elimination profiles of the hydrocarbon mixture were fit to the first-order kinetic model, and the first-order rate constant k (min^{-1}) was calculated and plotted (Figure 8). Generally speaking, the obtained results show that the order of hydrocarbon activity can be arranged as follows: propene > propane ≥ ethene >> ethane. However, the calculated rate constants for the elimination of ethene, propane, and propene over porous ceria were 48%, 32%, and 67% higher than those over commercial ceria. These results, again, reflect the high activity of porous (over commercial ceria).

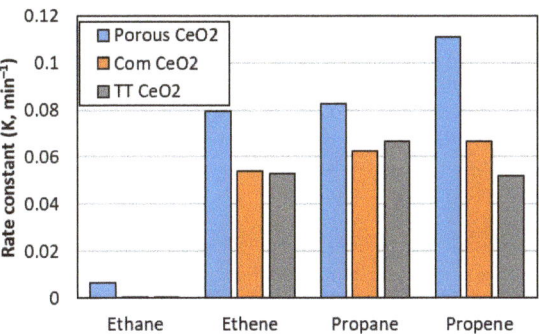

Figure 8. The first-order rate constant of the hydrocarbon elimination over porous, commercial, and thermally treated ceria.

3. Discussion

3.1. The Formation Mechanism of Porous Ceria

The synthesis of porous ceria was performed by using the citric–nitrate combustion process. The synthesis mechanism can be described in four main steps. Step (1) is the complex formation between citric acid and cerium cations. The citric acid ($C_6H_8O_7$) acts as a readily available, cost-effective, and effective chelating agent. Due to its high complexing ability, citric acid, when in an aqueous solution with cerium nitrate $Ce(NO_3)_3$, forms cerium-citric acid complex (Figure 9), as reported by K. Amalajyothi et al. [29].

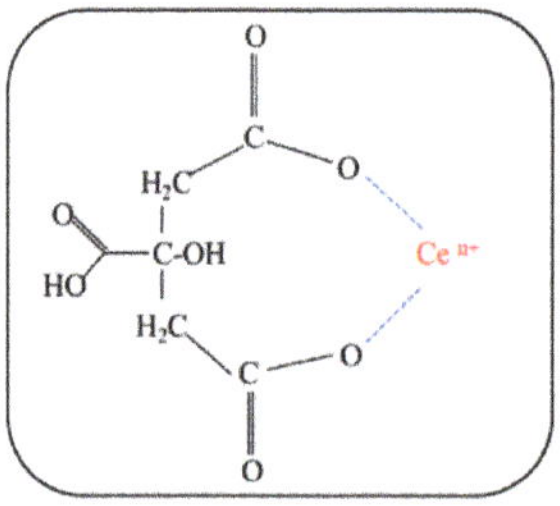

Figure 9. The formed complex between citric acid and cerium cation.

The gel formation is the second step of the synthesis mechanism. Under vigorous stirring, the citric–cerium complexes can be hydrolyzed to produce the colloidal sol or sometimes gel of cerium hydroxide nanoparticles, with extensive adsorption of citric acid on the surface of the formed nanoparticles. It was reported that the adsorption process of citric acid molecules on cerium hydroxide limits the growth of such crystals and produces nanoparticles [30].

In step (3), the obtained gel solution is subjected to thermal treatment. In this step, a great amount of gas (mainly H_2O, CO_2 and N_2) is released [31]; these gases are responsible for the formation of the voluminous foamy structure [32,33]. Finally, step four is the combustion process. At this stage, very small ceria particles (under 5 nm) are formed without agglomeration [30]. The calcined powders showed characteristic porous features due to the release of large amounts of gas during combustion. The combustion reaction can be expressed as follows:

$$Ce(NO_3)_3 \cdot 6H_2O + 5/6\, C_6H_8O_7 = CeO_2 + 5\, CO_2 + 28/3\, H_2O + 3/2\, N_2$$

As observed, there is no impurity content in the as-synthesized product, which can be attributed to the fact that the heat released during the combustion reaction is lower for the lean fuel composition, thereby yielding powders without any carbonaceous residue. Moreover, a good amount of oxygen will be available for combustion when fuel-deficient composition is used [32]. The combustion reactions with citric acid are less violent and more controllable compared to urea or glycine due to its weak exothermic nature [34]. Citric acid fuel can play dual functions: first, as a fuel, and second, as a chelating agent. So, the morphology of the sample synthesized by citric acid was porous with a sponge-like morphology [35].

3.2. The Higher Photocatalytic Activity of Porous Ceria

The photocatalytic performance of a certain material can be effect by many factors, such as light absorption, chemical composition, the presence of contamination, textural properties, OH surface density, reaction conditions, etc. In this section, the reason for the high photocatalytic performance of porous ceria, when compared to that of commercial ceria, is discussed. Chemical structure and the presence of contaminations factors can be neglected because the two materials have the same chemical composition. Moreover, the presence of contaminations was not observed. As obtained from the bandgap calculation, it was found that the bandgap of the commercial ceria is 3.14 eV, which is very close to earlier reports [36]. However, the bandgap of porous ceria was reduced to 3.01 eV as a result of particle size, which was reduced to a few nanometers. The relationship between bandgap and particle size was discussed earlier by Segets et al. [37,38]. As a result, this small red shift simply means that porous ceria can harvest more photons from the visible light region, and more electron/hole pairs can be formed. On the other hand, different opinions were reported about the role of surface area in photocatalytic activity. Laosiripojana reported a positive opinion about the effect of surface area on ceria [39]. In the current study, it was found that porous ceria is 3.1 times greater than commercial ceria. The obtained high surface area could play a positive role in the photocatalytic activity of porous ceria due to the high number of catalytic centers which are subject to light. Obviously, this can be compared with the high surface area of porous titania (UV-100) over Titania P-25 [40]. Finally, the morphology of the porous ceria may play a role in activity enhancement [41]. The presence of a three-dimensional pore system offers high diffusion of the dye molecules and hydrocarbon gases to and from the catalytic-active centers. A similar hypothesis was proposed earlier by Shen et al. [42,43]. Based on the obtained results, porous ceria seems a promising photocatalyst, either in liquid or gas phase applications.

4. Materials and Methods

4.1. Synthesis

The ceria sample was prepared by using a one-step thermal technique. In a typical synthesis, 5 g of cerium nitrate hexahydrate (97% Sigma) was dissolved in 5 g of demi water, then 1 g of citric acid (98% Aldrich) was added to the solution, and it was stirred until complete dissolution. The solution was moved into a porcelain crucible and dried at 90 °C for 24 h. Finally, the formed solid was calcined in a muffle furnace at 550 °C for 180 min by using a heating ramp of 18 °C/min. The obtained porous ceria was labeled (porous CeO_2). On the other hand, for comparison purposes, another sample was prepared by thermal treatment for the commercial ceria (TT CeO_2) at 550 °C for 180 min.

4.2. Characterization

A Shimadzu LabX-6000 diffractometer with CuKα radiation (λ = 1.54056 Å) was operated at 40/30 kV/mA at 2°/m between 20–70° angles for structural studies. A JEOL JSM 6310-SEM coupled with an EDX system operating at 20 kV was used to capture e-mapping, elemental composition, and morphology. BET surface area was calculated from nitrogen adsorption/desorption isotherms, which was recorded on a QuantaChromeNOVA2000e instrument. A Shimadzu UV-3600 diffused reflectance spectrophotometer (DRS) setup was used, employed to investigate the optical properties of the prepared samples. The diffuse reflectance spectra were converted into a Kubelka-Munk function F(R) by using the equation $F(R) = (1-R)^2/2R$. Raman spectra were recorded using a THERMO SCIENTIFIC DXRFT-Raman spectrometer with a laser source emitting at 532 nm and had a power of 2 mW.

4.3. Catalytic Activity

The photocatalytic activity of the prepared samples was investigated through the photocatalytic decolorization of methyl green (MG) dye under light illumination. In a typical experiment, 0.1 g of catalyst is dispersed into 50 mL of 0.02 g/L of the dye solution, and the overall mixture is stirred for 1 h to attain a uniform dispersion of catalytic material. Later, the suspension was kept in a photocatalytic chamber containing six Phillips light bulbs (18 W power), and the moment the light turns ON is considered the initial time of the reaction. Then, the samples were taken periodically at equal time intervals, and the absorption spectra were recorded to measure the catalytic activity parameters. After completing the experiments, the catalyst material was removed (by filtering) and was tested for reusability for up to 4 sequential runs. Moreover, the photocatalytic elimination of the hydrocarbon mixture was performed in a stainless-steel reactor with a maximum capacity of 35 mL. A total of 0.1 g of the applied catalyst was spread at the bottom of the reactor, and the reactor was closed tightly. Air was vacuumed from the reactor through the ultra-vacuum pump and then a hydrocarbon mixture was introduced into the reactor. The mixture contained 1% vol. of the five different gases: methane, ethane, ethene, propane, and propene balanced with argon as an inner gas. The applied light source (wavelength range from 300–650 nm) was introduced into the reactor through the glass window. Samples were withdrawn every 15 min through the automated valve and sent to on-line gas chromatography equipped with FID and TCD detectors.

5. Conclusions

Porous ceria nanoparticles were prepared by applying a one-step thermal technique. The prepared material exhibited a surface area eight times greater than commercial ceria. The photocatalytic activity of the porous ceria was higher than other commercial semiconductors under the illumination of visible light. The obtained higher photocatalytic activity of porous ceria can be related to its morphological, textural, and optical properties.

Author Contributions: Conceptualization, M.S.H. and N.S.A.; methodology, A.A.A.; formal analysis, F.A.I. and M.S.; investigation, A.A.A.; resources, N.S.A. and M.S.H.; writing—original draft preparation, A.A.A.; writing—review and editing, M.S.H.; supervision, M.S.H.; project administration, M.S.H.; funding acquisition, M.S.H. All authors have read and agreed to the published version of the manuscript.

Funding: The current research was funded by the Ministry of Education in KSA through project number KKU-IFP2-P-6.

Data Availability Statement: Data are available through direct contact with the corresponding author, Mohamed S. Hamdy: mhsaad@kku.edu.sa.

Acknowledgments: The authors extend their appreciation to the Ministry of Education in KSA for funding this research through project number KKU-IFP2-P-6.

Conflicts of Interest: The authors declare no conflict of interest.

References

1. Kusmierek, E. A CeO$_2$ semiconductor as a photocatalytic and photoelectrocatalytic material for the remediation of pollutants in industrial wastewater: A review. *Catalysts* **2020**, *10*, 1435. [CrossRef]
2. Manisalidis, I.; Stavropoulou, E.; Stavropoulos, A.; Bezirtzoglou, E. Environmental and health impacts of air pollution: A review. *Front. Public Health* **2020**, *8*, 14. [CrossRef]
3. Nadarajan, R.; Bakar, W.A.W.A.; Ali, R.; Ismail, R. Photocatalytic degradation of 1, 2-dichlorobenzene using immobilized TiO$_2$/SnO$_2$/WO$_3$ photocatalyst under visible light: Application of response surface methodology. *Arab. J. Chem.* **2018**, *11*, 34–47. [CrossRef]
4. Iervolino, G.; Zammit, I.; Vaiano, V.; Rizzo, L. Limitations and prospects for wastewater treatment by UV and visible-light-active heterogeneous photocatalysis: A critical review. In *Heterogeneous Photocatalysis*; Springer: Cham, Switzerland, 2020; pp. 225–264.
5. Kottappara, R.; Palantavida, S.; Vijayan, B.K. Enhancing semiconductor photocatalysis with carbon nanostructures for water/air purification and self-cleaning applications. In *Carbon Based Nanomaterials for Advanced Thermal and Electrochemical Energy Storage and Conversion*; Elsevier: Amsterdam, The Netherlands, 2019; pp. 139–172.
6. Cañón, A.M.R. Nanostructured ZnO Films for Water Treatment by Photocatalysis. Ph.D. Thesis, University of Bath, Bath, UK, 2014.
7. Amaterz, E.; Tara, A.; Bouddouch, A.; Taoufyq, A.; Bakiz, B.; Benlhachemi, A.; Jbara, O. Photo-electrochemical degradation of wastewaters containing organics catalysed by phosphate-based materials: A review. *Rev. Environ. Sci. Bio/Technol.* **2020**, *19*, 843–872. [CrossRef]
8. Karthikeyan, C.; Arunachalam, P.; Ramachandran, K.; Al-Mayouf, A.M.; Karuppuchamy, S. Recent advances in semiconductor metal oxides with enhanced methods for solar photocatalytic applications. *J. Alloys Compd.* **2020**, *828*, 154281. [CrossRef]
9. Mekonnen, T.B. An Overview on the Photocatalytic Degradation of Organic Pollutants in the Presence of Cerium Oxide (CeO$_2$) Based Nanoparticles: A Review. *Nanosci. Nanometrol* **2021**, *7*, 11648.
10. Rocha, L.S.R.; Amoresi, R.A.C.; Duarte, T.M.; Marana, N.L.; Sambrano, J.R.; Aldao, C.M.; Longo, E. Experimental and theoretical interpretation of the order/disorder clusters in CeO$_2$: La. *Appl. Surf. Sci.* **2020**, *510*, 145216. [CrossRef]
11. Arabaci, A. Ceria-based solid electrolytes for IT-SOFC applications. *Acta Phys. Pol. A* **2020**, *137*, 530–534. [CrossRef]

12. Bowen, M.S. Evaluation of Materials for Use in Open-Cycle Magnetohydrodynamic Power Generation. Master's Thesis, Oregon State University, Corvallis, OR, USA, 2021.
13. Chiu, F.C.; Lai, C.M. Optical and electrical characterizations of cerium oxide thin films. *J. Phys. D Appl. Phys.* **2010**, *43*, 075104. [CrossRef]
14. Wolski, L.; Sobańska, K.; Walkowiak, A.; Akhmetova, K.; Gryboś, J.; Frankowski, M.; Pietrzyk, P. Enhanced adsorption and degradation of methylene blue over mixed niobium-cerium oxide–Unraveling the synergy between Nb and Ce in advanced oxidation processes. *J. Hazard. Mater.* **2021**, *415*, 125665. [CrossRef]
15. Wang, F.; Zhang, L.; Zhu, J.; Han, B.; Zhao, L.; Yu, H.; Shi, W. Study on different CeO_2 structure stability during ethanol steam reforming reaction over Ir/CeO_2 nanocatalysts. *Appl. Catal. A Gen.* **2018**, *564*, 226–233. [CrossRef]
16. Rizzuti, A.; Dipalo, M.C.; Allegretta, I.; Terzano, R.; Cioffi, N.; Mastrorilli, P.; Mali, M.; Romanazzi, G.; Nacci, A.; Dell'Anna, M.M. Microwave-Assisted Solvothermal Synthesis of Fe_3O_4/CeO_2 Nanocomposites and Their Catalytic Activity in the Imine Formation from Benzyl Alcohol and Aniline. *Catalysts* **2020**, *10*, 1325. [CrossRef]
17. Murugadoss, G.; Ma, J.; Ning, X.; Kumar, M.R. Selective metal ions doped CeO_2 nanoparticles for excellent photocatalytic activity under sun light and supercapacitor application. *Inorg. Chem. Commun.* **2019**, *109*, 107577. [CrossRef]
18. Sayed, M.A.; Abo-Aly, M.M.; Aziz, A.A.A.; Hassan, A.; Salem, A.N.M. A facile hydrothermal synthesis of novel $CeO_2/CdSe$ and $CeO_2/CdTe$ Nanocomposites: Spectroscopic investigations for economically feasible photocatalytic degradation of Congo red dye. *Inorg. Chem. Commun.* **2021**, *130*, 108750. [CrossRef]
19. Yulizar, Y.; Juliyanto, S.; Apriandanu, D.O.B.; Surya, R.M. Novel sol-gel synthesis of CeO_2 nanoparticles using Morinda citrifolia L. fruit extracts: Structural and optical analysis. *J. Mol. Struct.* **2021**, *1231*, 129904. [CrossRef]
20. Safat, S.; Buazar, F.; Albukhaty, S.; Matroodi, S. Enhanced sunlight photocatalytic activity and biosafety of marine-driven synthesized cerium oxide nanoparticles. *Sci. Rep.* **2021**, *11*, 14734. [CrossRef] [PubMed]
21. El-Habib, A.; Addou, M.; Aouni, A.; Diani, M.; Nouneh, K.; Zimou, J.; El Jouad, Z. Effect of indium doping on the structural, optical and electrochemical behaviors of CeO_2 nanocrystalline thin films. *Opt. Mater.* **2022**, *127*, 112312. [CrossRef]
22. Shkir, M.; Khan, A.; Chandekar, K.V.; Sayed, M.A.; El-Toni, A.M.; Ansari, A.A.; AlFaify, S. Dielectric and electrical properties of La@ NiO SNPs for high-performance optoelectronic applications. *Ceram. Int.* **2021**, *47*, 15611–15621. [CrossRef]
23. Shkir, M.; Chandekar, K.V.; Khan, A.; Alshahrani, T.; El-Toni, A.M.; Sayed, M.A.; AlFaify, S. Tailoring the structure-morphology-vibrational-optical-dielectric and electrical characteristics of Ce@ NiO NPs produced by facile combustion route for optoelectronics. *Mater. Sci. Semicond. Process.* **2021**, *126*, 105647. [CrossRef]
24. Zimou, J.; Nouneh, K.; Hsissou, R.; El-Habib, A.; El Gana, L.; Talbi, A.; Addou, M. Structural, morphological, optical, and electrochemical properties of Co-doped CeO_2 thin films. *Mater. Sci. Semicond. Process.* **2021**, *135*, 106049. [CrossRef]
25. Villa-Aleman, E.; Houk, A.L.; Dick, D.D.; Hunyadi Murph, S.E. Hyper-Raman spectroscopy of CeO_2. *J. Raman Spectrosc.* **2020**, *51*, 1260–1263. [CrossRef]
26. Huang, X.; Zhang, K.; Peng, B.; Wang, G.; Muhler, M.; Wang, F. Ceria-based materials for thermocatalytic and photocatalytic organic synthesis. *Acs Catal.* **2021**, *11*, 9618–9678. [CrossRef]
27. Ponnar, M.; Sathya, M.; Pushpanathan, K. Enhanced UV emission and supercapacitor behavior of Zn doped CeO_2 quantum dots. *Chem. Phys. Lett.* **2020**, *761*, 138087. [CrossRef]
28. Han, K.; Wang, Y.; Wang, S.; Liu, Q.; Deng, Z.; Wang, F. Narrowing bandgap energy of CeO_2 in $(Ni/CeO_2)@SiO_2$ catalyst for photothermal methane dry reforming. *Chem. Eng. J.* **2021**, *421*, 129989. [CrossRef]
29. Amalajyothi, K.; Berchmans, L.J. Combustion synthesis of nanocrystalline cerium hexaboride using citric acid as a fuel. *Int. J. Self-Propagating High-Temp. Synth.* **2009**, *18*, 151–153. [CrossRef]
30. Masui, T.; Hirai, H.; Imanaka, N.; Adachi, G.; Sakata, T.; Mori, H. Synthesis of cerium oxide nanoparticles by hydrothermal crystallization with citric acid. *J. Mater. Sci. Lett.* **2002**, *21*, 489–491. [CrossRef]
31. Jamshidijam, M.; Mangalaraja, R.V.; Akbari-Fakhrabadi, A.; Ananthakumar, S.; Chan, S.H. Effect of rare earth dopants on structural characteristics of nanoceria synthesized by combustion method. *Powder Technol.* **2014**, *253*, 304–310. [CrossRef]
32. Manoharan, D.; Vishista, K. Optical properties of nano-crystalline cerium dioxide synthesized by single step aqueous citrate-nitrate gel combustion method. *Asian J. Chem.* **2013**, *25*, 9045. [CrossRef]
33. Deganello, F.; Marcì, G.; Deganello, G. Citrate–nitrate auto-combustion synthesis of perovskite-type nanopowders: A systematic approach. *J. Eur. Ceram. Soc.* **2009**, *29*, 439–450. [CrossRef]
34. Purohit, R.D.; Saha, S.; Tyagi, A.K. Powder characteristics and sinterability of ceria powders prepared through different routes. *Ceram. Int.* **2006**, *32*, 143–146. [CrossRef]
35. Ghahramani, Z.; Arabi, A.M.; Shafiee Afarani, M.; Mahdavian, M. Solution combustion synthesis of cerium oxide nanoparticles as corrosion inhibitor. *Int. J. Appl. Ceram. Technol.* **2020**, *17*, 1514–1521. [CrossRef]
36. Bhosale, A.K.; Shinde, P.S.; Tarwal, N.L.; Pawar, R.C.; Kadam, P.M.; Patil, P.S. Synthesis and characterization of highly stable optically passive CeO_2–ZrO_2 counter electrode. *Electrochim. Acta* **2010**, *55*, 1900–1906. [CrossRef]
37. Segets, D.; Lucas, J.M.; Klupp Taylor, R.N.; Scheele, M.; Zheng, H.; Alivisatos, A.P.; Peukert, W. Determination of the quantum dot bandgap dependence on particle size from optical absorbance and transmission electron microscopy measurements. *Acs Nano* **2012**, *6*, 9021–9032. [CrossRef] [PubMed]
38. Kouotou, P.M.; El Kasmi, A.; Wu, L.N.; Waqas, M.; Tian, Z.Y. Particle size-bandgap energy-catalytic properties relationship of PSE-CVD-derived Fe_3O_4 thin films. *J. Taiwan Inst. Chem. Eng.* **2018**, *93*, 427–435. [CrossRef]

39. Laosiripojana, N.; Assabumrungrat, S. Catalytic dry reforming of methane over high surface area ceria. *Appl. Catal. B Environ.* **2005**, *60*, 107–116. [CrossRef]
40. Alhanash, A.M.; Al-Namshah, K.S.; Hamdy, M.S. The effect of different physicochemical properties of titania on the photocatalytic decolourization of methyl orange. *Mater. Res. Express* **2019**, *6*, 075519. [CrossRef]
41. Huang, X.S.; Sun, H.; Wang, L.C.; Liu, Y.M.; Fan, K.N.; Cao, Y. Morphology effects of nanoscale ceria on the activity of Au/CeO_2 catalysts for low-temperature CO oxidation. *Appl. Catal. B Environ.* **2009**, *90*, 224–232. [CrossRef]
42. Shen, Y.; Fang, Q.; Chen, B. Environmental applications of three-dimensional graphene-based macrostructures: Adsorption, transformation, and detection. *Environ. Sci. Technol.* **2015**, *49*, 67–84. [CrossRef]
43. Yue, D.; Lei, J.; Peng, Y.; Li, J.; Du, X. Three-dimensional ordered phosphotungstic acid/TiO_2 with superior catalytic activity for oxidative desulfurization. *Fuel* **2018**, *226*, 148–155. [CrossRef]

Disclaimer/Publisher's Note: The statements, opinions and data contained in all publications are solely those of the individual author(s) and contributor(s) and not of MDPI and/or the editor(s). MDPI and/or the editor(s) disclaim responsibility for any injury to people or property resulting from any ideas, methods, instructions or products referred to in the content.

MDPI
St. Alban-Anlage 66
4052 Basel
Switzerland
www.mdpi.com

Catalysts Editorial Office
E-mail: catalysts@mdpi.com
www.mdpi.com/journal/catalysts

Disclaimer/Publisher's Note: The statements, opinions and data contained in all publications are solely those of the individual author(s) and contributor(s) and not of MDPI and/or the editor(s). MDPI and/or the editor(s) disclaim responsibility for any injury to people or property resulting from any ideas, methods, instructions or products referred to in the content.

www.ingramcontent.com/pod-product-compliance
Lightning Source LLC
LaVergne TN
LVHW070419100526
838202LV00014B/1488